# Hemicelluloses and Lignin in Biorefineries

# GREEN CHEMISTRY AND CHEMICAL ENGINEERING

### Series Editor: Sunggyu Lee
Ohio University, Athens, Ohio, USA

**Proton Exchange Membrane Fuel Cells: Contamination and Mitigation Strategies**
Hui Li, Shanna Knights, Zheng Shi, John W. Van Zee, and Jiujun Zhang

**Proton Exchange Membrane Fuel Cells: Materials Properties and Performance**
David P. Wilkinson, Jiujun Zhang, Rob Hui, Jeffrey Fergus, and Xianguo Li

**Solid Oxide Fuel Cells: Materials Properties and Performance**
Jeffrey Fergus, Rob Hui, Xianguo Li, David P. Wilkinson, and Jiujun Zhang

**Efficiency and Sustainability in the Energy and Chemical Industries:
Scientific Principles and Case Studies, Second Edition**
Krishnan Sankaranarayanan, Jakob de Swaan Arons, and Hedzer van der Kooi

**Nuclear Hydrogen Production Handbook**
Xing L. Yan and Ryutaro Hino

**Magneto Luminous Chemical Vapor Deposition**
Hirotsugu Yasuda

**Carbon-Neutral Fuels and Energy Carriers**
Nazim Z. Muradov and T. Nejat Veziroğlu

**Oxide Semiconductors for Solar Energy Conversion: Titanium Dioxide**
Janusz Nowotny

**Lithium-Ion Batteries: Advanced Materials and Technologies**
Xianxia Yuan, Hansan Liu, and Jiujun Zhang

**Process Integration for Resource Conservation**
Dominic C. Y. Foo

**Chemicals from Biomass: Integrating Bioprocesses into Chemical Production Complexes
for Sustainable Development**
Debalina Sengupta and Ralph W. Pike

**Hydrogen Safety**
Fotis Rigas and Paul Amyotte

**Biofuels and Bioenergy: Processes and Technologies**
Sunggyu Lee and Y. T. Shah

**Hydrogen Energy and Vehicle Systems**
Scott E. Grasman

**Integrated Biorefineries: Design, Analysis, and Optimization**
Paul R. Stuart and Mahmoud M. El-Halwagi

**Water for Energy and Fuel Production**
Yatish T. Shah

**Handbook of Alternative Fuel Technologies, Second Edition**
Sunggyu Lee, James G. Speight, and Sudarshan K. Loyalka

# Hemicelluloses and Lignin in Biorefineries

by

Jean-Luc Wertz

Magali Deleu

Séverine Coppée

Aurore Richel

## CRC Press
Taylor & Francis Group
Boca Raton  London  New York

CRC Press is an imprint of the
Taylor & Francis Group, an **informa** business

CRC Press
Taylor & Francis Group
6000 Broken Sound Parkway NW, Suite 300
Boca Raton, FL 33487-2742

First issued in paperback 2019

© 2018 by Taylor & Francis Group, LLC
CRC Press is an imprint of Taylor & Francis Group, an Informa business

No claim to original U.S. Government works

ISBN-13: 978-1-138-72098-5 (hbk)
ISBN-13: 978-0-367-88867-1 (pbk)

**Visit the Taylor & Francis Web site at**
**http://www.taylorandfrancis.com**

**and the CRC Press Web site at**
**http://www.crcpress.com**

# Contents

# Foreword

I have known Jean-Luc for more than 10 years, and have had the pleasure and honor to collaborate on several projects. We have worked together over that time for the promotion of the valorization of biomass at ValBiom, which is the Walloon Association dedicated to the promotion of biobased economy.

He was always passionate about his field and was instrumental in the development of the bioeconomy strategy for Wallonia, Belgium. As a published author, he has always impressed me with his expertise and zeal for promoting the biobased economy. His expertise in lignocellulosic biorefineries sciences and technologies is something he is continuously sharing with a range of audiences, as he is doing in this book.

The book focuses on the description of lignocellulosic biomass and its conversion through various processes in biorefineries. The authors also address strong socioeconomic and environmental messages in a context in which reducing the dependency of the European economy on fossil resources is of paramount importance. This is both in view of the increasing depletion of fossil resources, and their impact on climate change.

Through this book, readers will understand how biobased industries offer society the way forward to better resource efficiency. It describes in detail how we can expand the use of renewable biological resources as the possible substitutes for fossil fuels. It underlines the urgent need to do *more with less* and to succeed in *living well, within the limits of our planet*, which is also set in the recent policy initiatives.

Whether you are an engineer, scientist, decision-maker, or a citizen, this book provides key up-to-date information to understand the potential of lignocellulosic biomass for the construction of a sustainable biobased industry in Europe. Lignocellulosic biomass is made up primarily of cellulose, hemicelluloses, and lignin, with cellulose being the most abundant biopolymer on the Earth. It is estimated that around 100 billion tons of cellulose are synthesized annually by nature as a result of photosynthesis, which is by far more than the global crude oil production. The opportunity nature already provides for a plentiful supply of lignocellulosic biomass is evident, even if we refer to the nonedible part of plant such as agricultural and forestry residues, fractions of municipal and industrial wastes, and energy crops.

In the European Union, the biobased industries already account for 3.2 million jobs and provide €600 billion turn over. According to figures from 2013 provided by the bio-based Industries Consortium, this sector should create hundreds of thousands of skilled and nonskilled jobs, 80% of them being in the rural areas by 2030. This development will regenerate underdeveloped and/or abandoned regions and will grow and diversify farmers' income. It will enable the EU to reduce its dependency on the import of strategic raw material, such as fossil raw materials and expensive imported commodities such as chemicals and protein for animal feed. Again by 2030, with the right kind of development, 30% of fossil-based products could be replaced by biobased alternatives, leading to a reduction in greenhouse gas (GHG) emissions of 50%.

Addressing issues that are crucial to achieving these important goals, centre around ensuring the steady and sustainable supply of lignocellulosic biomass and innovating new technologies for second generation biorefineries to enable the transformation of the nonedible part of biomass. This book gives an overview of the current science and technology contributing to this ongoing biobased revolution, which is all about preparing Europe for the post-petroleum era.

In my role as an executive director of the Bio-Based Industries Joint Undertaking, which is the ambitious €3.7 billion joint initiative between biobased industry and the EU, I am delighted to see Jean-Luc and his coauthors to bring together their extensive knowledge and share it with the new generation of engineers and scientists, entrepreneurs, and innovators.

As its first reader I really welcome the initiative of Jean-Luc Wertz and his coauthors and recommend this book to you as a fascinating review of the current state of the art for biomass valorization.

**Philippe Mengal**
*Bio-Based Industries Joint Undertaking*
*Brussels, Belgium*

# Preface

The bioeconomy is a response to key environmental challenges that the world is facing already today. Bioeconomy aims to reduce the dependence on natural resources, transform manufacturing, promote sustainable production of renewable resources from land, fisheries and aquaculture and their conversion into food, feed, fiber, biobased products, and bioenergy, while developing new jobs and industries.

This book, titled *Hemicelluloses and Lignin in Biorefineries*, has been written to demonstrate why the bioeconomy is crucial for future generations and why hemicelluloses and lignins are the molecules of the future.

At the heart of this bioeconomy are biorefineries. These processing facilities transform biomass into energy and into a range of products in a sustainable way by combining plant chemistry and biotechnology. They will gradually replace oil refineries by using renewable resources in place of fossil fuels. A common goal for biorefineries is to use all parts of the biomass raw material as efficiently as possible, that is, maximizing the economic added value, whereas minimizing the environmental footprint. These biorefinery concepts include the application of advanced separation and conversion processes, which are the current hot topics in the research and development sector.

Our book deals with lignocellulosic biomass, the three main components of which are cellulose, hemicelluloses, and lignin, which form the cell wall of the plant. It not only presents new knowledge about the biosynthesis of these molecules but also about their enzymatic degradation by focusing particularly on hemicelluloses and lignin. Conversion of the latter is a topic of particular interest in recent years. Indeed, by taking an interest in its valorization, lignin is no longer considered as a waste, but a key product, which will be at the origin of most aromatic compounds.

This book is intended to serve as a comprehensive reference for any chemical company, research center, and university involved in or interested in the bioeconomy.

# Acknowledgments

We express our gratitude to the FRS–FNRS (Fonds national de la Recherche Scientifique, Belgium) who helped to get funding for this book.

We also thank Philippe Mengal, executive director of the Bio-Based Industries (BBI) Joint Undertaking, for accepting to write a foreword.

Jean-Luc Wertz dedicates this book to his family and, in particular, to the still unborn child of his daughter, and to Mathilda, Nicolas, and Carolina, the three children of his son.

Magali Deleu thanks the FRS–FNRS for her position as a senior research associate and dedicates this book to her family, particularly her husband and three children, Justine, Batiste, and Léane.

Aurore Richel acknowledges the University of Liege for supporting her research in the field of biomass and waste valorization, and development of green technologies. She also thanks the Walloon Region (Belgium).

Finally, we thank Teresita and Allison from the CRC Press for the huge support and assistance during the publication of this book.

# About the Authors

**Jean-Luc Wertz** holds degrees in chemical and civil engineering and in economic science from Catholic University of Louvain, Louvain-la-Neuve, Belgium, as well as a PhD from the same university in applied science, specializing in polymer chemistry. He has held various international positions in R&D, including Spontex, where he was the worldwide director of R&D. He holds several patents related to various products. Jean-Luc Wertz is now a project manager in biomass valorization at ValBiom and has worked for more than 8 years on lignocellulosic biorefineries and biobased products. He also wrote two books: *Cellulose Science and Technology* in 2010 and *Lignocellulosic Biorefineries* in 2013.

**Magali Deleu** holds a master's degree in chemical engineering and bioindustries from Gembloux Agro-Bio Tech (University of Liege), Belgium as well as a PhD degree from the same university in agricultural sciences and biological engineering. After a postdoc at Lund University in Sweden, Dr. Deleu got a permanent position since 2003 as research associate and since 2014 as senior research associate at Belgium National Funds for Scientific Research (FRS–FNRS) at the Laboratory of Molecular Biophysics at Interfaces, Gembloux Agro-Bio Tech, Belgium. She is engaged in research and education concerning the physicochemical and membrane properties of surfactants from biomass. Her main research interest is to explore the molecular mechanisms involved in the perception of structurally different amphiphilic molecules by biological membranes. She has published about 70 papers on this topic.

**Séverine Coppée** after completing her master's degree in chemical sciences earned her PhD degree in material sciences at the University of Mons, Mons, Belgium. She then worked as a postdoctoral fellow in the field of organic photovoltaics first in the United States and then for Materia Nova research center, Mons, Belgium. Since 2014, Dr. Coppée has been a project manager for GreenWin competitiveness cluster in Wallonia, Belgium and is supporting the cluster members with building up the biobased chemistry research projects through various funding opportunities, particularly in the framework of the Bio-based Industries Consortium (BIC) of which GreenWin is an effective member. She is also an life cycle assessment (LCA) advisor involved in the Life Cycle in Practice (LCiP), a European project (cofunded by the LIFE + Environmental Policy and Governance Programme of the EU) of which GreenWin is a partner. The LCiP project helps small and medium-sized enterprises (SMEs) in France, Belgium, Portugal, and Spain to reduce the environmental impacts of their products and services across the entire life cycle.

**Aurore Richel** holds a degree in chemical sciences and PhD in chemistry from the University of Liege, Liège, Belgium. Prof. Richel is currently professor and head of the Laboratory of Biological and Industrial Chemistry at the University of Liege. She is engaged in research and education in the fields of biological chemistry, biorefining, and industrial technologies. Prof. Richel is involved in numerous projects and industrial collaborations, specializing in the following areas: optimized use of vegetal biomass and waste for bioproducts and biofuels and development of new methodologies with low environmental footprints. She has published dozens of publications on this topic.

# 1 Introduction

## 1.1 BIOECONOMY AND CIRCULAR BIOECONOMY

This book is intended to accelerate the emergence of the bioeconomy defined as those parts of the economy that use biological renewable resources to produce food, feed, materials, chemicals, and energy. It will demonstrate why the bioeconomy is crucial for the future generations.

The bioeconomy is circular by nature because carbon is sequestered from the atmosphere by plants. After uses and reuses of products (or molecules) made from those plants, the carbon is cycled back as soil carbon or as atmospheric carbon once again (Figure 1.1).[1]

The circular economy focuses mainly on the efficient use of finite resources and ensures that those are used and recycled as long as possible.[2] The bioeconomy integrates the sustainable production and conversion of renewable resources. It is the renewable part of the circular economy. The principle of the circular economy is thus complementary to the renewable character of the bioeconomy and must facilitate the recycling of carbonaceous molecules after efficient uses (Figures 1.2 and 1.3). The huge benefits from the circular economy will be truly felt if the bioeconomy—the renewable part of the circular economy concept—is made to play its important and growing role.

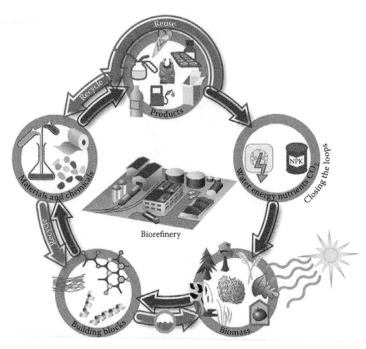

**FIGURE 1.1** Bioeconomy as part of the circular economy. (Courtesy of Patrick van Leeuwen; Reproduced from Bio-based Industries Consortium, *Bioeconomy: Circular by Nature*, http://biconsortium.eu/news/bioeconomy-circular-nature, 2015. With permission.)

**FIGURE 1.2** The left circle is the bioeconomy; the right circle is the technical economy. (Courtesy of Loes Deutekom; Reproduced from Partners4Innovation, *Towards a Biobased and Circular Economy*, https://www.youtube.com/watch?v=3ibJsOQdeoc, 2013. With permission.)

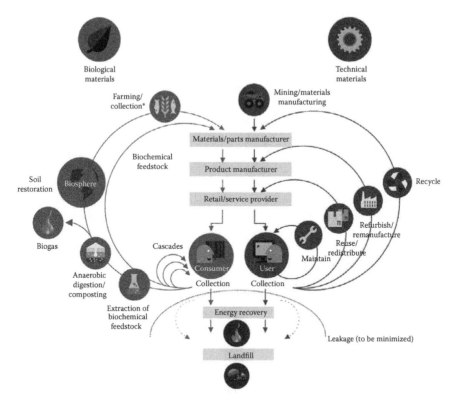

**FIGURE 1.3** Inspired by living systems, the circular economy concept is built around optimizing an entire system of resource or material flows. Like biological materials, technical materials can be part of a cycle built around reuse, remanufacture, and recycling. A circular economy advocates a shift away from the consumption of products to services. (Reprinted by permission from Macmillan Publishers Ltd. *Nat. Clim. Change*, Ellen MacArthur Foundation, 3, 180, copyright 2013.)

In the core of this economy are the biorefineries that sustainably transform biomass into food, feed, chemicals, materials, and bioenergy (fuels, heat, and power) generally through combination of plant chemistry and biotechnologies.

Lignocellulosic biomass consists of three major components—cellulose, hemicelluloses, and lignin—which form the plant cell wall. Although the cell walls of plants vary in both composition and organization, they are all constructed using a structural principle common to all fiber composites.[5] How the cell wall is biosynthesized is not yet completely understood today.

Historically, cellulose was investigated the first, about 20 years ago, which resulted in bioethanol fuel. Research on hemicelluloses followed resulting in valorization of 5 carbon (C5) sugars. Investigation on lignin is starting now, with the growing understanding that the whole plant should be valorized and not only its cellulose and hemicelluloses fractions.

## 1.2 BIOECONOMY

### 1.2.1 DEFINITION AND STAKES

Over the coming decades, the world will witness increased competition for limited and finite natural resources. A growing global population will need a safe and secure food supply. Climate change will have an impact on primary production systems such as agriculture, forestry, fisheries, and aquaculture.[6]

A transition is needed toward an optimal use of renewable biological resources. We must move toward sustainable primary production and processing systems that can optimally produce food, fiber, and other biobased products with fewer inputs, less environmental impact, and reduced greenhouse gas emissions. The optimal production refers to modern concepts such as Functional Economy. The Functional Economy is a concept that describes new economic dynamics based on providing solutions including the use of equipment and services from a perspective that reduces the mobilization of material resources while increasing the *servicial* value (the useful effects) of the solution.[7] The model proposes to exchange the material goods by functional goods.

In 1989, the World Commission on Environment and Development—created by the United Nations and chaired by the Norwegian Prime Minister Gro Harlem Brundtland—has defined sustainable development as "development which meets the needs of current generations without compromising the ability of future generations to meet their own needs."[8] Sustainability includes three pillars: environment, social, and economic. That definition explains without ambiguity that economic and social well-being cannot coexist with measures that impact over the environment (Figure 1.4).

According to the EU definition, the bioeconomy encompasses the sustainable production of renewable biological resources and their conversion into food, feed, biobased products, and bioenergy.[11] It includes agriculture, forestry, fisheries, food and pulp, and paper production, as well as parts of chemical, biotechnological, and energy industries. Its sectors and industries have a strong innovation potential due to an increasing demand for its goods (food, energy, other biobased products) and to

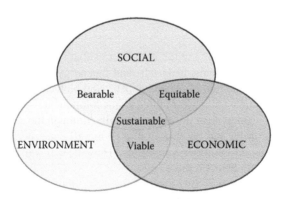

**FIGURE 1.4** The three pillars of sustainability. (Adapted from Green Planet Ethics, http://greenplanetethics.com/wordpress/sustainable-development-can-we-balance-sustainable-development-with-growth-to-help-protect-us-from-ourselves; http://en.wikipedia.org/wiki/File:Sustainable_development.svg.)

## TABLE 1.1
### The Bioeconomy in the European Union (2009)

| Sector | Annual Turnover (billion EUR) | Employment (thousands) |
| --- | :---: | :---: |
| Food | 965 | 4400 |
| Agriculture | 381 | 12000 |
| Paper/pulp | 375 | 1800 |
| Forestry/wood industries | 269 | 3000 |
| Fisheries and aquaculture | 32 | 500 |
| Biochemicals and plastics[a] | 50 | 150 |
| Enzymes[a] | 0.8 | 5 |
| Biofuels[a] | 6 | 150 |
| Total | 2078 | 22005 |

*Source:* Commission Staff Working Document of COM, 60 final. Innovation for Sustainable Growth. A Bioeconomy for Europe. http://ec.europa.eu/research/bioeconomy/pdf/bioeconomycommunication-strategy_b5_brochure_web.pdf, 2012; https://biobs.jrc.ec.europa.eu/sites/default/files/generated/files/policy/KBBE%20September%202010%20Report%20Achievements%20and%20Challenges.pdf

[a] Biobased industries.

their use of a wide range of sciences (from agronomy to process engineering or materials sciences) and industrial technologies, along with local and tacit knowledge.

Based on the available data from a wide range of sources, it is estimated that the European bioeconomy has an annual turnover of about EUR 2 trillion and employs more than 22 million people and approximately 9% of the total EU workforce (Table 1.1).[11]

### 1.2.2 SOCIETAL CHALLENGES

The bioeconomy's cross-cutting nature offers a unique opportunity to comprehensively address inter-connected societal challenges such as food security, natural resource scarcity, fossil resource dependence, and climate change, while achieving sustainable economic growth.[11]

Biological resources and ecosystems could be used in a more sustainable, efficient, and integrated manner. Food waste (including household waste, consumption waste, and agricultural and food processing waste) represents another serious concern. An estimated 30% of all food produced in developed countries is discarded.

The bioeconomy is an important element of reply to the challenges ahead. Managed in a sustainable manner, it can

- Sustain a wide range of public goods including biodiversity and ecosystem services.
- Reduce the environmental footprint of primary production and the supply chain as a whole.
- Increase competitiveness.
- Enhance Europe self-reliance.
- Provide job and business opportunities.[13]

## 1.3  BIOBASED ECONOMY

In the biobased economy, renewable biological resources instead of fossil resources are used as raw materials for the production of chemicals, materials, and fuels. The so-called biobased economy is the conversion of renewable feedstock (biomass and organic waste) into biobased products.[14] It can generally be regarded as the *nonfood pillar of the bioeconomy*.[15,16] Biotechnology plays a key role in the biobased economy.[17] Plant (green) biotechnology on the one hand is important for the primary production of biomass through the genetic improvement of crops. Industrial (white) biotechnology on the other hand is needed for the conversion of biomass into various products, using microorganisms (fermentation) and their enzymes (biocatalysis).

In the EU definition, the biobased economy is a low waste production chain starting from the use of land and sea, through the transformation and production of biobased products adapted to the requirements of end-users.[18]

## 1.4  LIFE-CYCLE ASSESSMENT

As environmental awareness increases, industries and businesses are assessing how their activities affect the environment.[19,20] The environmental performance of products and processes has become a critical issue. Companies have developed methods for evaluating the environmental impacts associated with a product, process, or activity. One such tool is Life-Cycle Assessment (LCA).

The LCA concept considers the entire life cycle of a product. It is a *cradle-to-grave* approach for assessing industrial systems.[19] *Cradle-to-grave* starts with the gathering of raw materials from the earth to create the product and ends at the point when all materials are returned to the earth. Therefore, LCA provides a comprehensive view of the environmental aspects of the product or process. In contrast, *cradle-to-cradle* refers to the optimum life cycle of the materials used in a product.[21]

The term life cycle refers to the major activities in the course of the product's lifespan from its manufacture, use, and maintenance, to its final disposal, including the raw material acquisition for its manufacture and transportation.[19] Figure 1.5 illustrates the typical life cycle stages that can be considered in a LCA and the typical inputs/outputs measured.

LCA is part of the ISO 14000 series of standards on environmental management.[22–24] The series provides principles, framework, and methodological requirements for conducting LCA studies.[25] The LCA framework consists of four components: goal and scope definition, inventory analysis, impact assessment, and interpretation (Figure 1.6).

Environmental considerations have been and will continue to be a strong driver for the development and introduction of biobased polymers including hemicelluloses and lignin. This calls for a comparison of their environmental performance with their petrochemical counterparts via, for example, LCA studies.

The available LCA studies indicate that most biobased polymers offer important environmental benefits today and for the future.[27] Compared to conventional polymers, the products studied generally contribute to the goals of saving energy resources

Inputs                           Outputs

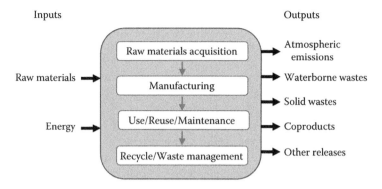

**FIGURE 1.5** Life cycle stages including inputs/outputs. (Adapted from Scientific Applications International Corporation (SAIC), *Life Cycle Assessment: Principles and Practice*, EPA/600/R-06/060, http://www.epa.gov/nrmrl/std/lca/pdfs/600r06060.pdf, 2006.)

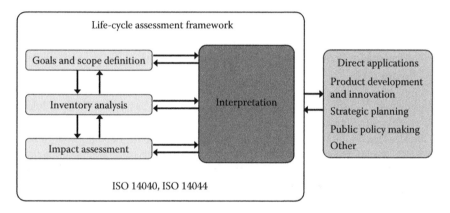

**FIGURE 1.6** The four phases of a LCA. (From ISO 14000, 1997; Adapted from Roes, A.L., Ex-ante life cycle engineering, Application to nanotechnology and white biotechnology, Thesis, Utrecht University, http://igitur-archive.library.uu.nl/dissertations/2011-0112-200331/roes.pdf, 2011.)

and mitigating greenhouse gas emissions. Overall, the available LCAs and environmental assessments strongly support the further development of biobased polymers.

## 1.5 LIGNOCELLULOSIC BIOMASS

*Biomass* has been defined as material of biological origin excluding material embedded in geologic formation and/or transformed to fossil.[28,29] This definition refers to the short carbon cycle, that is, the life cycle of biological materials such as plants, algae, marine organisms, forestry, microorganisms, animals and biological waste from households, agriculture, animals, and food/feed production.

Biomass is a renewable resource that can be used for the production of multiple energy and nonenergy products currently produced from fossil fuels and in particular petroleum. Currently, about 90% of fossil oil is used for production of energy (power, heat, and transport fuels), while ~10% is used for production of chemicals and materials.[30] Provided that biomass is harvested sustainably, biomass can also significantly reduce greenhouse gas emissions because primarily only the carbon fixed during photosynthesis can be released. Hence, biomass has the potential to resolve two of the main current challenges of humanity: energy security and environmental concerns. Furthermore, no resource is able to compete with biomass for generating the molecules currently produced from petroleum.

The Earth's biomass represents an enormous store of energy. It has been estimated that just one-eighth of the total biomass produced annually would provide all of humanity's current demand for energy (~5 $10^{20}$ J per year).[31,32] The world production of biomass is estimated at 146 billion metric tons a year, mostly wild plant growth.[33]

*Lignocellulosic biomass* refers to nonedible plant materials or nonedible parts of plants made up primarily of cellulose, hemicelluloses, and lignin. It represents the vast bulk of plant material. It includes in particular

- Agricultural residues.
- Forestry residues.
- A fraction of municipal and industrial (paper) wastes.
- Energy crops.

Such biomass resources typically contain on dry weight basis 40%–60% cellulose, 20%–40% hemicelluloses, and 10%–25% lignin (Figure 1.7).

Cellulose, hemicelluloses, and lignin are the main structural components of plant cell walls, the source of lignocellulosic biomass. Another structural component of cell walls is pectins.[34,35] Plant cell walls are usually divided into two categories: *primary walls* that surround growing cells and *secondary walls* that are thickened structures containing lignin and surrounding specialized cells.[36] Most of the lignocellulosic biomass comes from secondary walls. The cell walls of plants are the most abundant source of organic carbon on the planet.[37]

The chemical composition of plants varies widely. Factors that influence chemical composition include the plant variety, the part of the plant, the age of the plant, and the growing conditions such as soil, temperature, and water.[38] Variation in cellulose, hemicelluloses, and lignin content of some plants and plant residues is shown in Table 1.2.

Lignocellulosic biomass is expected to be used increasingly in the coming years for the industrial production of energy and nonenergy products.

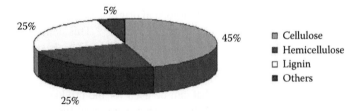

**FIGURE 1.7**   Average composition of lignocellulosic biomass.

## TABLE 1.2
## Chemical Composition of Selected Lignocellulosic Biomass
## (dry weight % basis)

| Lignocellulosic Biomass | Cellulose % | Hemicelluloses % | Lignin % | Reference |
|---|---|---|---|---|
| Corn stover | 37–42 | 20–28 | 18–22 | [28] |
| Sugarcane bagasse | 26–50 | 24–34 | 10–26 | [29–31] |
| Wheat straw | 31–44 | 22–24 | 16–24 | [32–34] |
| Hardwood stems | 40–45 | 18–40 | 18–28 | [31, 34, 35] |
| Softwood stems | 34–50 | 21–35 | 28–35 | [34–36] |
| Rice straw | 32–41 | 15–24 | 10–18 | [33, 37, 38] |
| Barley straw | 33–40 | 20–35 | 8–17 | [34, 39, 40] |
| Switch grass | 33–46 | 22–32 | 12–23 | [35, 41, 42] |
| Energy crops | 43–45 | 24–31 | 19–12 | [41] |
| Grasses (average)[a] | 25–40 | 25–50 | 10–30 | [31] |
| Manure solid fibers | 8–27 | 12–22 | 2–13 | [43] |
| Municipal organic waste | 21–64 | 5–22 | 3–28 | [44] |

*Source:* https://www.researchgate.net/publication/271447231_Wet_Explosion_a_Universal_and_Efficient_Pretreatment_Process_for_Lignocellulosic_Biorefineries#pfe\.

[a] e. g., reed canary grass, smooth brome grass, tall fescue, etc.

### 1.5.1 CELLULOSE

Cellulose (Figure 1.9) is the most abundant biopolymer on earth with an estimated 100 billion tons synthesized annually as a result of photosynthesis.[40] Cellulose is made of parallel unbranched β-1,4-linked *glucan* chains that form microfibrils.

The cellulose microfibrils with diameters of ~3 nm in higher plants consist of well-packed, long hydrogen-bonded stretches of crystalline cellulose and less ordered amorphous regions. They are the main load-bearing components of the cell wall.

### 1.5.2 HEMICELLULOSES

Hemicelluloses (Figure 1.9) are a heterogeneous group of matrix polymers of relatively low molecular weight, which are associated with cellulose and other polymers in plant cell walls. In primary cell walls, they constitute the matrix polysaccharides together with pectins, whereas in secondary cell walls, they constitute the matrix polymers with lignin. Hemicelluloses bind tightly to the surface of the cellulose microfibrils and to each other by hydrogen bonding and can be referred to as cross-linking *glycans*.[41] The most important biological role of hemicelluloses is their contribution to strengthening the cell wall by interaction with cellulose, and, in some walls, with lignin.[42]

Hemicelluloses are branched polysaccharides that have mostly β-1,4-linked backbones with an *equatorial* configuration.[42] In contrast to cellulose that contains only glucose units, they contain a variety of pentose and hexose sugars. Hemicelluloses include several polymers with a glucose, xylose, or mannose backbone.

Pectins form a gel matrix around the cellulose/hemicelluloses scaffold and contain α-1,4-linked-D-galacturonic acid residues in the backbone.[43,44]

### 1.5.3 Lignin

Lignin (Figure 1.9) is a complex phenolic heteropolymer that imparts strength, rigidity, and hydrophobicity to plant secondary cell walls.[45] It is the main natural polymer with an aromatic backbone.[46] Found in all vascular plants, particularly within the woody tissues, lignin makes up a substantial fraction of the total organic carbon in the biosphere. It is mainly deposited in terminally differentiated cells of supportive and water-conducting tissues.

Lignin is the generic term for a large group of aromatic polymers formed by the random condensation of *phenylpropanoid* precursors known as monolignols.[47–50] The three most abundant monolignols are *p*-coumaryl alcohol, coniferyl alcohol, and sinapyl alcohol, differing in their methoxyl substitution.

The units resulting from the monolignols, when incorporated into lignin, are called *p*-hydroxyphenyl (H), *guaiacyl* (G), and *syringyl* (S) units. Lignin monomer composition varies among plant species. With some exceptions, lignin from gymnosperms is composed of G units only (with minor amounts of H units), whereas lignin from dicots is primarily composed mainly of G and S units, and lignin from monocots (grasses) is a mixture of G, S, and H units.[40,51]

The units in the lignin polymer are linked by a variety of chemical bonds that have different chemical properties.[52]

## 1.6 PLANT CELL WALL

### 1.6.1 Structure of Cell Walls

Plant cell walls are complex structures composed mainly of polysaccharides and lignin (Figure 1.8).

Plant cell walls do not only provide rigidity and strength to the plant body but also protect the plant against environmental stresses.[54,55] The composition of the cell wall depends on the cell type. The cell walls of growing cells are called primary cell walls. They are relatively thin and extensible.

Once growth stops, secondary cell walls are deposited interior to the primary cell walls in certain cell types such as the fiber cells in wood. Secondary walls are thicker and stronger than primary walls. The primary walls are mainly composed of cellulose, hemicelluloses, pectins, and proteins, whereas the secondary wall typically contains cellulose, hemicelluloses, and lignin.

Probably, the biggest gap in our knowledge about cell walls relates to biosynthesis of the various wall components.[36] It has been estimated that more than 2000 genes are required for the synthesis and metabolism of cell wall components.

### 1.6.2 Diversity

All photosynthetic multicellular eukaryotes, including land plants and algae, have cells that are surrounded by a dynamic, complex, carbohydrate-rich cell wall.[56] All plant

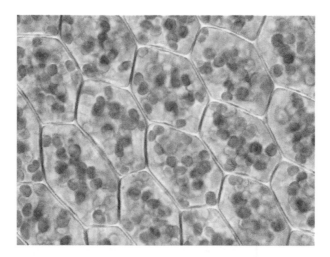

**FIGURE 1.8**   Plant cell walls that surround plant cells (size of plant cells varies from 10 to 100 μm). Green chloroplasts are visible in the cells. (Reproduced with permission of Kristian Peters, https://commons.wikimedia.org/wiki/File:Plagiomnium_affine_laminazellen.jpeg; Chloroplasts, https://en.wikipedia.org/wiki/Chloroplast.)

cell walls share several common features.[57] They are composed of cellulose microfibrils that form the mechanical framework of the wall, and a matrix phase that forms cross-links among the microfibrils and fills the space between the fibrillar framework.

Since the early 1970s, the chemistry of cell walls has been a prominent area of research, as cell walls are composed of a relatively small number of basic building blocks (Figure 1.9), and hence could be readily extracted and analyzed chemically.[57]

The cell wall exerts considerable biological and biomechanical controls over individual cells and organisms, thus playing a key role in their environmental interactions.[56] This has resulted in compositional variation that is dependent on developmental stage, cell type, and season. Further variation is evident that has a phylogenetic basis (Figure 1.9). Studies of a widespread group of organisms have shown the diversity of their cell walls and provide an insight into their evolutionary relationships.[57]

Figure 1.10 shows the simplified cell walls of six representative groups of plants:

- Nongrass angiosperms (A) have high amount of hemicelluloses and structural proteins. The primary walls have high amount of pectins, while the secondary walls have high amounts of lignin with G and S units.
- Grasses (B) have mixed-linkage glucans instead.
- Gymnosperms (C) have wall composition similar to nongrass angiosperms except they have higher amounts of glucomannans and their lignin is consisting primarily of G units.
- Leptosporangiates (group of ferns) (D) have high amounts of xylans, mannans, uronic acids, 3-*O*-methyl rhamnose, and lignins.
- In eusporangiates (other group of ferns) and bryophytes (mosses) (E) and charophytes (green algae) (F), the primary and secondary cell walls are not clearly differentiated.[57]

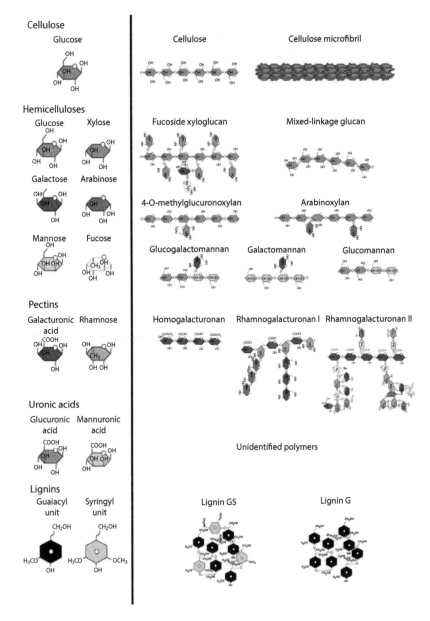

**FIGURE 1.9** Chemical structure of the predominant building blocks of plant cell walls. Left panel: monomers. Right panel: subunit of the respective polymers. (From Sarkar, P. et al., *J. Exp. Bot.*, 2009, 60, 3615–3635, by permission of Oxford University Press.)

During the course of evolution, plants have repeatedly adapted to their respective niche, which is reflected in the changes of their body plan and the specific design of cell walls.[57] Cell walls not only changed throughout evolution but also are constantly remodeled and reconstructed during the development of an individual plant, and in

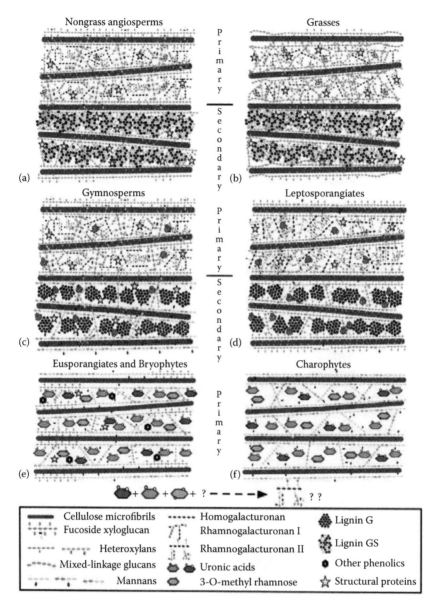

**FIGURE 1.10** Simplified 2D representation of general cell wall composition in the different groups of Kingdom Plantae. All groups have cellulose microfibrils. (From Sarkar, P. et al., *J. Exp. Bot.*, 2009, 60, 3615–3635, by permission of Oxford University Press.)

response to environmental stress or pathogen attacks. Carbohydrate-rich cell walls display complex designs, which together with the presence of phenolic polymers constitute a barrier for microbes, fungi, and animals.

Throughout evolution microbes have coevolved strategies for efficient breakdown of cell walls.[57] Our current understanding of cell walls and their evolutionary changes is limited. Comprehensive plant cell wall models will aid in the redesign of plant cell walls for the purpose of commercially viable lignocellulosic biofuel production as well as for many industries. Such knowledge will also be interesting for agriculture and plant biologists. It is expected that detailed plant cell wall models will require integrated correlative imaging and modeling approaches.

While cell walls are a characteristic feature of all plants, they are not exclusive to plants, with most bacterial and algal cells as well as all fungal cells also being surrounded by extracellular macromolecular barriers. The macromolecular composition, however, is characteristically different among the major evolutionary lineages of the living world (Figure 1.11).

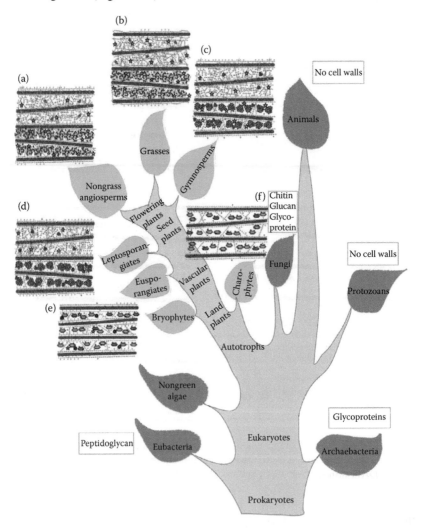

**FIGURE 1.11** Diagram showing changes in cell wall composition during the course of evolution (see Figure 1.9 for letters A to F). (From Sarkar, P. et al., *J. Exp. Bot.*, 2009, 60, 3615–3635, by permission of Oxford University Press.)

## 1.7  LIGNOCELLULOSIC BIOREFINERIES

### 1.7.1  Biorefinery Concept

Simultaneously resolving energy security and environmental concerns is a key challenge for policy makers today.[58] The world is facing a fast-growing human population and the consequent growing demand for food, water, and energy.[59,60] Furthermore, anthropogenic climate change is a severe threat to mankind and requires a significant reduction of our current greenhouse gas emissions to avoid detrimental consequences for the planet. Only the use of new technologies will allow us to progressively bridge the gap between economic growth and environmental sustainability. These technologies will transform the world economy from one based on fossil resources to one based on renewable resources such as biomass, while improving the sustainable production of energy, fuels, chemicals, and materials.[61–63]

The major drivers for a bioeconomy are independence from fossil energy sources, climate change abatement, and economic growth through new value chains.[64]

> *Biorefineries* that convert biomass and/or organic waste into chemicals and fuels were identified as a potential solution to mitigate the threat of climate change and to meet the growing demand for energy and nonenergy products.[59] Generally, one can make a distinction between first- and second-generation biorefineries.[65]
>
> *First-generation biorefineries*, which are already well implemented around the world, convert edible biomass into energy and nonenergy products. They have generated the *food versus fuel* debate about increased biofuel production and rising food prices, as well as the displacement of food crops.[66] Furthermore, direct and indirect land-use change affects the biodiversity as well as releasing carbon stored in the land.[67,68] As a consequence, the sustainability of many first-generation biorefineries has been increasingly questioned.
>
> *Second-generation biorefineries* use lignocellulosic feedstocks, such as organic wastes and residues, and convert them into energy and nonenergy products. Their objective is to optimize the valorization of all plant components. Therefore, their economic viability depends on optimal valorization of hemicelluloses and lignin in addition to cellulose. The advantages of such biorefineries are multiple as follows:
>   - No competition with food production
>   - Much higher amounts available
>   - Use of waste and residue materials that do not find applications today
>   - Use of abandoned land
>   - Diversion of waste materials from landfills
>   - In general, lower environmental impacts[68]

In this way, the lignocellulosic biorefineries can offer considerable potential to promote rural development and improve economic conditions worldwide. However, lignocellulosic biorefineries are not sustainable by themselves. While second-generation technologies are more efficient than first-generation, they could become unsustainable if, for instance, they compete with food crops for available land or with wildlife habitats. Thus, their sustainability will depend on whether producers comply with criteria such as minimum life cycle, greenhouse gas (GHG) reductions, including land-use change and social standards. In other words, global changes are necessary to reach sustainability.

Most lignocellulosic biorefineries are expected to be ready for large-scale commercial production in a few years. The landscape of active players is rather scattered and fragmented with many relatively small technology players, but there is an ever increasing number of large players starting to invest.

Two of the main industry drivers, in addition to energy security and environmental concerns, are mandates and policies.[59] Fuel regulations are examples of potential industry drivers.

### 1.7.2 FEEDSTOCK

Lignocellulosic biorefineries utilize biomass consisting of the residual nonfood parts of current crops or other nonfood sources, the goal being to maximize the value of the products obtained from the whole lignocellulosic content of the plant.[59,69]

#### 1.7.2.1 Biomass Residues and Wastes

The constraints related to the availability of additional land suggest that lignocellulosic biorefineries should focus on currently available feedstock sources in the initial phase of the industry's development.[65] Agricultural and forestry residues form a readily available source of biomass and can provide feedstock from current harvesting activities without the need for additional land cultivation. Biomass residues in general include agricultural residues (corn stover, corn cob, straws from wheat, rice, and other grain crops, bagasse, molasses); wood residues resulting from lumber, furniture, and fiber production; forest residues (tops and limps from harvest for wood products, material from fuel reduction treatments); black liquors from pulp production; and animal wastes and clean wood from trimmings and construction demolition.[70] Municipal solid waste (MSW) constitutes an alternative promising source of lignocellulosic biomass. Green waste can also be envisaged as feedstock.

#### 1.7.2.2 Dedicated Crops

Miscanthus, switchgrass (Figure 1.12), and other perennial grasses are also considered to be promising sources of lignocellulosic biomass.[71] Perennial grasses are less expensive to produce because they do not have to be replanted each year. Fast-growing trees, such as willow and poplar, are also attractive options because of harvesting and storage advantages.

Overall, second-generation biorefineries convert a variety of feedstocks leading to a variety of hemicelluloses and lignins in addition to cellulose. These hemicellulose and lignin molecules can lead to building blocks which introduced into adequate value chains are responsible for a high variety of final products.

(a)  (b)

**FIGURE 1.12** (a) *Miscanthus giganteus* (Courtesy of Jean-Michel DEPLANQUE, Tournai, Belgium.) and (b) switchgrass. (Courtesy of Warren Gretz from the National Renewable Energy Laboratory [NREL], U.S. Department of Energy, https://en.wikipedia.org/wiki/File:Panicum_virgatum.jpg.)

### 1.7.3 CONVERSION PROCESSES

Depending on the nature of the feedstock and the desired output, lignocellulosic biorefineries employ a variety of conversion technologies.[60]

There are basically two primary pathways for producing biobased products from lignocellulosic biomass: *biochemical* (including chemical) and *thermochemical*. Biochemical conversion basically involves hydrolysis of the polysaccharides in the biomass, and *fermentation* of the resulting sugars into ethanol. Thermochemical conversion typically involves *gasification* or other thermal treatment of the biomass followed by catalytic synthesis or fermentation of the resulting gas or liquid into ethanol, diesel, and other fluids.

#### 1.7.3.1 Biochemical Processes

Before fermentation, lignocellulosic feedstock processing needs to separate the cellulose and hemicelluloses from the nonfermentable lignin. This is usually performed by a pretreatment (see Chapter 8). While the lignin is currently mostly burned to generate energy, cellulose and hemicelluloses are hydrolyzed enzymatically with use of cellulases and hemicellulases (see Chapter 2) to deliver sugar solutions for subsequent fermentation (Figure 1.13).

##### 1.7.3.1.1 Thermochemical Processes

Primary routes for biomass thermal conversion are schematized in Figure 1.14.[74,75]

Combustion of biomass mainly leads to heat.

Gasification of biomass converts carbonaceous materials into synthesis gas, namely $H_2$ and CO, a gas mixture known as syngas. Gasification is achieved at high temperatures in the presence of a limited amount of oxygen. The resulting syngas can then be used in the Fischer–Tropsch process to produce hydrocarbons. When applied to biomass and biofuel production, the process is commonly referred to biomass to liquids (BtL).

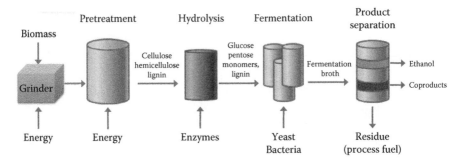

**FIGURE 1.13** A schematic of the biochemical conversion process, from lignocellulosic biomass to biofuels. (Adapted from IFP Energies nouvelles, http://www.ifpenergiesnouvelles.com/Research-themes/New-energies/Producing-fuels-from-biomass/Biocatalysts-one-of-IFPENs-expertise-field-Questions-to-Frederic-Monot-Head-of-the-Biotechnology-Department-at-IFPEN, 2016.)

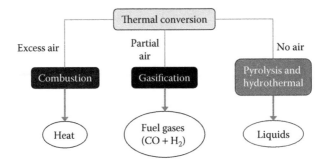

**FIGURE 1.14** Primary routes for biomass thermal conversion. (Adapted from Bain, R.L., *An Introduction to Biomass Thermochemical Conversion*, NREL. www.nrel.gov/docs/gen/fy04/36831e.pdf, 2004.)

*Pyrolysis* is the thermal decomposition of the biomass in the absence of oxygen to produce char, gas, and a liquid product (bio-oil) rich in oxygenated hydrocarbons.[76] It can be upgraded to lower the oxygen content and transported using the same infrastructure used by the oil industry. Bio-oil and upgraded oil can be employed in applications ranging from value-added chemicals to transportation fuels. *Hydrothermal treatments* which involve the processing of biomass in subcritical and supercritical water medium are very similar to pyrolysis except that it occurs at much higher feedstock moisture contents.[77]

### 1.7.4 VALORIZATION OF HEMICELLULOSES AND LIGNIN DERIVED FROM BIOMASS

An essential feature of the viable development of most second-generation biorefineries will be the coproduction of high-value chemicals from hemicellulose and lignin components.[78,79]

Biomass pretreatment technologies prior to enzymatic hydrolysis have the potential to extract hemicelluloses and lignin from biomass (Figure 1.15). The isolation of these macromolecules makes feasible their valorization.

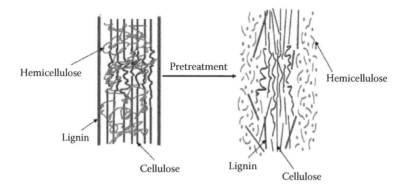

**FIGURE 1.15**   Schematic of the role of pretreatment in removing hemicelluloses and lignin from biomass prior to enzymatic hydrolysis. (From G. Brodeur, G. et al., *Enzyme Res.*, 2011, 787532, 2011.)

Furthermore, several routes for chemical conversion of hemicelluloses and lignin have been developed over the years by the pulp and paper industry and some of these can be applied for the production of valuable biobased products.[79] For lignin products, thermochemical, chemical pulping, and paper bleaching methods for production of monomeric and polymeric products can be considered. For hemicellulose products, preextraction of hemicelluloses from biomass is important and influences the mixture of solubilized material obtained.

## 1.8   STRUCTURE OF THE BOOK

*Chapter 1* of the book starts with a brief review of bioeconomy, circular economy, and lignocellulosic biomass. Life-Cycle Assessment is then briefly described. The plant cell wall, source of lignocellulosic biomass, is presented in terms of structure and diversity. Lignocellulosic biorefineries that convert biomass into energy and products are then discussed. The focus of the book is on biological conversion rather than chemical conversion. Indeed, the biological conversion appears generally simpler, more effective, and more efficient than conventional chemical systems.

*Chapter 2* is devoted to carbohydrate-active enzymes including glycosyltransferases, glycoside hydrolases, and esterases. These enzymes degrade, modify, or create glycosidic bonds. They are vital in polysaccharide biosynthesis and degradation.

*Chapter 3* is devoted to cellulose, its structure, its biosynthesis, and its enzymatic hydrolysis. The book does not concentrate on chemical hydrolysis of cellulose but concentrates on enzymatic hydrolysis, which is more effective and is used at the commercial scale today. Cellulose is the main component of the plant cell wall and forms a network with hemicelluloses and lignin.

*Chapter 4* investigates the structure and biosynthesis of hemicelluloses, with a detailed structure of specially xyloglucans, xylans, and mannans.

*Chapter 5* describes the enzymatic degradation of hemicelluloses including specially xyloglucans, xylans, and mannans.

*Chapter 6* deals with the structure and biosynthesis of lignin with a focus of monolignols and polymerization.

*Chapter 7* deals with the enzymatic degradation of lignin with a focus on laccases, lignin peroxidases, manganese peroxidases, and versatile peroxidases.

*Chapter 8* is about the biomass pretreatment technologies that are required to mainly separate cellulose, hemicelluloses, and lignin.

*Chapter 9* is devoted to the valorization of hemicelluloses into valuable products, especially through biorefineries. Valorization is considered in papermaking processes as well as in the production of building blocks from sugars. Case studies close the chapter.

*Chapter 10* is devoted to the valorization of lignin into valuable products, especially through biorefineries. Valorization is considered in papermaking processes as well as in the production of present, medium and long-term applications from lignin. Case studies close the chapter.

Finally, *Chapter 11* gives the perspectives regarding the potential of hemicelluloses and lignin to generate biobased products as an alternative to fossil-based products.

## REFERENCES

1. Bio-based Industries Consortium, *Bioeconomy: Circular by Nature*, 2015. http://biconsortium.eu/news/bioeconomy-circular-nature.
2. D. Carrez and P. van Leeuwen, *Bioeconomy: Circular by Nature*, 2015. http://biconsortium.eu/sites/biconsortium.eu/files/downloads/European_Files_september2015_38.pdf.
3. Partners4Innovation, *Towards a Biobased and Circular Economy*, 2013. https://www.youtube.com/watch?v=3ibJsOQdeoc.
4. Ellen MacArthur Foundation, *Nat. Clim. Change.*, 3, 180, 2013. http://www.nature.com/nclimate/journal/v3/n3/fig_tab/nclimate1842_F1.html
5. B. Alberts, A. Johnson, J. Lewis et al, *Molecular Biology of the Cell*, New York, Garland Science, 2002. http://www.ncbi.nlm.nih.gov/books/NBK26928/
6. http://ec.europa.eu/research/bioeconomy/policy/bioeconomy_en.htm
7. C. Du Tertre, S. Le Pochat and S. Boucq, *Proceedings 2nd Conference LCA*, Lille, 2012. http://www.avnir.org/documentation/book/LCAconf_dutertre_2012_en.pdf.
8. United Nations Economic Commission for Europe, UNECE in 2004–2005, *Sustainable development-concept and action* in http://www.unece.org/oes/nutshell/2004-2005/focus_sustainable_development.html
9. Green Planet Ethics. http://greenplanetethics.com/wordpress/sustainable-development-can-we-balance-sustainable-development-with-growth-to-help-protect-us-from-ourselves
10. http://en.wikipedia.org/wiki/File:Sustainable_development.svg
11. Commission Staff Working Document of COM(2012) 60 final. Innovation for Sustainable Growth. A Bioeconomy for Europe. http://ec.europa.eu/research/bioeconomy/pdf/bioeconomycommunicationstrategy_b5_brochure_web.pdf.
12. https://biobs.jrc.ec.europa.eu/sites/default/files/generated/files/policy/KBBE%20September%202010%20Report%20Achievements%20and%20Challenges.pdf
13. http://ec.europa.eu/programmes/horizon2020/en/h2020-section/bioeconomy
14. http://www.essenscia.be/Upload/Docs/Bio.be-strategybiobasedeconomy.pdf
15. http://www.besustainablemagazine.com/cms2/sustainable-use-of-and-value-creation-from-renewable-resources-in-a-biobased-economy-in-flanders/

16. http://www.era-ib.net/events/cinbios-forum-industrial-biotechnology-and-biobased-economy
17. http://www.gbesummerschool.be/
18. http://ec.europa.eu/research/consultations/bioeconomy/bio-based-economy-for-europe-part2.pdf
19. Scientific Applications International Corporation (SAIC), *Life Cycle Assessment: Principles and Practice*, EPA/600/R-06/060, 2006. http://www.epa.gov/nrmrl/std/lca/pdfs/600r06060.pdf.
20. Australian Life Cycle Assessment Society in http://www.alcas.asn.au/intro-to-lca.
21. MAERSK, *Cradle to cradle*, 2014. http://www.maersk.com/en/hardware/triple-e/the-hard-facts/cradle-to-cradle.
22. United States Environment Protection Agency in http://www.epa.gov/nrmrl/std/lca/lca.html.
23. ISO, ISO 14000 essentials, 2011. http://www.iso.org/iso/iso_14000_essentials.
24. ISO, ISO 14040: 2006. http://www.iso.org/iso/catalogue_detail?csnumber=37456.
25. DANTES website, *More about LCA*. http://www.dantes.info/Tools&Methods/Environmentalassessment/enviro_asse_lca_detail.html.
26. A.L. Roes, Ex-ante life cycle engineering, Application to nanotechnology and white biotechnology, Thesis, Utrecht University, 2011. http://igitur-archive.library.uu.nl/dissertations/2011-0112-200331/roes.pdf.
27. M. Patel, C. Bastioli, L. Marinu and E. Wurdinger, *Environmental assessment of bio-based polymers and natural fibers*, 2005. http://onlinelibrary.wiley.com/doi/10.1002/3527600035.bpola014/abstract.
28. http://www.fao.org/bioenergy/52184/en
29. ftp://ftp.cen.eu/CEN/Sectors/List/bio_basedproducts/DefinitionsEN16575.pdf
30. R. Hatti-Kaul, Lund University, *Biorefineries – A path to Sustainability*, 2009. www.scienceforum2009.nl/Portals/11/6Kaul-pres.pdf.
31. Nova: Science in the news, Biomass–the growing energy resource, Australian Academy of Science, 2009. http://www.science.org.au/nova/039/039print.htm.
32. Massachusetts Agriculture in the Classroom, *Winter 2008 Newsletter*, 2008. www.umass.edu/umext/mac/Newsletters/winter2008.htm
33. M. Balat and G. Ayar, *Energy Sources*, 27, 931, 2005. http://www.reventurepark.com/uploads/1_WTE_ART_18.pdf.
34. R. Kumar, S. Singh and O.V. Singh, *J. Ind. Microbiol. Biotechnol.*, **35**, 377, 2008. http://www.springerlink.com/content/n358853567w53245.
35. D. Mohnen, *Curr. Opin. Plant Biol.* 11, 266, 2008. http://www.sciencedirect.com/science/article/pii/S1369526608000630.
36. K. Keegstra, *Plant Physiol.* 154, 483, 2010. http://www.plantphysiol.org/content/154/2/483.full.
37. http://www.plantphysiol.org/content/153/2/444.full
38. https://www.sharecare.com/health/herbal-supplements/how-herbal-products-standardized?logref=Helpful_Button&regref=Helpful_Button
39. https://www.researchgate.net/publication/271447231_Wet_Explosion_a_Universal_and_Efficient_Pretreatment_Process_for_Lignocellulosic_Biorefineries#pfe\.
40. M.M.R. Ambavaram, A. Krishnan, K.R. Trijatmiko and A. Pereira, *Plant Physiol.* 155, 916, 2011. http://www.plantphysiol.org/content/155/2/916.full.
41. D. Hildebrand, *Plant Biochemistry*, BCH/PPA/PLS 609, 2010. www.uky.edu/~dhild/biochem/11B/lect11B.html.
42. H.V. Scheller and P. Ulvskov, *Ann. Rev. Plant Biol.* 61, 263, 2010. http://arjournals.annualreviews.org/doi/full/10.1146/annurev-arplant-042809-112315?amp;searchHistoryKey=%24%7BsearchHistoryKey%7D.
43. Complex Carbohydrate Research Center, The University of Georgia, *Plant Cell Walls, Galacturonans*. www.ccrc.uga.edu/~mao/galact/gala.htm.

44. D.J. Cosgrove, *Nat. Rev. Mol. Cell Biol.* 6, 850, 2005. https://homes.bio.psu.edu/expansins/reprints/CosgroveNatureRevMCB2005.pdf.

45. N.D. Bonawitz and C. Chapple, *Annu. Rev. Genet.* 44, 337, 2010. http://www.annualreviews.org/doi/pdf/10.1146/annurev-genet-102209-163508.

46. M. Dashtban, H. Schraft, T.A. Syed and W. Qin, *Int. J. Biochem. Mol. Biol.* 1, 36, 2010. http://www.ijbmb.org/files/IJBMB1004005.pdf.

47. M.M. Haggblom and I.D. Bossert, (Eds,) *Dehalogenation: Microbial Processes and Environmental Applications*, Boston, MA: Kluwer Academic Publishers, 2003.

48. R. Vanholme, B. Demedts, K. Morreel, J. Ralph and W. Boerjan, *Plant Physiol.* 153, 895, 2010. http://www.plantphysiol.org/cgi/content/full/153/3/895.

49. J. Ralph, K. Lundquist, G. Brunow, F. Lu, H. Kim, P.F. Schatz, J.M. Marita, R.D. Hatfield, S.A. Ralph, J.H. Christensen and W. Boerjan, Lignins: Natural polymers from oxidative coupling of 4-hydroxyphenyl-propanoids. *Phytochem. Rev.* 3, 29, 2004. http://www.springerlink.com/content/lx20h1488802t565.

50. J. RALPH, *Lignin Structure: Recent Developments*, Proceedings of the 6th Brazilian *Symposium Chemistry of Lignins and Other Wood Components*, 1999. http://www.dfrc.wisc.edu/DFRCWebPDFs/JR_Brazil99_Paper.pdf.

51. W. Boerjan, J. Ralph and M. Baucher, *Annu. Rev. Plant Biol.*, 54, 519, 2003. http://www.dfrc.wisc.edu/DFRCWebPDFs/2003-Boerjan-ARPB-54-519.pdf.

52. F.R.D. Van Parijs, K. Moreel, J. Ralph, W. Boerjan and R.M.H. Merks, *Plant Physiol.* 153, 1332, 2010. http://www.plantphysiol.org/cgi/content/full/153/3/1332.

53. Chloroplasts. https://en.wikipedia.org/wiki/Chloroplast.

54. N.C. Carpita and M.C. Mccann, *Trends Plant Sci.* 13, 415, 2008. http://www.sciencedirect.com/science/article/pii/S1360138508001817.

55. A. Endler and S. Persson, *Mol. Plant.*, 4, 199, 2011. http://mplant.oxfordjournals.org/content/4/2/199.full#ref-112.

56. Z.A. Popper, G. Michel, C. Herve, D.S. Domozych, W.G.T. Willats, M.G. Tuohy, B. Kloareg and D.B. Stengel, *Annu. Rev. Plant Biol.* 62, 567, 2011. http://www.annualreviews.org/doi/abs/10.1146/annurev-arplant-042110-103809.

57. P. Sarkar, E. Bosneaga and M. Auer, *J. Exp. Bot.* 60, 3615, 2009. http://jxb.oxfordjournals.org/content/60/13/3615.full.

58. International Energy Agency, *Energy Security and Climate Policy*, 2007. http://www.iea.org/publications/free_new_Desc.asp?PUBS_ID=1883.

59. World Economic Forum, *The Future of Industrial Biorefineries*, 2010. http://www3.weforum.org/docs/WEF_FutureIndustrialBiorefineries_Report_2010.pdf.

60. OECD Green Growth Studies, Energy, 2011. http://www.oecd.org/dataoecd/37/42/49157219.pdf.

61. U.R. Wageningen *Biobased Economy across the board*, 2011. http://www.themabiobasedeconomy.wur.nl/UK/newsagenda/news/Biobased_Economy_across_the_board.htm.

62. M. Bonaccorso, *Inside the World Bioeconomy,* 2014, Il Bioeconomista. http://www.lulu.com/shop/mario-bonaccorso/inside-the-world-bioeconomy/paperback/product-21878056.html.

63. United Nations, RIO+20, Corporate sustainability forum, 2012, http://csf.compact4rio.org/events/rio-20-corporate-sustainability-forum/custom-114-.251b87a2deaa4e56a3e00ca1d66e5bfd.aspx.

64. European Biogas Association, EBA's position on bioeconomy, http://european-biogas.eu/wp-content/uploads/2013/08/2014-08-EBA-position_bio-economy.pdf.

65. International Energy Agency, Sustainable production of second-generation biofuels, 2010. http://www.iea.org/papers/2010/second_generation_biofuels.pdf.

66. International Centre for Trade and Sustainable Development, Biofuel production, trade and sustainable development, 2008.

67. BirdLife International, Environmental impacts of current biofuels, 2012. http://www.birdlife.org/eu/EU_policy/Biofuels/eu_biofuels2.html.
68. R. Zah, TA-Swiss, Future perspectives of 2nd generation biofuels, 2010.
69. SupraBio, Newsletter, January 2011. http://www.suprabio.eu/klanten/supra/media/documenten/SUPRABIO%20newsletter%20number%201.pdf.
70. US Climate Change Technology Program, *Biomass Residues*, 2003. http://www.climatetechnology.gov/library/2003/tech-options/tech-options-2-3-7.pdf.
71. Energy Future Coalition, 2007. http://www.energyfuturecoalition.org/biofuels/fact_ethanol_cellulose.htm.
72. http://en.wikipedia.com/wiki/Switchgrass
73. IFP Energies nouvelles. 2016 http://www.ifpenergiesnouvelles.com/Research-themes/New-energies/Producing-fuels-from-biomass/Biocatalysts-one-of-IFPEN-s-expertise-field-Questions-to-Frederic-Monot-Head-of-the-Biotechnology-Department-at-IFPEN
74. R.L. Bain, *An Introduction to Biomass Thermochemical Conversion*, NREL, 2004. www.nrel.gov/docs/gen/fy04/36831e.pdf.
75. R.L. Bain, *Biomass Gasification*, NREL, 2006. http://www.ars.usda.gov/sp2UserFiles/Program/307/biomasstoDiesel/RichBainpresentationslides.pdf.
76. http://www.pnl.gov/main/publications/external/technical_reports/PNNL-18284.pdf.
77. S. Kumar, Hydrothermal treatment for biofuels: Lignocellulosic biomass to bioethanol, biocrude and biochar, University of Auburn, Auburn University, 2011. http://etd.auburn.edu/etd/handle/10415/2055.
78. J.L. Wertz and O. Bedue, *Lignocellulosic Biorefineries*, EPFL Press, 2013.
79. X. Zhang, M. Tu and M.G. Paice, *Bioenerg. Res.* 4, 246, 2011. http://www.springerlink.com/content/h800224105045774.
80. G. Brodeur, E. Yau, K. Badal, J. Collier, K.B. Ramachandran and S. Ramakrishnan, *Enzyme Res.*, 2011; 2011:787532.

# 2 Carbohydrate-Active Enzymes

## 2.1 INTRODUCTION

Carbohydrate-active enzymes are vital in an abundance of cellular processes. They create, modify, and degrade glycosidic bonds in monosaccharides, oligo-, or polysaccharides.[1]

For the production of any product based on lignocellulose, polysaccharides present in plant raw materials are of great interest as feedstock for further conversions.[2] The biochemical biorefinery developed for the production of biobased products basically involves the following steps:

- First, the cellulose and hemicellulose components of the biomass must be broken down (hydrolyzed), enzymatically or chemically, into sugars (mostly into monosaccharides or disaccharides); a variety of pretreatments are required to carry out this saccharification step in an efficient and low-cost manner.
- Second, some of these sugars, which are a complex mixture of 5-carbon and 6-carbon sugars, are typically fermented.

Feedstock, processes, and enzymes are the key trio in any second-generation project.[3] Recalcitrance of lignocellulose to deconstruction and high cost of enzymatic conversion form the major bottlenecks in this biochemical technology. Significant research, therefore, has to be directed toward the identification of efficient (hemi) cellulase systems and process conditions, besides those aimed at the biochemical and genetic improvement of existing organisms utilized in the process.[4]

Lignin, the most abundant aromatic biopolymer on Earth, is highly recalcitrant to enzymatic degradation.[5] By linking to both hemicelluloses and cellulose, it creates a barrier to any solutions or enzymes and prevents the penetration of lignocellulolytic enzymes into the interior lignocellulosic structure. Of the biomass components, lignin is the most resistant to degradation. Unlike most microorganisms, white-rot fungi are able to degrade lignin efficiently using extracellular ligninolytic enzymes. These ligninolytic enzymes include phenol oxidase (laccase) and heme peroxidases.

Carbohydrate-active enzymes include glycosyltransferases (GTs), glycoside hydrolases (GHs, also called glycosidases or glycosyl hydrolases), and esterases. They often display a modular structure with catalytic and noncatalytic modules.[6,7] An essential noncatalytic module is the *carbohydrate-binding module.*

*Catalytic modules* are subdivided into various families that catalyze the breakdown, biosynthesis, and/or modification of carbohydrates and *glycoconjugates.* Carbohydrate-binding modules are also subdivided in families of modules found attached to the catalytic modules. Protein clans are groups of glycoside families sharing a protein fold and catalytic machinery.

This review will help the reader to understand the recent improvements in the cocktails of enzymes proposed by enzyme suppliers and to conceive new cocktails.

## 2.2 GLYCOSYLTRANSFERASES

### 2.2.1 GLYCOSYLTRANSFERASE ACTIVITIES

GTs are enzymes (EC 2.4, EC for enzyme classification) that catalyze glycosyl group transfer from glycosyl donors to nucleophilic acceptors to form *glycosides* (Equation 2.1).[8,9] The acceptor substrates are most commonly oligosaccharides, but they can be monosaccharides.[10]

$$\text{Acceptor} + \text{Glycosyl donor} \xrightarrow{\text{GT}} \text{Glycosylated acceptor} \tag{2.1}$$

$$+ \text{ Nucleotide or isoprenoid} - P$$

It is frequent that GTs utilize an activated donor sugar substrate that contains a (substituted) phosphate-leaving group.[11] Donor sugar substrates are most commonly activated in the form of nucleoside diphosphate sugars. A classical nucleoside diphosphate sugar (nucleotide sugar) is UDP-glucose (Figure 2.1).

However, nucleoside monophosphate sugars, lipid phosphate sugars, and phosphate sugars can also act as donor molecules (Figure 2.2).

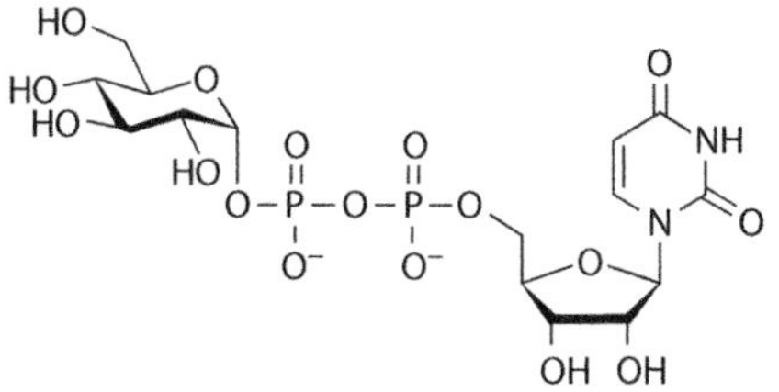

**FIGURE 2.1** Uridine diphosphoglucose (UDPGlc).

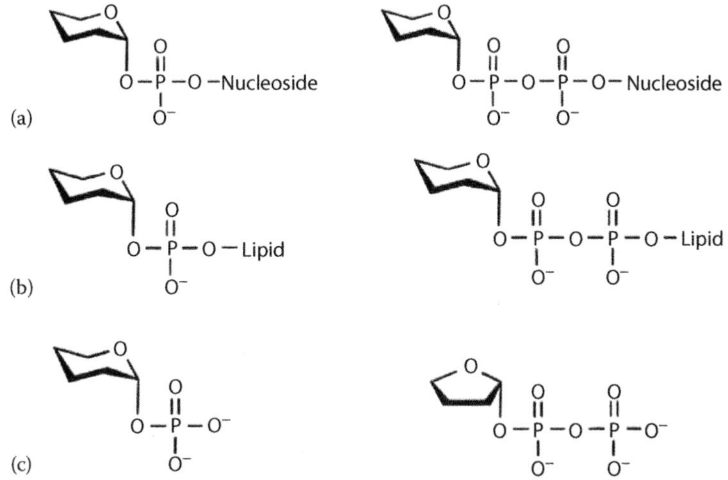

**FIGURE 2.2** Sugar donor substrates. (a) Nucleotide sugars and (b,c) Nonnucleotide sugars. (From Withers, S.G. and Williams, S.J., *Glycosyl Transferases* in CAZypedia, http://www. cazypedia.org/, accessed 2016; http://www.cazypedia.org/index.php/Glycosyltransferases.)

Nucleotide sugar-dependent GTs are often referred to as Leloir enzymes, in honor of Luis F. Leloir who discovered the first sugar nucleotide and was awarded the Nobel Prize in chemistry in 1970.[8]

Glycosyltransferases that utilize nonnucleotide donors are termed non-Leloir glycosyltransferases.

## 2.2.2 GLYCOSYLTRANSFERASE FOLDS

As for other classes of carbohydrate-active enzymes, GTs are classified into families based on amino acid sequence similarities.[12,13] The number of distinct GT families was 97 in June 2015. In striking contrast with glycosidases, which exhibit a wide variety of overall folds, nucleotide sugar-dependent (Leloir) GTs have been found to possess only two general folds termed GT-A and GT-B.[11] This finding may indicate that the majority of GTs has evolved from a small number of progenitor sequences. However, it also reflects the requirement for at least one nucleotide-binding domain of the *Rossmann fold*[14] type (structural motif found in proteins that bind nucleotides, especially the cofactor NAD). Other structural folds have been observed recently in non-Leloir GTs.

The GT-A fold was first described for the inverting GT SpsA from *Bacillus subtilis*.[15] Consisting of an open twisted β-sheet surrounded by α-helices on both sides, the overall architecture of the GT-A fold is reminiscent of two abutting Rossmann-like folds, typical of nucleotide-binding proteins.[11] Two tightly associated β/α/β domains, the sizes of which vary, abut closely, lead to the formation of a continuous central β-sheet. For this reason, the GT-A fold is sometimes described as a single-domain fold. However, distinct nucleotide- and acceptor-binding domains are present. Most GT-A enzymes possess an Asp-X-Asp (DXD; D = Asp = aspartic acid; X = any amino acid) motif in which the carboxylates coordinate a divalent cation and/or a ribose.

The GT-B fold was first described for the DNA-modifying β-GT from bacteriophage T4.[16] Like the GT-A fold, the architecture of GT-B enzymes consists of two β/α/β Rossmann-like domains; however, in this case, the two domains are less tightly associated and face each other with the active site lying within the resulting cleft.[11] As with the GT-A fold, these domains are associated with the donor and acceptor substrate-binding sites. Unlike GT-A enzymes, GT-B enzymes are metal-ion independent and do not possess a DXD motif.[9]

### 2.2.3 GLYCOSYLTRANSFERASE CLASSIFICATION

It has been shown that GTs catalyze the transfer of glycosyl groups to a nucleophilic acceptor with either retention or inversion of configuration at the anomeric carbon of the donor substrate.[8] This allows the classification of GTs as either retaining or inverting enzymes (Figure 2.3).

GT families can be classified into clans depending on their fold and the stereochemical outcome of the reactions that they catalyze (Figure 2.4).[11] Among GT-A and GT-B superfamilies, the overall fold of the enzyme does not dictate the stereochemical

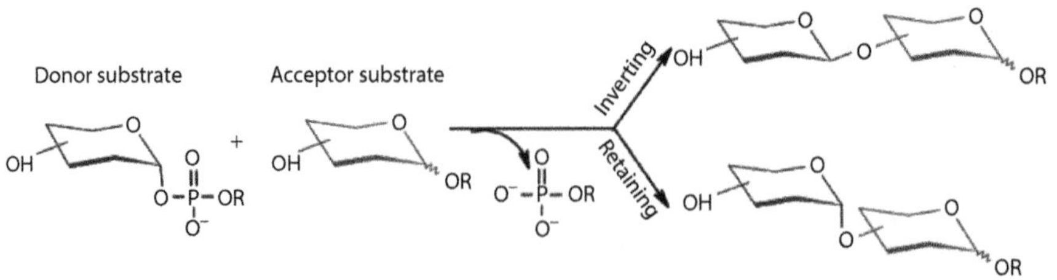

**FIGURE 2.3**  Like glycosidases, glycosyltransferases catalyze glycosyl group transfer with either inversion or retention of the anomeric stereochemistry with respect to the donor sugar. (Adapted from http://www.ncbi.nlm.nih.gov/books/NBK1921/?report=reader#!po=8.33333.)

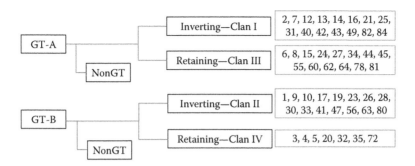

**FIGURE 2.4** GT classification system proposed by Coutinho et al.[12] Families are classified into clans on the basis of their fold (GT-A or GT-B) and activity. The GT-A and GT-B folds are also shared by non-GT enzymes. GT family numbers belonging to each clan are indicated on the far right. The remaining families are those predicted to adopt either the GT-A or GT-B fold. This classification system does not include 39 of the 90 GTs. (Adapted from Coutinho, P.M. et al., *J. Mol. Biol.*, 328, 307, 2003.)

outcome of the reaction that it catalyzes, as examples of both inverting and retaiing GTs have been identified within both the GT-A and GT-B fold classes.[11]

### 2.2.4 MECHANISM OF INVERTING GLYCOSYLTRANSFERASES

The mechanism utilized by inverting GTs is that of a *direct displacement* $S_N2$-like (substitution, nucleophilic, and bimolecular) reaction.[11,17] Inversion of stereochemistry results from this $S_N2$-like reaction where an acceptor hydroxyl group performs a nucleophilic attack at carbon C1 of the sugar donor from one side and the (substituted) phosphate-leaving group leaves from the other side.[9] The transition state is believed to possess substantial oxocarbenium ion character.[8]

An active-site side chain serves as a base catalyst that deprotonates the incoming nucleophile of the acceptor, facilitating direct $S_N2$-like displacement of the activated (substituted) phosphate-leaving group (Figure 2.5a). Typically, enzymes of this type possess an aspartic acid or glutamic acid residues whose side chains serve to partially deprotonate the incoming acceptor hydroxyl group, rendering it a better nucleophile.[9]

The $S_N2$-like reaction is also facilitated by Lewis acid activation of the departing (substituted) phosphate-leaving group.[11] A vast majority of enzymes from GT Clan I (inverting GT-A enzymes) that have been subjected to biochemical analysis use an essential divalent cation (usually $Mn^{2+}$ or $Mg^{2+}$), coordinated by the so-called DXD (aspartate–any amino acid–aspartate) motif, to facilitate departure of the nucleoside diphosphate-leaving group by electrostatically stabilizing the developing negative charge. Most inverting GT-B enzymes appear to use metal ion-independent methods for stabilizing the departure of nucleoside diphosphate-leaving groups.

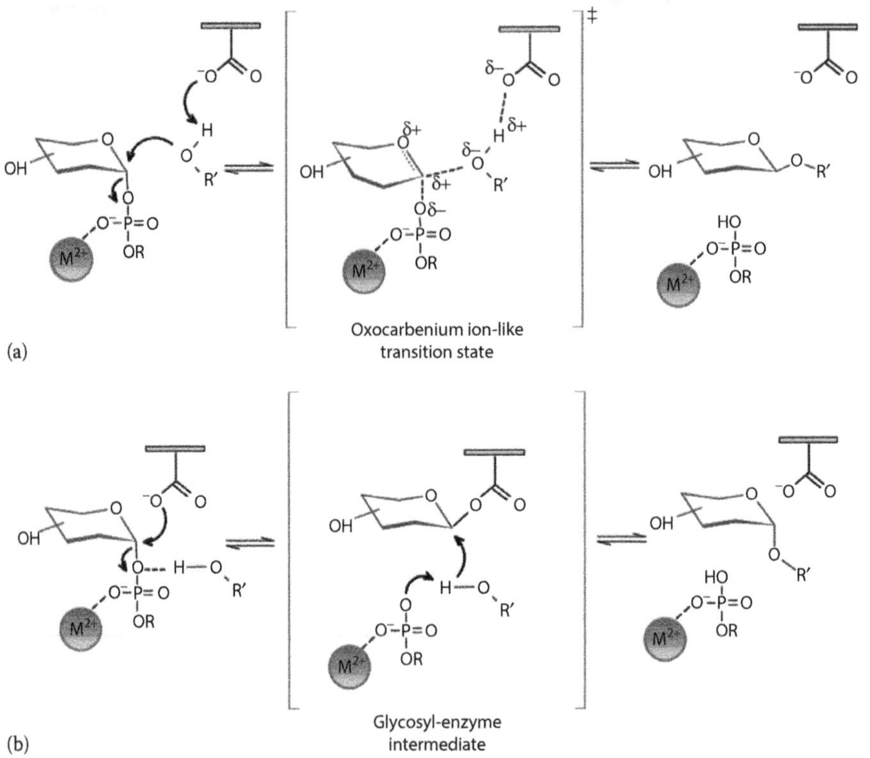

(a)

Oxocarbenium ion-like
transition state

(b)

Glycosyl-enzyme
intermediate

**FIGURE 2.5**   (a) Inverting GTs utilize a direct-displacement $S_N2$-like reaction that results in an inverted anomeric configuration via a single oxocarbenium ion-like transition state. (b) A proposed double displacement mechanism for retaining GTs involves the formation of a covalently bound glycosyl-enzyme intermediate. R, a nucleoside, a nucleoside monophosphate, a lipid phosphate, or phosphate; and R′OH, an acceptor group. (Adapted from http://www.ncbi. nlm.nih.gov/books/NBK1921/?report=reader#!po=8.33333.)

The key parameters in investigating the catalytic mechanism of inverting GTs are, therefore, the identity of the base catalyst and the method used to facilitate departure of the (substituted) phosphate-leaving group.[11]

Inverting GT-A GTs, notably those in family GT2, are involved in the formation of many β-linked polysaccharides such as cellulose, chitin, and hyaluronan. These enzymes exemplify a controversy in the polysaccharide field: Is the UDP monosaccharide the donor (in which case the growing chain extends by addition to its nonreducing end) or is a UDP-growing chain the donor with a UDP monosaccharide acting as the acceptor (which would result in a growing chain extending by addition at its reducing end)? Labeling experiments on cellulose synthase (Figure 2.6)[18] as well as on chitin synthase[19] support nonreducing end elongation, with the reducing end of the growing chains pointing away from the site of synthesis.

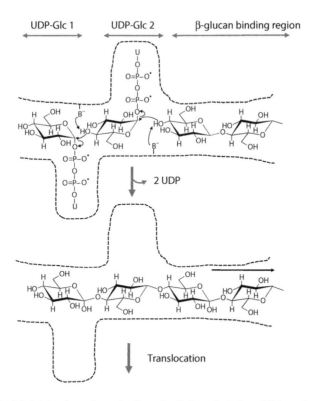

**FIGURE 2.6** Model for the polymerization of cellulose chain by addition of sugar residues at the nonreducing end. (Adapted from Breton, C. et al., *GlycoBiology*, 16, 29R, http://glycob.oxfordjournals.org/content/16/2/29R.full.pdf+html, 2006.)

## 2.2.5 MECHANISM OF RETAINING GLYCOSYLTRANSFERASES

By direct comparison to inverting GHs, the mechanism of retaining GTs has been proposed to be that of a *double displacement* mechanism involving a covalently bound glycosyl-enzyme intermediate (Figure 2.5b), demanding the existence of an appropriately positioned nucleophile within the active site.[11] A divalent cation or suitably positioned positively charged side chains or helix dipoles would presumably play the role of the Lewis acid as was described earlier for the inverting GTs. The leaving diphosphate group itself probably plays the role of a base catalyst activating the incoming acceptor hydroxyl group for nucleophilic attack. Typically, in the double displacement mechanism, an aspartic acid or glutamic acid side chain in the active site reacts with the glycosyl donor to generate an inverted glycosyl-enzyme intermediate which is in turn attacked by the glycosyl acceptor with a second inversion of stereochemistry to give overall retention of stereochemistry.[9,20]

## 2.3  GLYCOSIDE HYDROLASES

### 2.3.1  Glycoside Hydrolase Activities

Carbohydrates show wide stereochemical variation and can be assembled in many different ways.[21] Living organisms take advantage of this diversity by using oligosaccharides and polysaccharides for a multitude of biological functions, from storage and structure to highly specific signaling roles. Selective hydrolysis of glycosidic bonds is therefore crucial for energy uptake, cell wall expansion and degradation, and turnover of signaling molecules.[22,23]

GHs (EC 3.2.1) are a widespread group of enzymes that hydrolyze the glycosidic bond between two or more carbohydrates or between a carbohydrate and a noncarbohydrate moiety.[24] These enzymes can catalyze the hydrolysis of O-, N-, and S-linked glycosides (Equation 2.2).[25] They are extremely common enzymes with roles in nature including degradation of cellulose and hemicelluloses.

$$
\underset{OR}{\overset{O}{\diagdown}} + H_2O \xrightarrow{\text{Glycoside hydrolase}} \underset{OH}{\overset{O}{\diagdown}} + HOR \qquad (2.2)
$$

GHs typically can act either on α- or on β-glycosidic bonds, but not on both.[26] β-Glycosidases act on β-glycosidic bonds and α-glycosidases act on α-glycosidic bonds.

### 2.3.2  Endo- and Exo-Action

GHs have been divided into two groups, *exoglucanases* (including cellobiohydrolase and β-glucosidase activities) and *endoglucanases*, according to their respective capacity to cleave a substrate at the end (most frequently, but not always the nonreducing end) or within the middle of a chain (Figure 2.7).[25]

### 2.3.3  Active-Site Topologies

Although many protein folds are represented in glycoside hydrolase families, the overall topologies of the *active sites* fall into only three general classes.[21,27] These

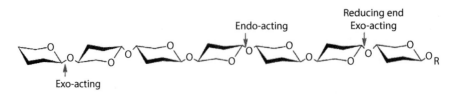

**FIGURE 2.7**  Exo- and endo-acting GH. (From Withers, S.G. and Williams, S.J., *Glycoside Hydrolases* in CAZypedia, http://www.cazypedia.org/, accessed 2016; http://www.cazypedia.org/index.php/Glycoside_hydrolases.)

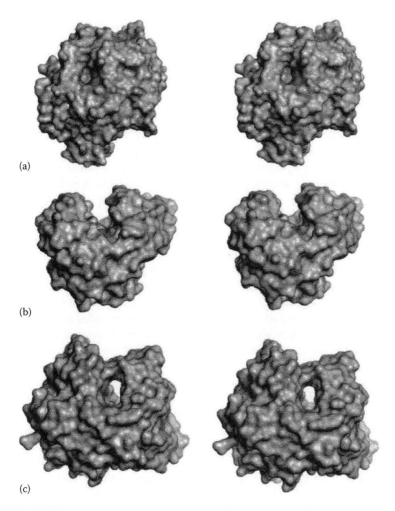

**FIGURE 2.8** The three types of active site found in glycoside hydrolases: (a) the pocket, (b) the cleft, and (c) the tunnel. (Reprinted from *PLoS One*, 8, Schuman, B. et al., e71077, 2013, Copyright 2017 with permission from Elsevier.)

three topologies (Figure 2.8) can, in principle, be built on the same fold, with the same catalytic residues. They are as follows:

. • The pocket or crater. It is optimal for the recognition of a saccharide nonreducing end and is encountered in monosaccharidases such as β-galactosidase and β-glucosidase, and in exopolysaccharidases such as glucoamylase and β-amylase.
  • The cleft or groove. It has an open structure which allows a random binding of several sugar units in polymer substrates and is commonly found in endo-acting polysaccharidases.

- The tunnel. It arises from the previous one when the protein evolves long loops that cover part of the cleft. The resulting tunnel enables a polysaccharide chain to be threaded through it.

The endoglycanases are commonly characterized by a groove into which a polysaccharide chain can fit in a random manner.[28] Usually, exoglycanases possess tunnel-like active sites, which can only accept a substrate chain via its terminal regions. The exo-acting enzymes act by threading the polymer chain through the tunnel, where successive sugar units are removed in a sequential manner. The sequential cleavage of a polymer chain is termed *processivity*. Although this division based on substrate specificity is instructive, some GHs show both modes of action.

### 2.3.4 Catalytic Mechanism

In most cases, enzymatic hydrolysis of the glycosidic bond takes place via general acid catalysis that requires two critical carboxylic acid residues of the enzyme: a general acid (proton donor) and a nucleophile/base.[21,24,29] The pair of carboxylic acid groups located in the active site of the enzyme is typically provided by either aspartic acid or glutamic acid. Depending on the spatial position of these catalytic residues and hence on the sequence, hydrolysis occurs via overall retention or overall inversion of the anomeric configuration (Figure 2.9).[30,31] This involves the conservation of the mechanism within each sequence-based family.

In both the retaining and the inverting mechanisms, the position of the proton donor is identical; in other words, it is within hydrogen bonding of the glycosidic oxygen.[21] In retaining enzymes, the nucleophilic catalytic base is in close vicinity of the sugar anomeric carbon. This base, however, is more distant in inverting enzymes which must accommodate a water molecule between the base and the sugar. This difference results in an average distance between the two catalytic residues of ~5.5 Å in remaining enzymes as opposed to 10 Å in inverting enzymes.

In the retaining mechanism, the two critical amino acid residues are involved in a two-step catalysis in which a covalent enzyme substrate is formed and then hydrolyzed (Figure 2.10).[32] Each step passes through an oxocarbenium-like transition state.[25] In the first step, one of the residues performs a nucleophilic attack at the sugar anomeric carbon, while the other residue functions as an acid/base and assists aglycon departure by protonation of the glycosidic oxygen. The result of this first step is the formation of a covalent glycosyl-enzyme intermediate. In the second step, the deprotonated acid/base residue abstracts a proton from a water molecule, which attacks the glycosyl-enzyme to release a sugar with a stereochemistry identical to that of the substrate.

Hydrolysis of a GH[33] with net inversion of anomeric configuration is generally achieved via a one-step, single-displacement mechanism involving oxocarbenium

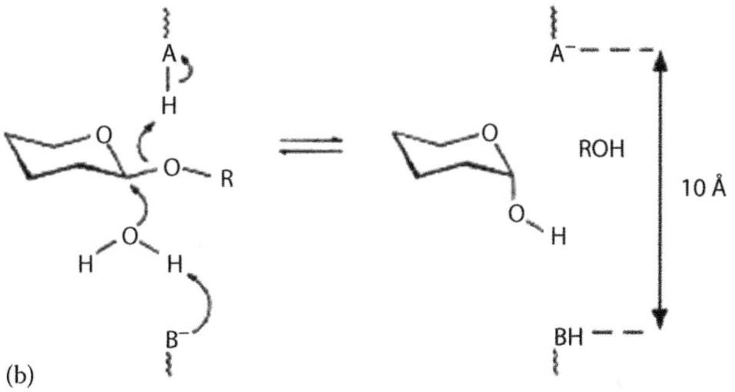

**FIGURE 2.9** The two major mechanisms of enzymatic glycosidic bond hydrolysis. (a) The *retaining* mechanism, in which the glycosidic oxygen is protonated by the acid catalyst (AH) and nucleophilic assistance to aglycone departure is provided by the base B⁻. The resulting glycosyl enzyme is hydrolyzed by a water molecule and this second nucleophilic substitution at the anomeric carbon generates a product with the same stereochemistry as the substrate. (b) The *inverting* mechanism, in which protonation of the glycosidic oxygen and aglycon departure are accompanied by a concomitant attack of a water molecule that is activated by the base residue (B⁻). This single nucleophilic substitution yields a product with opposite stereochemistry to the substrate. (Reprinted from *PLoS One*, 8, Schuman, B. et al., e71077, 2013, Copyright 2017 with permission from Elsevier.)

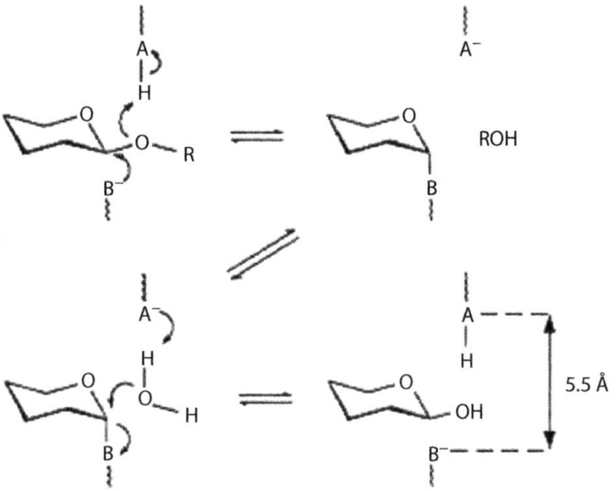

ion-like transition states (Figure 2.11).[34] The reaction typically occurs with general acid and general base assistance from two amino acid side chains. In this mechanism, the proton transfer from the catalyst promotes the leaving group departure, while the catalytic base deprotonates a water molecule for nucleophilic substitution at the anomeric center.

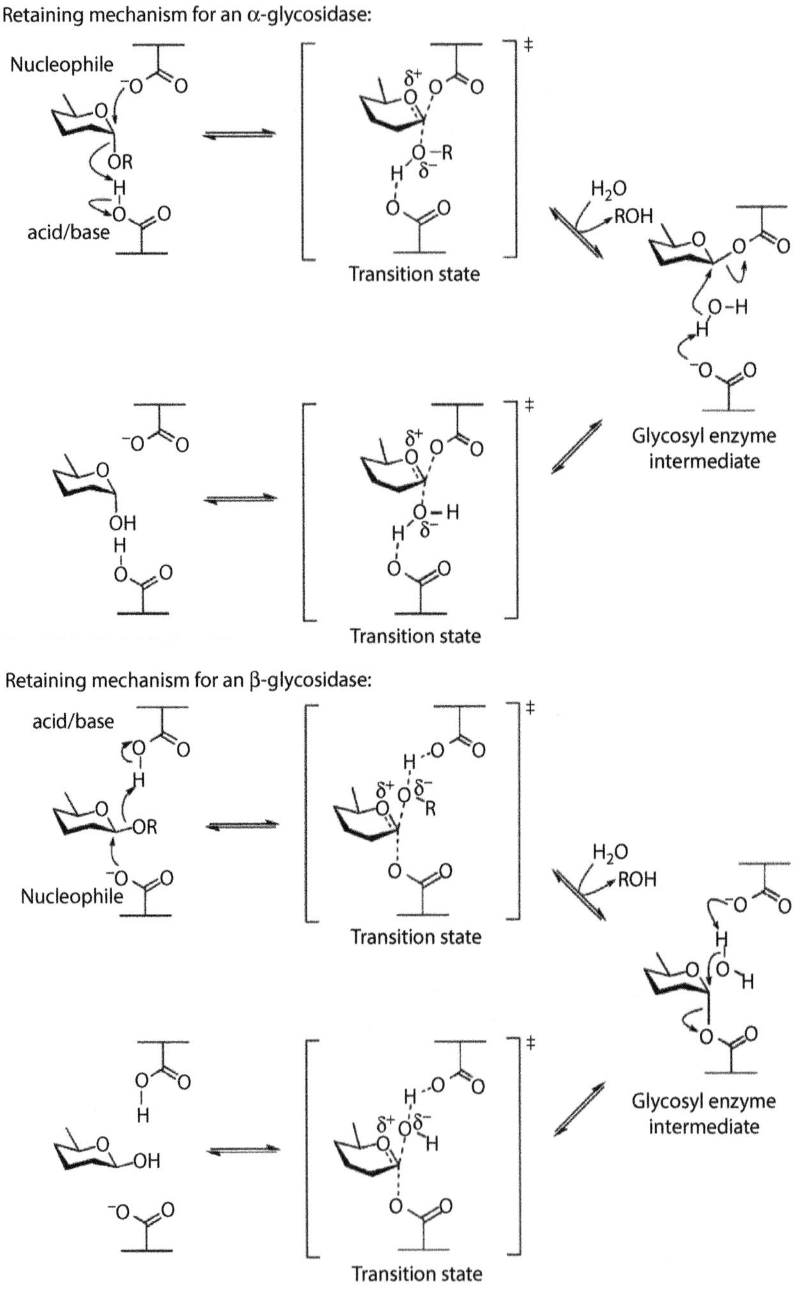

Retaining mechanism for an α-glycosidase:

FIGURE 2.10  Retaining mechanism for an α-glycosidase and for a β-glycosidase. (From Withers, S.G. and Williams, S.J., *Glycoside Hydrolases* in CAZypedia, http://www.cazypedia. org/, accessed 2016; https://www.cazypedia.org/index.php/Glycoside_hydrolases#Overview.)

Inverting mechanism for an α-glycosidase:

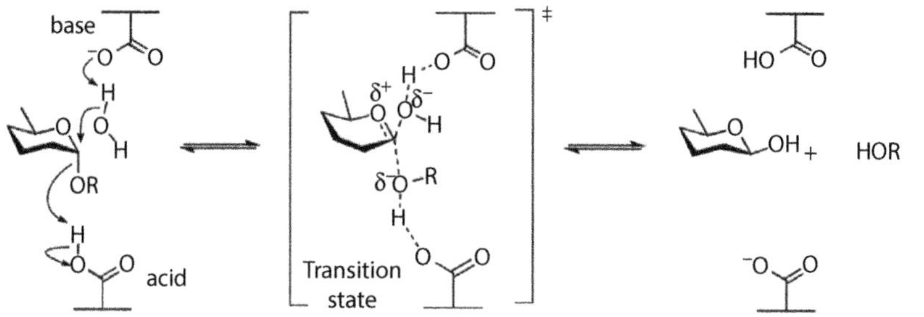

Inverting mechanism for an β-glycosidase:

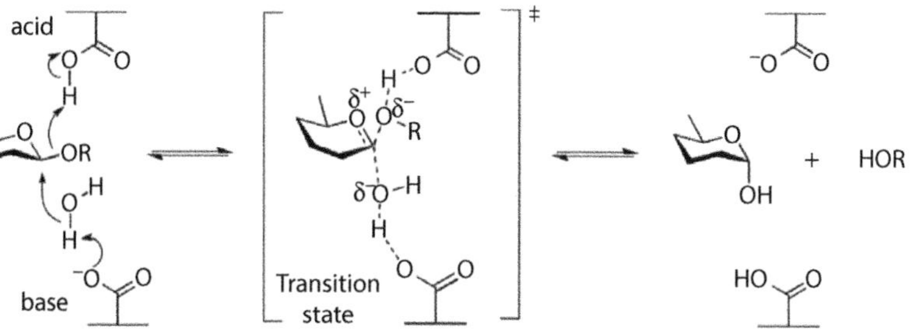

**FIGURE 2.11** Inverting mechanism for an α-glycosidase and for a β-glycosidase. (From Withers, S.G. and Williams, S.J., *Glycoside Hydrolases in CAZypedia*, http://www.cazypedia. org/, accessed 2016; http://www.cazy.org/Glycoside-Hydrolases.html.)

## 2.3.5 Catalytic Modules

### 2.3.5.1 Classification of Various Glycoside Hydrolase Families

GHs can be classified into families according to the amino acid sequence similarities of their catalytic modules (Table 2.1).[21] Because there is a direct relationship between sequence and folding similarities, such a classification reflects the structural features of these enzymes better than their sole substrate specificity.[24] Table 2.1 reports the various GH families, their clan characterized by the fold of proteins, the fold of the catalytic module, the type of enzymes, the organism source, and the mechanism are known.[24,35]

In this classification, enzymes with different substrate specificities are sometimes found in the same family, indicating an evolutionary divergence to acquire new specificities.[36] On the other hand, enzymes that hydrolyze the same substrate are sometimes found in different families.

Cellobiohydrolases (CBHs) display loops that cause their catalytic centers to lie within enclosed tunnels. This topology allows these enzymes to release the product while remaining bound to the polysaccharide chain, thereby creating the conditions for processivity.

It can be seen from the table that family GH7 contains only fungal enzymes, whereas family GH8 contains only bacterial enzymes. CBHs are in GH families 6, 7, and 44.

In November 2015, the CAZy database included 14 clans and 135 GH families (Table 2.2).[24] Main fold of representative GH families is represented in Figure 2.12.

## TABLE 2.1
## Fold of the Catalytic Module, Enzymes, Organism Source and Mechanism for Various GH Families (clans)

| GH Family (clan) | Fold | Enzymes | Organism Source | Mechanism[b] |
|---|---|---|---|---|
| 5 (A) | $(\beta/\alpha)_8$ barrel | Mainly endoglucanases | Bacteria, fungi | Retaining |
| 6 | Modified $\beta/\alpha$ barrel | Endoglucanases and CBHs[a] | Bacteria, fungi | Inverting |
| 7 (B) | $\beta$-jelly roll | Endoglucanases and CBHs | Fungi | Retaining |
| 8 (M) | $(\alpha/\alpha)_6$ barrel | Mainly endoglucanases | Bacteria | Inverting |
| 9 | $(\alpha/\alpha)_6$ barrel | Mainly endoglucanases | Bacteria, fungi | Inverting |
| 10 (A) | $(\beta/\alpha)_8$ barrel | Xylanases | Bacteria, Eukaryota | Retaining |
| 12 (C) | $\beta$-jelly roll | Endoglucanases | Bacteria, fungi | Retaining |
| 45 | $\beta$ barrel | Endoglucanases | Bacteria, fungi | Inverting |
| 48 (M) | $(\alpha/\alpha)_6$ barrel | Processive endoglucanases and/or CBH | Bacteria | Inverting |

*Source:*  Wertz, J.L. and Bedue, O., *Lignocellulosic Biorefineries*, EPFL Press, p. 192, Table 5.1, 2013; Bayer, E.A. et al., *Curr. Opin. Struct. Biol.*, 8, 548, 1998.

[a]  CBH: cellobiohydrolase.

[b]  retention or inversion of the configuration of the anomeric carbon.

## TABLE 2.2
## GH Clans of Related Families

| Clan | Fold | GH Families |
|---|---|---|
| GH-A | $(\beta/\alpha)_8$ barrel | 1, 2, 5, 10, 17, 26, 30, 35, 39, 42, 50, 51, 53, 59, 72, 79, 86, 113, 128 |
| GH-B | $\beta$-jelly roll | 7, 16 |
| GH-C | $\beta$-jelly roll | 11, 12 |
| GH-D | $(\beta/\alpha)_8$ Barrel | 27, 31, 36 |
| GH-E | 6-fold $\beta$-propeller | 33, 34, 83, 93 |
| GH-F | 5-fold $\beta$-propeller | 43, 62 |
| GH-G | $(\alpha/\alpha)_6$ barrel | 37, 63 |
| GH-H | $(\beta/\alpha)_8$ barrel | 13, 70, 77 |
| GH-I | $\alpha + \beta$ | 24, 46, 80 |
| GH-J | 5-fold $\beta$-propeller | 32, 68 |
| GH-K | $(\beta/\alpha)_8$ barrel | 18, 20, 85 |
| GH-L | $(\alpha/\alpha)_6$ barrel | 15, 65, 125 |
| GH-M | $(\alpha/\alpha)_6$ barrel | 8, 48 |
| GH-N | $\beta$-helix | 28, 49 |

*Source:*  http://www.cazy.org/Glycoside-Hydrolases.html

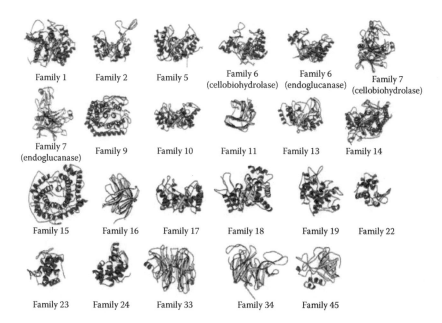

**FIGURE 2.12** Ribbon representation of the main fold of the catalytic domain in various glycosyl hydrolase families (see Table 2.2). β strands are shown in cyan and α-helices in red. (From Schuman, B. et al., *PLoS One*, 8, e71077, 2013.)

The clan GH-A is the biggest clan superfamily and includes 19 GH families.[24] It is defined by three features: (1) $(\beta/\alpha)_8$ barrel fold, (2) two catalytic residues that are located at the ends of the fourth and seventh β strands of the barrel, and (3) hydrolysis through a retaining mechanism.[38] It is also termed as the 4/7 superfamily because of the positions of catalytic residues.

### 2.3.5.2 Folds of the Catalytic Module

Catalytic modules of GH families display various folds including[39–42]

1. $(\alpha/\alpha)_6$ *barrels*, consisting of six inner and six outer *α-helices* forming a barrel-like structure, such as in clan GH-G, L, and M.[43]
2. $(\beta/\alpha)_8$ *barrels*, or TIM (triose-phosphate isomerase) barrels, consisting of eight repeating units of β/α module, in which the eight β *strands* form an inner parallel *β-sheet* arranged in a barrel structure, which is surrounded by the eight α-helices, such as in clan GH-A.[44]
3. β *barrels*, consisting of a β-sheet that twists and coils to form a closed structure, such as in GH45.
4. β *sandwiches* in which two β-sheets pack together, face-to-face, in a layered arrangement, such as in GH domains forming a β sandwich with a jelly roll topology.[45]
5. *Greek key* topologies, in which typically three anti-parallel β strands connected by hairpins are followed by a longer connection to the fourth strand, which lies adjacent to the first.[46]

6. *Jelly rolls*, variants of Greek key topology with both ends of a β sandwich or a β barrel fold being crossed by two interstrand connections, such as in clans GH-B and C.[46–48]

7. β-*propellers*, a type of all β protein architecture characterized by 4 to 8 blade-shaped β-sheets arranged toroidally around a central axis, such as in clans GH-E, F, and J.

8. β-*helix*, a protein structure formed by the association of parallel β strands in a helical pattern with either two or three faces, such as in clan GH-N.

### 2.3.6 CARBOHYDRATE-BINDING MODULE

A carbohydrate-binding module (CBM) is defined as a contiguous amino acid sequence within a carbohydrate-active enzyme with a discreet fold having carbohydrate-binding activity.[49] A few exceptions are CBMs of cellulosomal *scaffoldin* proteins and rare instances of independent putative CBMs. Most cellulases have a catalytic module and a CBM joined by a highly glycosylated and presumably flexible linker peptide.[50] Removal of the CBM results in a significantly reduced enzymatic activity on crystalline cellulose probably due to a decreased binding capacity, but the activity on soluble cellulose oligomers is retained. Like the catalytic modules, the CBMs also form distinct families of related amino acid sequences (Table 2.3).[35] Representative structures for each CBM family have been elucidated either by crystallography or by NMR spectroscopy.[37]

---

**TABLE 2.3**
**Representative Carbohydrate-Binding Modules (CBMs) of Families 1, 2, and 3**

| Family | 1 | 2a[a] | 3a | 3c |
|---|---|---|---|---|
| Organism | *Trichoderma reesei* | *Cellulomonas fimi* | *Clostridium thermocellum* | *Thermomonospora fusca* |
| Protein | Cel7A (GH7)[b] | Exoglucanase Cex (GH10)[b] | Cellulosomal scaffoldin CipA | Endoglucanase E4 (GH9)[b] |
| Size[c] | 36 | 110 | 155 | 143 |
| Dominant fold | Three-stranded anti-parallel β-sheet | Nine-stranded β barrel | Nine-stranded β sandwich | Nine-stranded β sandwich |
| Preferred substrate | Crystalline cellulose | Crystalline cellulose | Crystalline cellulose | Single cellulose chain |
| Proposed binding | Three coplanar aromatic residues[51] | Three coplanar aromatic residues[52,53] | Coplanar aromatic and polar residues[54,55] | Coplanar aromatic and polar residues[55–57] |

*Source:* Bayer, E.A. et al., *Curr. Opin. Struct. Biol.*, 8, 548, 1998; CAZy Carbohydrate-Binding Module Family Server. www.cazy.org/fam/acc_CBM.html.

[a] CBM2a binds cellulose, whereas CBM2b binds xylan.[52,53]

[b] glycoside hydrolase family.

[c] number of amino acid residues.

---

The family 1 CBMs (CBM1s) of ~40 residues are found almost exclusively in fungi, whereas the CBM2s of ~100 residues, as well as the CBM3s of ~150 residues, are found in bacteria. CMB1 is found either at the N-terminal or at the C-terminal extremity of the enzymes known to contain such a domain.[52] The CBM2s can be classified in two subfamilies according to substrate specificities: CMB2a, which binds cellulose, and CMB2b, which interacts specifically with xylans.[52] CMB2 is found either at the N-terminal or at the C-terminal extremity of cellulases and *xylanases*. The CBM3s are separated into two functionally different types. One type (family 3a and 3b) binds strongly to crystalline cellulose, whereas another (family 3c) fails to bind crystalline cellulose but serves in a helper role in the hydrolysis of a single cellulose chain by feeding it into the active site of the neighboring catalytic module.[51,56–58] The families 3a (cellulosomal scaffoldins) and 3b (mainly free enzymes) are closely similar in their primary structures. The CMB3 domain is mainly found in C-terminal to the catalytic domain, which corresponds to a wide range of bacterial GH-like families 9, 5, and 10.[56]

The major function of the CBMs is to deliver the catalytic domain to the crystalline substrate. The binding appears extremely stable, although the enzyme may undergo a lateral diffusion on the substrate. Some CBMs also appear to disrupt the noncovalent interactions between the chains of the crystalline substrate, while others bind preferentially to noncrystalline substrates.[37]

In July 2017, the CAZY database collated 81 CBM families.[49]

### 2.3.7 Nomenclature

A nomenclature scheme has been developed for designating glycoside hydrolases.[59,60] In accordance with standard practice in bacterial genetics, the genes and their products are designated by three letters. If the enzymes are named according to their preferred substrates, then the designations would be as given in Table 2.4. Note that there must be strict correspondence between the letters for gene and protein.

**TABLE 2.4**
**Acronyms for GH Genes and Encoded Enzymes**

| Enzyme | Gene | Protein | EC Designation |
| --- | --- | --- | --- |
| Cellulase | *Cel* | Cel | EC 3.2.1.4; EC 3.2.1.91 |
| Xylanase | *xyn* | Xyn | EC 3.2.1.8 |
| Mannanase | *Man* | Man | EC 3.2.1.78 |
| Lichenase | *lic* | Lic | EC 3.2.1.73; EC 3.2.1.58 |
| Laminarinase | *lam* | Lam | EC 3.2.1.39 |

*Source:* Henrissat, B., Teeri, T.T., and Warren, R.A.J., *FEBS Letters*, 425, 352, http://www.sciencedirect.com/science/article/pii/S0014579398002658, 1998.

These three letters are followed by the family number of the GH enzyme (e.g., Cel5). If an organism produces multiple enzymes from the same family, these will be designated by, for example, Cel5A, Cel5B, with the final capital letters indicating the order in which the enzymes were first reported.

As the scheme does not distinguish between endo- and exo-action, particular enzymes will be referred to as endoglucanase CelA, cellobiohydrolase Cel6A, and so on. Furthermore, two similar enzymes from different organisms will be differentiated by indicating the organism of origin (e.g., CfCel9B from *Cellulomonas fimi* and TfCel9B from *Thermomonospora fusca*).

A nomenclature scheme has been also developed for designating sugar-binding subsites in GHs.[61] It enables to demarcate substrate binding in GH active sites. Subsites are labeled from −n to +n (where n is an integer). −n represents the nonreducing end and +n represents the reducing end, with cleavage taking place between the −1 and +1 subsites. The scissile bond itself belongs to the ring in subsite −1.[62] Thus, enzyme subsites toward the reducing end of the substrate are labeled +1, +2, +3, and so on, and those toward the nonreducing end, away from the point of cleavage, are −1, −2, −3, and so on. In this system, −1, −2, −3, and so on are the *glycone* (sugar moiety of a glycoside) subsites and +1, +2, +3, and so on are the *aglycone* (nonsugar moiety of a glycoside) subsites on each side of the bond being hydrolyzed (Figure 2.13).[61]

Subsite −1 is the most important, as it holds the sugar ring that bears the scissile bond and contains the anomeric carbon atom that is the center of the hydrolysis reaction.[62] The positive subsites are the leaving group (the alcohol that departs with one electron pair) and the negative subsites are the reducing-end group.[63]

The hydrolysis of a glycosidic bond can be seen as a nucleophilic substitution (Equation 2.3) at the anomeric carbon where water is the nucleophile (Nuc), and the leaving group (LG) is an alcohol (HOR).[64] The electron pair (:) from the nucleophile attacks the substrate (R-LG) forming a new bond, while the leaving group departs with an electron pair.[65]

$$\text{Nuc:} + \text{R-LG} \rightarrow \text{R-Nuc} + \text{LG:} \hspace{3cm} (2.3)$$

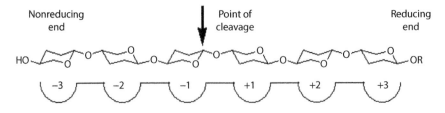

**FIGURE 2.13** Subsite nomenclature. (From Withers, S.G. and Williams, S.J., Glycoside Hydrolases in CAZypedia, http://www.cazypedia.org/, accessed 2016; Petersen, L., Catalytic strategies of glycoside hydrolases, Thesis, Iowa State University, http://lib.dr.iastate.edu/cgi/viewcontent.cgi?article=2086&context=etd, 2010.)

## 2.4  ESTERASES

### 2.4.1  OVERVIEW

A hydrolase (*EC 3*) is an enzyme that catalyzes the hydrolysis of a chemical bond such as C–O, C–N, C–C and some other specific bonds, (Equation 2.4).[23,66,67]

$$A\text{-}B + H_2O \rightarrow A\text{-}OH + B\text{-}H \tag{2.4}$$

The second figure in the code number of the hydrolases indicates the nature of the bond hydrolyzed; *EC 3.1* are the esterases, *EC 3.2* are the glycosylases, and so on. The third figure normally specifies the nature of the substrate, for example, in the esterases, the carboxylic ester (RCOOR) hydrolases (*EC 3.1.1*) and thiol ester (RCOSR) hydrolases (*EC 3.1.2*).

Carboxylesterase (*EC 3.1.1.1*; systematic name: carboxylic ester hydrolase) catalyzes the following reaction (Equations 2.5/2.6):[68]

$$\text{A carboxylic ester} + H_2O \rightarrow \text{an alcohol} + \text{a carboxylate} \tag{2.5}$$

$$RCOOR + H_2O \rightarrow ROH + RCOOH \tag{2.6}$$

Arylesterase (*EC 3.1.1.2*; systematic name: aryl-ester hydrolase) catalyzes the following reaction (Equations 2.7/2.8):[69,70]

$$\text{A phenyl acetate} + H_2O \rightarrow \text{a phenol} + \text{acetate} \tag{2.7}$$

$$C_6H_5COOCH_3 + H_2O \rightarrow C_6H_5OH + CH_3COOH \tag{2.8}$$

Triacylglycerol lipase, also called triglyceride lipase or lipase (*EC 3.1.1.3*; systematic name: triacylglycerol acylhydrolase), catalyzes the hydrolysis of triesters of fatty acids (Equations 2.9/2.10):[71–73]

$$\text{Triacylglycerol (triglycerides)} + H_2O \rightarrow \text{diacylglycerol}$$
$$+ \text{a carboxylate (a fatty acid)} \tag{2.9}$$

$$CH_2COOR\text{-}CHCOOR\text{-}CH_2COOR$$
$$+ H_2O \rightarrow CH_2COOR\text{-}CHOH\text{-}CH_2COOR + RCOOH \tag{2.10}$$

Acetylesterase (*EC 3.1.1.6*; systematic name: acetic ester acetylhydrolase) catalyzes the following hydrolysis (Equations 2.11/2.12):[74]

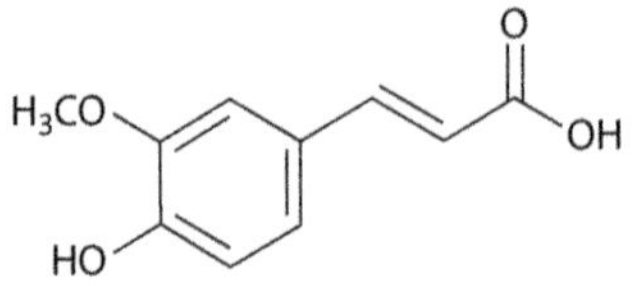

**FIGURE 2.14**   Ferulic acid.

$$\text{Acetic ester} + H_2O \rightarrow \text{alcohol} + \text{acetate} \tag{2.11}$$

$$CH_3COOR + H_2O \rightarrow ROH + CH_3COOH \tag{2.12}$$

Acetylxylan esterase (*EC 3.1.1.72*; systematic name: acetylxylan esterase) catalyzes the deacetylation of xylans and xylooligosaccharides.[75–77]

Feruloyl esterase (*EC 3.1.1.73*; systematic name: 4-hydroxy-3-methoxycinnamoyl-sugar hydrolase) catalyzes the hydrolysis of feruloyl-polysaccharide, releasing ferulate (Figure 2.14) and polysaccharide (Equation 2.13):[78,79]

$$\text{Feruloyl-polysaccharide} + H_2O \rightarrow \text{ferulate} + \text{polysaccharide} \tag{2.13}$$

### 2.4.2   CARBOXYLESTERASES

Carboxyl ester hydrolases are ubiquitous enzymes.[80] In the presence of water, they catalyze the hydrolysis of an ester bond resulting in the formation of an alcohol and a carboxylic acid. However, in an organic solvent, they can catalyze the reverse reaction or a transesterification reaction.

Most carboxylic ester hydrolases, which are cofactor-independent enzymes, conform to a common structural organization: the α/β-hydrolase fold, which is also present in many other hydrolytic enzymes such as proteases, dehalogenases, peroxidases, and epoxide hydrolases.[80] The canonical α/β-hydrolase fold consists of an eight-stranded mostly parallel β-sheet, with the second strand anti-parallel. The parallel strands, β3 to β8, are connected by helices, which pack on either side of the central β-sheet. The sheet is highly twisted and bent so that it forms a half-barrel. The active site contains a conserved catalytic triad. The catalytic triad is usually composed of a nucleophilic serine in a GXSXG pentapeptide motif (where X is any residue), and an acidic residue (aspartate or glutamate) that is hydrogen bonded to a histidine residue. The substrate-binding site is located inside a pocket on top of the central β-sheet that is typical of this fold. The size and shape of the substrate-binding cleft have been related to substrate specificity.

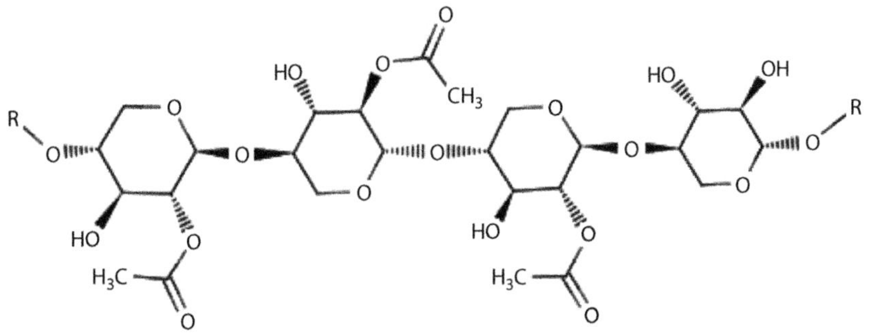

**FIGURE 2.15** Acetylated xylan. (From Brenda, the Comprehensive Enzyme Information System. http://www.brenda-enzymes.org/Mol/Mol.php4?n=207724&compound=acetylate d&s_type=5&back=1&limit_start=20.)

### 2.4.3 CARBOHYDRATE ESTERASES

Carbohydrate esterases promote the de-$O$ or de-$N$-acylation of substituted saccharides.[81,82] Since a carboxylic ester is formed by an acid and an alcohol, two classes of substrates for carbohydrate hydrolases might be considered: those in which sugar plays the role of the acid, such as pectin methyl esters, and those in which sugar behaves as the alcohol, such as in acetylated xylan (Figure 2.15).[83]

A number of possible reaction mechanisms could be involved: the most common is a serine (S)–histidine (H)–aspartate (D) catalytic triad catalyzed deacetylation analogous to the action of classical lipase and serine proteases.[81] A catalytic triad usually refers to the three amino acid residues that function together at the center of the active site of certain hydrolase and transferase enzymes.[85] A common method for generating a nucleophilic residue for covalent catalysis is by using an acid–base–nucleophile triad. The residues form a charge-relay network to polarize and activate the nucleophile, which attacks the substrate, forming a covalent intermediate which is then hydrolyzed to regenerate free enzyme. Other mechanisms such as a $Zn^{2+}$ catalyzed deacetylation might also be considered for some families.[81,86]

Through evolution, enzymes have acquired the ability to attack ester linkages between hydroxycinnamic acids and carbohydrates in the process of biodegradation of plant cell walls (Figures 2.16 and 2.17).[83]

Cinnamic acid derivatives such as ferulic or coumaric acids are covalently bound to the polysaccharides through ester linkages. The action of depolymerases is often limited by the presence of esterifications, which need to be removed prior to depolymerization.[88] Thus, esterases are considered important accessory enzymes. Ferulic acid is the major phenolic acid esterified to carbohydrates in the plant cell wall.

The CAZy database contains 16 carbohydrate esterase families.[81] Known activities for each family are summarized in Table 2.5.[89]

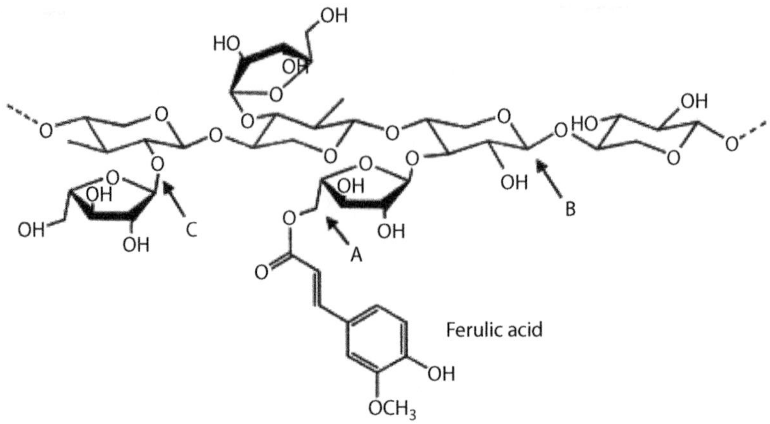

**FIGURE 2.16** Structure of ferulic acid (FA) esterified to arabinoxylan. (A) FA linked to O-5 of arabinose chain of arabinoxylan. (B) β-1,4-linked xylan backbone. (C) α-1,2-linked L-arabinose. (Reprinted from *Mol. Plant*, 2, De M.M. and Buanafina, O., 861, 2009, Copyright 2017 with permission from Elsevier.)

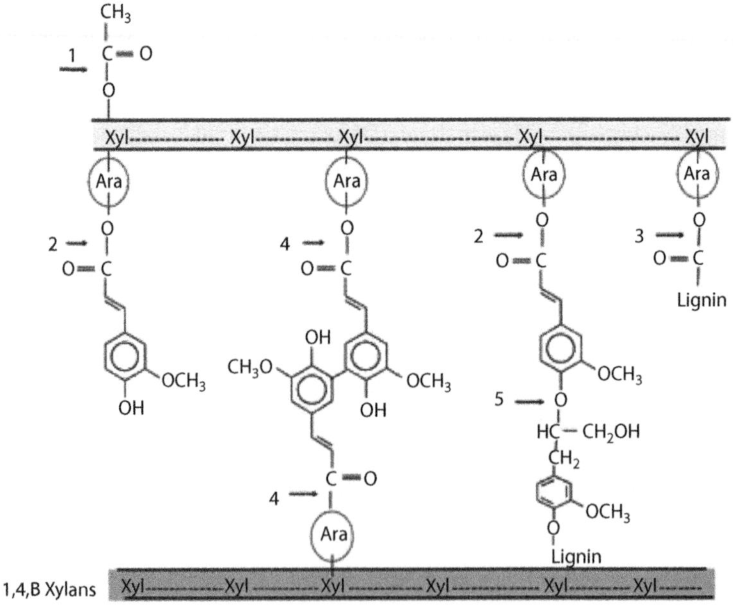

**FIGURE 2.17** Simplified structure of ferulic acid (FA) cross-linking arabinoxylan (AX) in grass cell walls. The 1,4-β-linked xylan backbone is represented by dotted lines and the side sugars, arabinose (Ara), are shown in circles. Ester and ether bonds are shown by arrows as follows: (1) acetyl group, (2) ester linked FA to AX, (3) arabinose-lignin, (4) 5–5 –ester linked FA dimer cross-linking AX chains, and (5) FA ether linked to lignin. (Modified from Williamson et al. (1998). Reprinted from *General introduction: catalysis by Ser-His-Asp catalytic triads*, Thesis, Chapter 1, http://dissertations.ub.rug.nl/FILES/faculties/science/2004/t.r.m.barends/c1.pdf. Copyright 2017 with permission from Elsevier.)

## TABLE 2.5
## Carbohydrate Esterase (CE) Family Classification

| CE Family | Some Known Activities | 3D Structure |
|---|---|---|
| CE1 | Acetyl xylan esterase. Cinnamoyl esterase. Feruloyl esterase. Carboxyl esterase | $\alpha/\beta/\alpha$ sandwich |
| CE2 | Acetyl xylan esterase | |
| CE3 | Acetyl xylan esterase | $\alpha/\beta/\alpha$ sandwich |
| CE4 | Acetyl xylan esterase. Chitin deacetylase | $(\beta/\alpha)_7$ barrel |
| CE5 | Acetyl xylan esterase. Cutinase | $\alpha/\beta/\alpha$ sandwich |
| CE6 | Acetyl xylan esterase | $\alpha/\beta/\alpha$ sandwich |
| CE7 | Acetyl xylan esterase Cephalosporin-C deacetylase | $\alpha/\beta/\alpha$ sandwich |
| CE8 | Pectin methylesterase | $\beta$-helix |
| CE9 | N-acetylglucosamine-6-phosphate deacetylase | $(\beta/\alpha)_8$ barrel |
| CE10 | Arylesterase. Carboxyl esterase. Acetylcholinesterase. Cholinesterase. Sterol esterase | $\alpha/\beta/\alpha$ sandwich |
| CE11 | UDP-3-o-acyl N-acetylglucosamine deacetylase | 2-layer-sandwich |
| CE12 | Pectin acetylesterase. Rhamnogalacturonan acetylesterase. Acetyl xylan esterase | $\alpha/\beta/\alpha$ sandwich |
| CE13 | Pectin acetylesterase | $\alpha/\beta/\alpha$ sandwich |
| CE14 | N-acetyl-1-D-myo-inosityl-2-amino-2-deoxy-$\alpha$-D-glucopyranoside deacetylase. Diacetylchitobiose deactylase. Mycothiol S-conjugate amidase | $\alpha/\beta$ fold |
| CE15 | 4-O-methyl-glucuronoyl methylesterase | |
| CE16 | Acetylesterase active on various carbohydrate acetyl esters | |

*Source:* http://www.cazy.org/Carbohydrate-Esterases.html

## REFERENCES

1. http://www.cazy.org/Welcome-to-the-Carbohydrate-Active.html
2. S. Mohanram, D. Amat, J. Choudhary, A. Arora, and L. Nain, Sustainable Chemical Processes 2013, 1:15. http://www.sustainablechemicalprocesses.com/content/1/1/15
3. Global Biobusiness <newsletter@globalbiobusiness.com>
4. Novozymes in http://www.bioenergy.novozymes.com/en/cellulosic-ethanol/Cellic-HTec3/Documents/CE_APP_Cellic_Ctec3.pdf
5. M. Dashtban, H. Schraft, T.A. Syed, and W. Qin, *Int. J. Biochem. Mol. Biol.*, 1, 36, 2010. http://www.ncbi.nlm.nih.gov/pmc/articles/PMC3180040/.
6. http://www.cazy.org/
7. B. Henrissat and G.J. Davies, Plant Physiol. 124, 1515, 2000. http://www.plantphysiol.org/content/124/4/1515.full
8. http://www.cazypedia.org/index.php/Glycosyltransferases

9. J. Rini, J. Esko and A. Varki, *Essentials of Glycobiology*, 2nd edition, 2009. http://www.ncbi.nlm.nih.gov/books/NBK1921/

10. http://www.ncbi.nlm.nih.gov/books/NBK1921/?report=reader#!po=8.33333

11. L.L. Lairson, B. Henrissat, G.J. Davies, and S.G. Whiters, *Annu. Rev. Biochem.*, 77, 521, 2008. http://www.ncbi.nlm.nih.gov/pubmed/18518825.

12. P.M. Coutinho, E. Deleury, G.J. Davies, and B. Henrissat, *J. Mol. Biol.* 328, 307, 2003. http://www.ncbi.nlm.nih.gov/pubmed/12691742

13. http://www.cazy.org/GlycosylTransferases.html

14. http://en.wikipedia.org/wiki/Rossmann_fold

15. S.J. Charnock and G.J. Davies, *Biochemistry* 38, 6380, 1999. http://pubs.acs.org/doi/abs/10.1021/bi990270y.

16. A. Vrielink, W. Ruger, D.H.P. Driessen, and P.S. Freemont, *EMBO J.* 13, 3413, 1994. http://www.ncbi.nlm.nih.gov/pubmed/8062817?dopt=Abstract

17. C. Breton, L. Snajdrova, C. Jeanneau, J. Koca, and A. Imberty, *GlycoBiology* 16, 29R, 2006. http://glycob.oxfordjournals.org/content/16/2/29R.full.pdf+html

18. M. Koyama, W. Helbert, T. Imai, J. Sugiyama, and B. Henrissat, *Proc. Natl. Acad. Sci. USA* 94, 9091, 1997. http://www.ncbi.nlm.nih.gov/pmc/articles/PMC23045/

19. T. Imai, T. Watanabe, T. Yui, and J. Sugiyama, *Biochem. J.* 374, 755, 2003. http://www.ncbi.nlm.nih.gov/pmc/articles/PMC1223643/pdf/12816541.pdf

20. B. Schuman, S.V. Evans, and T.M. Fyles, *PLoS One* 8(8), e71077, 2013. doi:10.1371/journal.pone.0071077. http://www.plosone.org/article/info:doi/10.1371/journal.pone.0071077

21. G. Davies and B. Henrissat, *Structure* 3, 853, 1995. http://www.sciencedirect.com/science/article/pii/S0969212601002209

22. http://www.chem.qmul.ac.uk/iubmb/enzyme/

23. http://www.chem.qmul.ac.uk/iubmb/enzyme/rules.html

24. http://www.cazy.org/Glycoside-Hydrolases.html

25. http://www.cazypedia.org/index.php/Glycoside_hydrolases

26. http://en.wikipedia.org/wiki/Glycosidic_bond

27. F. Cuskin, Mechanisms by which Glycoside Hydrolases Recognize Plant, Bacterial and yeast Polysaccharides, Thesis, Newcastle University, 2012.

28. A. Grassick, P.G. Murray, R. Thompson, C.M. Collins, L. Byrnes, G. Birrane, T.M. Higgins, and M.G. Tuohy, *Eur. J. Biochem.* 271, 4495, 2004. http://www.ncbi.nlm.nih.gov/pubmed/15560790

29. B. Henrissat, I. Callebaut, S. Fabrega, P. Lehn, J.P. Mornon, and G. Davies, *Proc. Natl. Acad. Sci. USA* 92, 7090, 1995. http://www.pnas.org/content/92/15/7090.long

30. D.E. Koshland, *Biol. Rev. Camb. Philos. Soc.* 28, 416, 1953.

31. B. Henrissat and G. J. Davies, *Plant Physiol.* 124, 1515, 2000.

32. D.A. Brooks, S. Fabrega, L.K. Hein, E.J. Parkinson, P. Durand, G. Yogalingam, U. Matte et al. *Glycobiol.* 11, 741, 2001. http://glycob.oxfordjournals.org/content/11/9/741.long

33. Nomenclature Committee of the International Union of Biochemistry and Molecular Biology (NC-IUBMB) http://www.chem.qmul.ac.uk/iubmb/enzyme/EC3/2/1/.

34. CAZypedia, *Glycoside Hydrolases.* http://www.cazypedia.org/index.php/Glycoside_hydrolases.

35. J.L. Wertz and O. Bedue, *Lignocellulosic Biorefineries*, EPFL Press, 2013.

36. B. Henrissat and G.J. Davies, *Plant Physiol.* 124, 1515, 2000.

37. E. A. Bayer, H. Chanzy, R. Lamed, and Y. Shoham, *Curr. Opin. Struct. Biol.* 8, 548, 1998.

38. Y. Kitago, S. Karita, N. Watanabe, M. Kamiya, T. Aizawa, K. Sakka, and I. Tanaka, *J. Biol. Chem.* 282, 35703, 2007. http://www.jbc.org/content/282/49/35703.full.pdf

39. G.J. Davies and B. Henrissat, *Biochem. Soc. Trans.* 30, 291, 2002. http://www.biochemsoctrans.org/bst/030/0291/0300291.pdf

40. Glossary of terms used in the brief descriptions of protein folds, 2009. http://scop.mrc-lmb.cam.ac.uk/scop/gloss.html
41. http://en.wikipedia.org/wiki/Alpha_helix
42. http://en.wikipedia.org/wiki/Beta_sheet
43. P.M. Alzari, H. Souchon, and R. Dominguez, *Structure*, 4, 265, 1996. http://www.ncbi.nlm.nih.gov/pubmed/8805535
44. C. Jurgens, A. Strom, D. Wegener, S. Hettwer, M. Wilmanns, and R. Sterner, *Proc. Natl. Acad. Sci. USA* 97, 9925, 2000. http://www.pnas.org/content/97/18/9925.full.pdf+html
45. W.H. Freeman and Company, *Biochemistry*, 5th edition, 2002. http://www.ncbi.nlm.nih.gov/books/NBK22461/
46. http://swissmodel.expasy.org/course/text/chapter4.htm
47. A. Williams, D.R. Gilbert, and D.R. Westhead, *Protein Eng.* 16, 913, 2003. http://peds.oxfordjournals.org/content/16/12/913.full
48. http://kinemage.biochem.duke.edu/teaching/anatax/html/anatax.3d.html
49. CAZy Carbohydrate-Binding Module Family Server in www.cazy.org/fam/acc_CBM.html
50. V. Receveur, M. Czjzek, M. Schülein, P. Patine, and B. Henrissat, *J. Biol. Chem.* 277, 40887, 2002.
51. ExPASy (Expert Protein Analysis System) server of the Swiss Institute of Bioinformatics, Prosite Documentation PDOC00486, CBM1 Domain Signature and Profile. http://au.expasy.org/cgi-bin/prosite-search-ac?PDOC00486.
52. P.J. Simpson, H. Xie, D.N. Bolam, H.J. Gilbert, and M.P. Williamson, *J. Biol. Chem.* 275, 41137, 2000.
53. ExPASy, Prosite Documentation PDOC00485, CBM2 Domain Signature and Profile. http://kr.expasy.org/cgi-bin/nicedoc.pl?PS51173
54. J. Tormo, R. Lamed, A.J. Chirino, E. Morag, E.A. Bayer, Y. Shoham, and T.A. Steitz, *EMBO J.* 15, 5739, 1996.
55. ExPASy, Prosite Documentation PDOC51172, CBM3 Domain Profile. http://au.expasy.org/cgi-bin/prosite-search-ac?PDOC51172
56. J. Sakon, D. Irwin, D.B. Wilson, and P.A. Karplus, *Nat. Struct. Biol.* 4, 810, 1997.
57. D. Irwin, D.H. Shin, S. Zhang, B.K. Barr, J. Sakon, P.A. Karplus, and D.B. Wilson, *J. Bacteriol.* 180, 1709, 1998.
58. S. Y. Ding, E.A. Bayer, D. Steiner, Y. Shoham, and R. Lamed, *J. Bacteriol.* 181, 6720, 1999.
59. B. Henrissat, T.T. Teeri, and R.A.J. Warren, *FEBS Letters* 425, 352, 1998. http://www.sciencedirect.com/science/article/pii/S0014579398002658
60. B.R. Urbanowicz, A.B. Bennett, E. del Campillo, C. Catalá, T. Hayashi, B. Henrissat, H. Höfte, S.J. McQueen-Mason et al. *Plant Physiol.* 144, 1693, 2007. http://www.ncbi.nlm.nih.gov/pmc/articles/PMC1949884/.
61. G.J. Davies, K.S. Wilson, and B. Henrissat, *Biochem. J.* 321, 557, 1997. http://www.ncbi.nlm.nih.gov/pmc/articles/PMC1218105/pdf/9020895.pdf.
62. L. Petersen, Catalytic strategies of glycoside hydrolases, Thesis, Iowa State University, 2010. http://lib.dr.iastate.edu/cgi/viewcontent.cgi?article=2086&context=etd.
63. CAZY, 2010. http://www.cazypedia.org/index.php/Sub-site_nomenclature.
64. U.C. Davies, ChemWiki by University of California, Section 9.2 Digestion of carbohydrate by glycosidase–an SN1 reaction, 2010. http://chemwiki.ucdavis.edu/Organic_Chemistry/Organic_Chemistry_With_a_Biological_Emphasis/Chapter__9%3A_Nucleophilic_substitution_reactions_II/Section_9.2%3A_Digestion_of_carbohydrate_by_glycosidase_-_an_SN1_reaction
65. http://en.wikipedia.org/wiki/Nucleophilic_substitution
66. http://en.wikipedia.org/wiki/Hydrolase
67. http://www.chem.qmul.ac.uk/iubmb/enzyme/EC3/1/1/

68. http://www.ebi.ac.uk/thornton-srv/databases/cgi-bin/enzymes/GetPage.pl?ec_number= 3.1.1.1

69. http://www.ebi.ac.uk/thornton-srv/databases/cgi-bin/enzymes/GetPage.pl?ec_number= 3.1.1.2

70. http://www.chem.qmul.ac.uk/iubmb/enzyme/EC3/1/1/2.html

71. http://www.chem.qmul.ac.uk/iubmb/enzyme/EC3/1/1/3.html

72. http://www.ebi.ac.uk/thornton-srv/databases/cgi-bin/enzymes/GetPage.pl?ec_number= 3.1.1.3

73. http://en.wikipedia.org/wiki/Lipase

74. http://www.ebi.ac.uk/thornton-srv/databases/cgi-bin/enzymes/GetPage.pl?ec_number= 3.1.1.6

75. http://www.ebi.ac.uk/thornton-srv/databases/cgi-bin/enzymes/GetPage.pl?ec_number= 3.1.1.72

76. http://www.ebi.ac.uk/pdbsum/1bs9

77. X. L. Li, C.D. Skory, M.A. Cotta, V. Puchart, and P. Bieey, *Appl. Environ. Microbiol.* 74, 7482, 2008. http://aem.asm.org/content/74/24/7482.full

78. http://www.ebi.ac.uk/thornton-srv/databases/cgi-bin/enzymes/GetPage.pl?ec_number= 3.1.1.73

79. http://www.chem.qmul.ac.uk/iubmb/enzyme/EC3/1/1/73.html

80. M. Levisson, J. Van der Oost, and S.W.M. Kengen, 13, 567, 2009. http://www.ncbi.nlm. nih.gov/pmc/articles/PMC2706381/

81. http://www.cazy.org/Carbohydrate-Esterases.html

82. J.F. Robyt, *Essentials of Carbohydrate Chemistry*, Springer, New York, 1998. http:// books.google.be/books/about/Essentials_of_Carbohydrate_Chemistry.html?id= l4NfU7_sAZoC&redir_esc=y

83. A.E. Fazari and Y.H. Ju, *Acta Biochim Biophys. Sin.* 39, 811, 2007. http://www.abbs. info/pdf/39-11/1-811-07120.pdf

84. Brenda, the Comprehensive Enzyme Information System. http://www.brenda-enzymes.org/Mol/Mol.php4?n=207724&compound=acetylated&s_type=5&back= 1&limit_start=20

85. http://en.wikipedia.org/wiki/Catalytic_triad

86. *General introduction: catalysis by Ser-His-Asp catalytic triads,* Thesis (Chapter 1). http://dissertations.ub.rug.nl/FILES/faculties/science/2004/t.r.m.barends/c1.pdf.

87. M.M. De and O. Buanafina, *Mol. Plant* 2, 861, 2009. http://mplant.oxfordjournals.org/ content/2/5/861.full.

88. S. Hassan and N. Hugouvieux-Cotte-Pattat, *J. Bacteriol.* 193, 963, 2011. http://jb.asm. org/content/193/4/963.full

89. D.B.R.K. Dupta Udatha, Towards Classification and Functional Description of Enzymes, A case study of feruloyl esterases, Thesis for the degree of Doctor of Philosophy, Chalmers University of Technology, Sweden, 2013. http://publications.lib. chalmers.se/records/fulltext/169680/169680.pdf

# 3 Cellulose, the Main Component of Biomass

Before dealing with hemicelluloses and lignin, it seems useful to give an update on cellulose, the major cell wall component, which interacts strongly with the two other major cell wall components. Some of the features of cellulose will serve for polysaccharides in general and their sugar monomers. The chapter will be devoted to its structure, its biosynthesis, and its enzymatic hydrolysis. After an introduction on bioeconomy and biorefineries, a chapter devoted to carbohydrate-active enzymes and one devoted to cellulose, the reader will be able to discuss fruitfully hemicelluloses and lignin.

As early as in 1838, Anselme Payen[1] established that the fibrous component of all higher plant cells had a unique chemical structure, which he named cellulose.[2] Cellulose is the main molecule in cell walls of higher plants. Cellulose is also produced by some algae, bacteria, fungi, protozoans, and animal tunicates. There is more cellulose in the biosphere than in any other substance. The macromolecular nature of cellulose was demonstrated toward 1930. It was so established that cellulose is a polymer of glucose units. The chemical composition and conformation of cellulose chains combined with their hydrogen bonding system are responsible for their tendency to form crystalline aggregates.

## 3.1  STRUCTURE

### 3.1.1  MOLECULAR STRUCTURE

The cellulose molecule is a linear polymer of D-*anhydroglucopyranose units* linked together by β-1,4-*glucosidic bonds* (Figure 3.1). In other words, it is a β-1,4-D-*glucan*.

The numbering of the carbon atoms in the ring is shown in Figure 3.1, with two attached oxygen atoms at C1, hydroxyl substituents at C2 and C3, one attached oxygen atom at C4, and one hydroxymethyl group at C5. The positions 1 and 4 are involved in the interunit linkage. C1 is an acetal center along the whole chain except for the right-hand end where it is a hemiacetal center with inherent reducing properties. Thus, cellulose, as all 1,4-linked glucans, has one *reducing end* containing an unsubstituted hemiacetal, and one non-reducing end containing an additional hydroxyl group at C4.

In most crystal structures of cellulose, the molecule has a twofold helical (two monomers per turn of the helix) *conformation*,[3] meaning that adjacent glucose units are oriented with their mean planes at an angle of 180° to each other. This conformation gives each molecule a flat ribbon-like structure that is stabilized by intramolecular hydrogen bonds, among which the O3–H···O5 bond between two consecutive glucosyl units is relatively strong and present in all crystalline forms of cellulose.[4] Other hydrogen bonds between adjacent molecules are themselves responsible for the aggregations of the molecules into crystals.

### 3.1.2  SUPRAMOLECULAR STRUCTURE

Cellulose is a semicrystalline polymer, with a nonuniform repartition of crystalline zones and noncrystalline areas defined as *amorphous zones*. The relative amount of crystalline polymer in native cellulose varies widely with the source. Native celluloses have usually a higher crystallinity than man-made celluloses such as viscose. Degrees of crystallinity can reach very high values, particularly in *Valonia* algae and in animal *tunicates*.

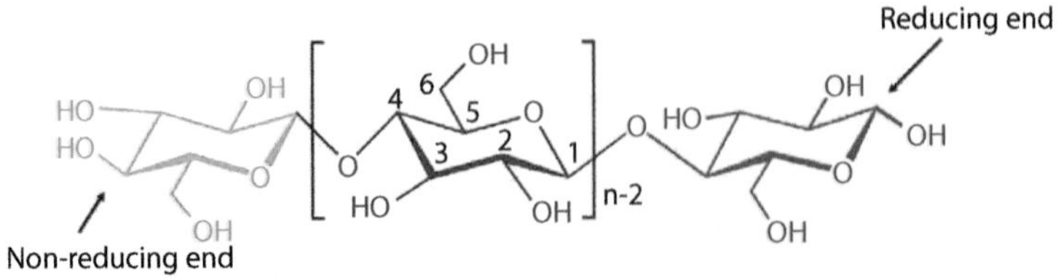

**FIGURE 3.1**  Molecular structure of cellulose showing the numbering of the carbon atoms, the reducing end in red with a hemiacetal, and the non-reducing end in green with a free hydroxyl at C4.

Cellulose can crystallize into several different crystalline *polymorphs*. Each polymorph is identified by its unit cell parameters.[2] Among all the cellulose polymorphs, cellulose I and cellulose II are the most important types. The chain sense is defined as *parallel up* when the z-coordinate of O5 is greater than that of C5. In this case, the reducing end of the chain is oriented in the same direction with respect to c-axis. The reverse direction is defined as *parallel down*.

### 3.1.2.1   Cellulose I

With only a few exceptions, all native celluloses consist of the structure called cellulose I. However, cellulose I is actually a composite of two crystalline *allomorphs*, labeled Iα and Iβ.[5,6] The relative amount of each depends on the native cellulose source. Algal and bacterial celluloses are rich in Iα, whereas celluloses from higher plants and tunicates are rich in Iβ. The two phases can coexist within native cellulose I.[7] Furthermore, it has been shown that the Iα phase is metastable and can be converted into the thermodynamically more stable Iβ phase by annealing or by treatments with various solvents.

Cellulose Iα has a triclinic one-chain unit cell where parallel cellulose chains stack, via van der Waals interactions, with progressive shear parallel to the chain axis (*P*1 space group).[8,9] There is only one chain with two glucose residues in the triclinic unit cell. The packing of the molecule in the unit cell is shown in Figure 3.2.

Cellulose Iβ has a monoclinic two-chain unit cell, which means that parallel cellulose chains are stacked with alternating shear.[8] There are two parallel-up cellulose chains, each with two glucose residues, located at the corners and center, in the monoclinic unit cell (Figure 3.2).

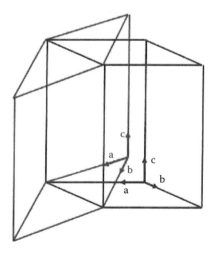

**FIGURE 3.2**   Relative orientation of the one-chain triclinic unit cell (colored in red, cellulose Iα) with respect to the two-chain monoclinic unit cell (colored in blue, cellulose Iβ). (Reprinted with permission from Sugiyama, J. et al., *Macromolecules*, 24, 4168. Copyright 1991 American Chemical Society.)

**TABLE 3.1**

**Comparison of the Crystal Structures Proposed for Most Common Cellulose Allomorphs**

| Type | Unit Cell | Chains | Repeat Distance (Å) | Asymmetric Unit |
|------|-----------|--------|---------------------|-----------------|
| Iα | One-chain triclinic | Parallel | ~10.35 | Two glucosyl |
| Iβ | Two-chain monoclinic | Parallel | ~10.35 | Two glucosyl |
| II | Two-chain monoclinic | Antiparallel | ~10.35 | Two glucosyl |

### 3.1.2.2 Cellulose II

Cellulose II is another cellulose polymorph, which results from the treatment of cellulose I by alkali solution, mercerization, or recrystallization from a solution.[8] The structure of cellulose II is believed to consist of a two-chain monoclinic cell unit where cellulose chains are stacked with opposite polarity, the so-called antiparallel structure. The antiparallel center and corner chains have different conformations.[10] Cellulose II forms a three-dimensional hydrogen-bonding network with hydrogen bonding between corner chains, center chains, and corner and center chains. This network appears stronger than the two-dimensional network of cellulose I. Cellulose II is thermodynamically more stable than cellulose I; therefore, the transformation from cellulose I to cellulose II is irreversible.

The crystal structures proposed for cellulose Iα, cellulose Iβ, and cellulose II are compared in Table 3.1.

In most crystal structures, glucose repeats as a dimer, with one set of conformational angles corresponding to a twofold helical symmetry (cellulose Iβ and cellulose II) or two sets corresponding to a onefold symmetry (cellulose Iα).

### 3.1.3 MORPHOLOGICAL STRUCTURE

The morphological structure of native cellulose is intended to describe the organization of crystals into microfibrils, layers (or lamellae), cell walls, fibers, tissues, or other cellulose morphologies (Figure 3.3).

Native cellulose from higher plants, algae, fungi, bacteria, amoebae, and tunicates is synthesized by the coordinated action of enzymatic polymerization associated with crystallization into nascent cellulose microfibrils.[12] This mechanism allows the generation of highly extended chains that can crystallize into microfibrils including cellulose Iα and Iβ. The microfibrils are then assembled to form higher order structures such as layers, cell walls, and fibers.[13]

### 3.1.3.1 Microfibrils

Electron microscopy has revealed the existence of very thin, crystalline fibrils, called microfibrils, in native cellulose. These microfibrils, or microcrystals, are biosynthesized by enzymatic terminal complexes located at the cell plasma membranes. It is believed that each native microfibril is in fact a single crystal.

The microfibrils are ~2 to 50 nm across depending on the synthesizing organism (Table 3.2)[14–16] and their lengths can reach several μm.[17] Therefore, they are examples

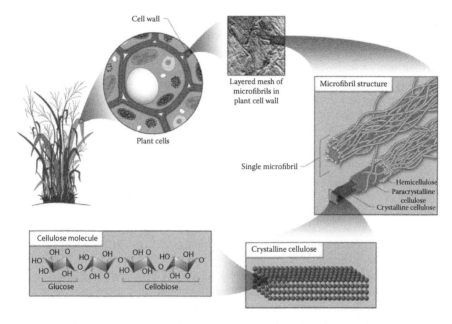

**FIGURE 3.3**  Cellulose: from macromolecules to plants. US DOE. 2005. *Genomics: GTL Roadmap*, DOE/SC-0090, U.S. Department of Energy Office of Science. (p. 204) Prepared by the Biological and Environmental Research Information System, Oak Ridge National Laboratory, genomicscience.energy.gov/ and genomics.energy.gov/. (From Cellulose structure and hydrolysis challenges, https://public.ornl.gov/site/gallery/originals/Fig2_Cellulose_Structure_a.jpg.)

## TABLE 3.2
## Width and Cross Section of Microfibrils of Various Natural Sources and Calculated Number of Glucan Chains per Microfibril

| Organism | Width (nm) | Cross Section | Glucan Chains per Microfibril |
|---|---|---|---|
| Alga *Micrasterias* | Up to 60 | Rectangular | 600–700 |
| Alga *Valonia* | 20 | Squarish | ~1000 |
| Tunicate *Halocynthia* | 10 | Parallelogram or truncated parallelogram | 600 |
| Ramie, cotton, flax | 5 | Not known | ~80 |
| Wood | 3–4 | Not known | 30–40 |
| Plant primary cell wall | 2–3 | Not known | ~30–36 |

*Source:*  Chanzy, H., *Cellulose Sources and Exploitation*, Ellis Horwood, Chichester, 1990; Chanzy, H., Personal communication, 2005, 2008; Kim, N.H. et al., *J. Struct. Biol.*, 117, 195, 1996; Vietor, R.J. et al., *Plant J.*, 30, 721, 2002; Somerville, C. et al., *Science*, 306, 2206, 2004; Delmer, D.P., *Annu. Rev. Plant Physiol. Plant Mol. Biol.*, 50, 245, 1999.

of bionanofibers. Algae such as *Micrasterias* and *Valonia*, and tunicates such as *Halocynthia* synthesize relatively large microfibrils. Ramie, wood, and primary wall synthesize relatively small microfibrils 5, 3–4, and 2–3 nm wide, respectively. Microfibrils are usually straight with extended cellulose molecules running parallel to the long axis of the microfibril.[14] An approximate number of glucan chains in the various microfibrils can be deduced from their lateral dimensions.

The demonstration of the parallel-up packing in cellulose Iα, and Iβ allowed the identification of the directionality of cellulose chains in microfibrils.[23] It was proven that the reducing ends of the growing chains point away from the synthesizing organism and therefore that polymerization by the cellulose synthase takes place at the non-reducing ends of the growing chains.

### 3.1.3.2  Plant Cell Walls

The plant cell wall is an extracellular matrix that encloses each cell in a plant.[24] The formation and differentiation of the cell wall play a crucial role in plant morphogenesis.[25] The walls of plant cells are generally thicker, stronger, and more rigid than the extracellular matrix produced by animal cells. In evolving relatively rigid walls, which vary from 0.1 to many micrometers in thickness, early plant cells adopted a sedentary life-style that has persisted in all current plants.

Plant cell walls do not only provide rigidity and strength to the plant body but also protect the plant against environmental stresses.[26,27] The composition of the cell wall depends on the cell type. The cell walls of growing cells are called primary cell walls. They are relatively thin and extensible to accommodate subsequent cell growth. In primary cell walls, organized cellulose microfibrils are embedded in an amorphous matrix.

Once growth stops, the cell wall no longer needs to be extensible: sometimes the primary wall is retained without major modifications, but more commonly a thicker, rigid, secondary cell wall is constructed by depositing new layers between the primary wall and the plasma membrane. The primary walls are mainly composed of cellulose, hemicelluloses, pectins, and proteins, whereas the secondary wall typically contains cellulose, hemicelluloses, and lignin.

Secondary wall formation is restricted to specialized cells and provides mechanical strength and rigidity to support aerial structures and hydrophobicity for transport functions.[25] Secondary walls may have a composition similar to that of the primary wall or be notably different. The most common additional polymer in secondary wall is *lignin*, found in the walls of the xylem vessels and fiber cells of woody tissues. When the plant cell dies, the wall remains.

Even if the plant cell walls differ in composition and organization, their underlying structure is remarkably consistent: tough microfibrils of cellulose are embedded in a highly cross-linked amorphous matrix. The microfibrils provide tensile strength to the wall, while the matrix provides resistance to compression, as in all fiber composites.[24]

### 3.1.3.3  Primary Cell Walls

In higher plants, the dominant structural features of primary cell walls are cellulose microfibrils with diameters of ~3 nm, which are cross-linked by single-chain polysaccharides.[21]

Structural analysis of primary cell wall polysaccharides has revealed the presence of three major classes of polysaccharides: cellulose, hemicelluloses, and *pectins*.[21] Hemicelluloses are branched polysaccharides that can form hydrogen bonds to the surface of cellulose microfibrils. The predominant hemicelluloses in many primary walls are *xyloglucans* (XyGs). Pectins are branched or unbranched polysaccharides containing D-galacturonic acid residues.[21]

Sets of microfibrils are arranged in layers or lamellae with each microfibril about 20–40 nm from its neighbors.[24] The primary cell wall consists of several such lamellae arranged in a plywood-like network.

The final shape of a growing plant cell is determined by controlled cell expansion.[24] Expansion occurs in response to *turgor* pressure in a direction that depends in part on the arrangement of the cellulose microfibrils in the wall. Cells anticipate their future morphology by controlling the orientation of microfibrils that they deposit in their walls. During their synthesis, nascent cellulose chains assemble spontaneously into microfibrils that form a lamella, in which all the microfibrils are approximately aligned. Each new lamella forms internally to the previous one, resulting in a wall composed of concentrically arranged lamellae, with the oldest on the outside. The most recently deposited microfibrils in elongating cells commonly lie perpendicular to the axis of cell elongation (*transverse microfibrils*). Although the orientation in the outer lamellae may be different, it is the orientation in these inner lamellae that have presumably a dominant influence on the direction of cell expansion.

The mechanism that dictates this orientation is connected with the *microtubules* in plant cells. The microtubules are arranged in the cortical cytoplasm with the same orientation as those of the cellulose microfibrils which are being deposited in the cell wall in that region.[24]

The organization of cellulose microfibrils and their complex interaction with other cell wall components results in an extensive and dynamic network that can be modified by the action of several *cell wall proteins*, such as endoglucanases, and expansins.[25,28]

### 3.1.3.4   Primary and Secondary Walls

In a typical wood cell such as a tracheid (xylem cell), the cell wall is composed of the lignified middle lamella (ML), which is shared by adjacent cells and ensures the adhesion between cells, the primary wall (P), the secondary wall composed of the S1, S2, and S3 layers and a lumen (Figure 3.4).[29–31]

The primary wall is ~0.1 μm thick.[32] It is built up of several layers of cellulose microfibrils.[32] The three layers of the secondary wall are built up by lamellae formed by ordered, parallel microfibrils embedded in lignin and hemicelluloses.[30,32]

The S2 layer, which is 1–10 μm thick, accounts for 75%–85% of the total thickness of the cell wall. The S1 and S3 layers are thin, 0.1–0.35 μm and 0.5–1.10 μm thick, respectively. The S1 layer, which represents 5%–10% of the total thickness of the cell wall, is considered as an intermediate between the primary cell wall and the S2 and S3 layers. The ordered parallel arrangement of microfibrils differs between the layers. In the S1 layer, the microfibrils are tightly wound in helices at a nearly transverse orientation (60°–80° with respect to the cell axis). In the S2 layer, the microfibrils are arranged in parallel, extended helices at a relatively small angle (5°–30°) with respect

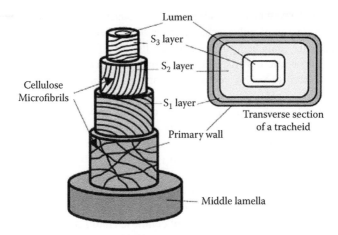

**FIGURE 3.4** Three-dimensional structure of the cell wall of a tracheid (xylem cell). The cell wall is divided into different layers, each layer having its own particular arrangement of cellulose microfibrils, which determine the mechanical and physical properties of the wood in that cell. These microfibrils may be aligned irregularly or at a particular angle to the cell axis. The middle lamella ensures the adhesion between cells. (Reprinted by permission from Macmillan Publishers Ltd: *Nature Reviews Genetics*, Sticklen, M.B., 2008, copyright 2008.)

to the cell axis. In the S3 layer, they are arranged in tightly wound helices at a nearly transverse orientation (60°–90° with respect to the cell axis).

Understanding how cell wall components are synthetized and integrated is essential to both plant biology, and efforts to utilize this renewable energy source in an efficient manner.[33,34] More than 2,000 genes are predicted to be involved in the synthesis and metabolism of cell wall components.[34,35] Identification of the genes responsible for wall biosynthesis and characterization of the biochemical and biological functions of the gene products that mediate wall biosynthesis are important areas of current research activity. Finally, once the process of wall biosynthesis is revealed, it will be important to understand how these processes are regulated, at both the biochemical and the transcriptional level.

## 3.2   BIOSYNTHESIS

Biosynthesis of cellulose is a multistep process involving *terminal complexes* (TCs) containing *cellulose synthase* enzymes organized in spinnerets at the cell membrane. Cellulose synthases use α-linked uridine diphosphoglucose (UDPGlc) as a substrate to polymerize glucose (Figure 3.5).

Each terminal complex spins a crystalline microfibril consisting of parallel, hydrogen-bonded glucan chains.

### 3.2.1   CELLULOSE-SYNTHESIZING TERMINAL COMPLEXES

Structures responsible for cellulose synthesis have been identified in freeze-fractured plasma membranes of many organisms.[36,37] Linearly arranged TCs in single or multiple rows are observed in bacteria, the cellular slime mold *Dictyostelium discoideum*

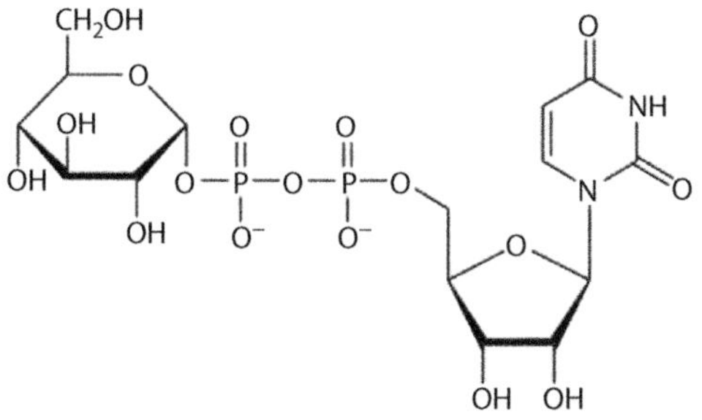

**FIGURE 3.5**  Uridine 5′-diphospho-glucose (UDPGlc).

and some algae, or hexagonal structures, with sixfold symmetry, termed rosettes, are observed in mosses, ferns, algae such as *Micrasterias*, and vascular plants.

The TC geometry correlates with microfibril size and shape.[38]

## 3.2.2  GLUCOSE POLYMERIZATION BY CELLULOSE SYNTHASES

### 3.2.2.1  Cellulose Synthase Substrate

Cellulose synthases are *glycosyltransferases* that catalyze the transfer of sugar moieties from activated donor molecules (glycones) to specific acceptor molecules (aglycones), forming glycosidic bonds.[22,38,39] These activated molecules are referred to as *nucleotide sugars* or sugar nucleotides. Nucleotide sugars act as glycosyl donors in glycosylation reactions.

The only cellulose synthase substrate is uridine 5′-diphospho-glucose (UDPGlc), which is a nucleotide sugar located in the cytoplasm of many plant and bacterial cells. UDPGlc is thought to bind to an active site of the enzyme on the cytoplasmic face of the plasma membrane.[40]

UDPGlc can be formed along a pathway in which:

1. α-D-glucose is phosphorylated to α-D-glucose-6-phosphate by the enzyme glucokinase.
2. α-D-glucose-6-phosphate is isomerized to α-D-glucose-1-phosphate by the enzyme phosphoglucomutase.
3. α-D-glucose-1-phosphate is converted to α-linked UDPGlc by the enzyme UDPGlc pyrophosphorylase.[41,42]

An alternative pathway for plant synthesis of UDPGlc is catalyzed by sucrose synthase (SuSy) which converts UDP and sucrose into UDPGlc and fructose. Membrane-bound forms of SuSy have been identified and some SuSys have been localized to the plasma membrane and cell walls, where cellulose biosynthesis takes place.[27]

### 3.2.2.2  Cellulose Synthases and Other Plant Glycosyltransferases

We refer to cellulose synthases as the proteins responsible, each, for the polymerization of one β-1,4-glucan chain. Cellulose synthases are also called catalytic subunits of the multimeric enzyme complex (TC).

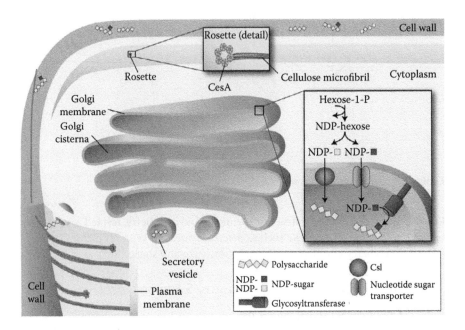

**FIGURE 3.6** Schematic representation of the key events in cell wall biosynthesis. Cellulose biosynthesis occurs at the plasma membrane in large complexes known as rosettes. The synthesis of matrix polysaccharides and glycoproteins occurs in the Golgi where the products accumulate in the lumen before transport in the cell wall via vesicles. CesA, cellulose synthase proteins that form the rosette; NDP-sugar, nucleotide sugars that act as donors for the sugars that go into polysaccharides; Csl, cellulose synthase-like proteins that are known to be involved in hemicellulose synthesis. (Reproduced from Keegstra, K., *Plant Physiol.*, 154, 483, http://www.plantphysiol.org/content/154/2/483.full, 2010. With permission of American Society of Plant Biologists.)

Wall polysaccharides are synthesized by glycosyltransferases (GTs) using diverse *nucleotide sugars*. (NDP-sugars).[43] GTs can be *processive* (catalyzing the addition of multiple sugar residues) or *nonprocessive* (catalyzing the addition of only one sugar residue).[44]

One important feature of cell wall biosynthesis is that it involves multiple cellular compartments (Figure 3.6).[35] Specifically, cellulose is synthesized at the plasma membrane with the insoluble cellulose microfibrils being deposited directly into the extracellular matrix.

On the one hand, cellulose is synthesized at the plasma membrane with the insoluble cellulose microfibrils being deposited directly into the extracellular matrix.[35] It is synthesized by large hexameric cellulose synthase complexes, ~25 nm in diameter, called rosettes, tracking along cortical microtubules at the plasma membrane.[27]

On the other hand, matrix polysaccharides and various glycoproteins are synthesized in the Golgi membranes, with the polymers being delivered to the wall via secretory vesicles.[35] Components synthesized in different locations must be assembled into a functional wall matrix.

Cellulose synthases are the active proteins in cellulose biosynthesis. They are processive β-glycosyltransferases. They have specificities for both donor and acceptor molecules, and for the type of linkage they form (α- or β-linkage). Cellulose synthases, which use α-linked UDPGlc as substrate, catalyze the formation of cellulose chains with β-1,4-linkages, involving a conformational inversion of the anomeric carbon. Therefore, they are *inverting* glycosyltransferases classified as family 2 glycosyltransferases (GT2).[45,46] Processive β-glycosyltransferases represent a large family of enzymes present in organisms from all domains of life.[47] Cellulose synthases are membrane proteins with a number of transmembrane regions. The globular fragment of the protein is probably involved in the catalysis.

Processive glycosyltransferases are characterized by two *conserved domains*, termed domains A and B.[48] Two aspartic acid residues in domain A, and a single aspartic acid residue and the QxxRW motif in domain B were found to be strictly conserved. The aspartic acid residues are believed to be involved in the catalytic reaction, while the QxxRW motif is thought to be responsible for the processive mechanism.

### 3.2.2.3 Models of Glucose Polymerization

Models of glucose polymerization have been proposed. In the model with single substrate-binding site, the A domain of cellulose synthases binds the UDP-sugar, and the B domain binds the acceptor molecule, the aspartates (D) in the A and B domains forming a single center for glycosyl transfer.[46] Inverting glycosyltransferases, including cellulose synthases, are assumed to use a single displacement mechanism with nucleophilic attack by the acceptor molecule at the anomeric carbon of the donor sugar.[49] Such a mechanism is presumed to require a base (aspartate residue) to activate the sugar acceptor for nucleophilic attack by deprotonation. For most enzymes, the reaction also involves additional carboxylate(s) (aspartate residue[s]) to coordinate a divalent metal ion on the phosphate group(s) of the nucleotide (Figure 3.7).

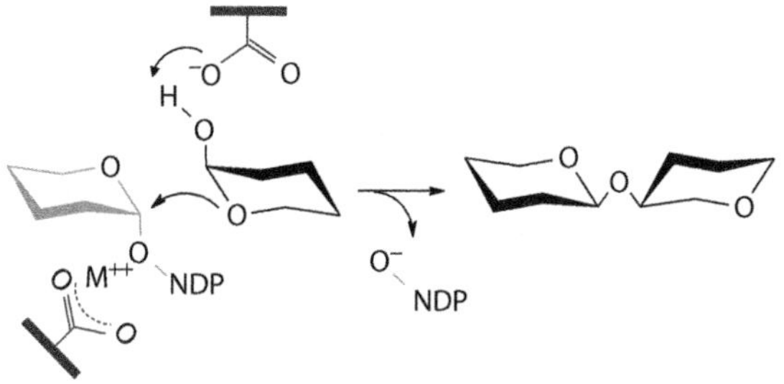

**FIGURE 3.7** Putative mechanism for an inverting nucleotide-sugar glycosyltransferase such as cellulose synthases. The mechanism is believed to require at least two catalytic carboxylates, one to activate the sugar acceptor and at least one to coordinate a divalent metal ion associated with the NDP-sugar (nucleoside diphosphate sugar). For cellulose synthases, it is presumed that at least two aspartates (D) in the A domain bind UDPGlc and one aspartate (D) in the B domain binds the sugar acceptor. (Adapted from Ross, J. et al., *Genome Biol.*, 2, 3004, http://genomebiology.com/2001/2/2/reviews/3004, 2001.)

### 3.2.2.4 Genes Encoding Cellulose Synthases

*3.2.2.4.1 Polymerization of Cellulose in Bacteria*

Due to the difficulties of characterizing cellulose synthase in higher plants, researchers looked toward simpler systems, notably the bacterium *Acetobacter xylinum*, to gain insight into the mechanism of cellulose synthesis.[41] Saxena, Lin, and Brown identified a *catalytic subunit* and the gene encoding it in the bacterium in 1990.[50] Furthermore, Wong et al. cloned an operon of four genes involved in *A. xylinum* in 1990 also.[51] Characterization of these genes indicates that the first gene termed *AxCesA* (first letters for Genus and species, followed by Ces for cellulose synthase, followed by A for genes encoding a presumed catalytic subunit, with a final number indicating the timing of the first report of the gene) codes for the 83-kDa subunit of the cellulose synthase that binds UDPGlc and presumably catalyzes the polymerization of glucose residues. It was in *A. xylinum* that UDPGlc was first implicated in cellulose biosynthesis.[52]

As mentioned before, the reducing end of the growing chains points away from the bacterium, so that polymerization by cellulose synthase takes place at the non-reducing end of the growing chains.[23] This mechanism is likely to be also valid for a number of processive glycosyltransferases that belong to the GT2 family. In 2003, the mechanism in β-chitin was confirmed.[53]

In 2012, Morgan et al.[54] presented the crystal structure of a complex of BcsA and BcsB from the gram-negative bacterium[55] *Rhodobacter sphaeroides* containing a translocating polysaccharide (Figure 3.8). Cellulose synthesis and transport across the inner bacterial membrane are mediated by a complex of the multispanning catalytic BcsA subunit (B for bacterial) and the membrane-anchored, periplasmic BcsB protein.

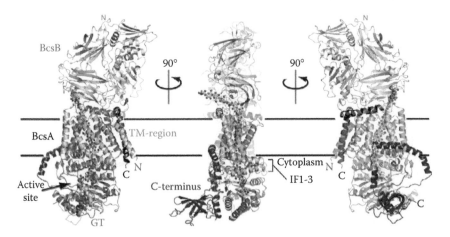

**FIGURE 3.8** Architecture of the BscA–BscB complex. BcsA and BcsB form an elongated complex with large cytosolic and periplasmic domains. BcsA's TM helices are colored green, the glycosyltransferase (GT) domain sand and the C-terminal domain red. BcsB is shown in light and dark blue for its periplasmic and membrane-associated regions, respectively. The N- and C-termini of both subunits are indicated and the translocating glucan and UDP are shown as cyan and violet spheres. Horizontal bars indicate the membrane boundaries. IF: Amphipathic interface helices of BcsA. (Reprinted by permission from Macmillan Publishers Ltd. *Nature*, Morgan, J.L.W. et al., 2013, copyright 2013.)

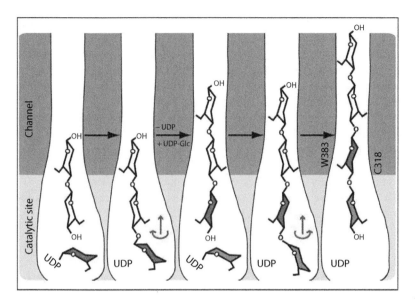

**FIGURE 3.9** Proposed model for cellulose synthesis and translocation. Following glycosyl transfer, the newly added glucose could rotate around the acetal linkage into the plane of the polymer. The rotation direction would be determined by steric interactions and formation of the β-1,4 glucan characteristic intramolecular O3–H...O5 hydrogen bond. The glucan might translocate into the channel during this relaxation. This process would be repeated with a second UDP-Glc but the rotation direction after glycosyl transfer would be in the opposite direction due to steric constraints. Alternatively, the glucan might not translocate into the channel until UDP is replaced by UDP-Glc. (Reprinted by permission from Macmillan Publishers Ltd. *Nature*, Morgan, J.L.W. et al., 2013, copyright 2013.)

The structure of the BcsA-B translocation intermediate shows the architecture of the cellulose synthase, demonstrates how BscA forms a cellulose-conducting channel, and suggests a model for the coupling of cellulose synthesis and translocation in which the nascent cellulose chain is extended by one glucose molecule at a time (Figure 3.9).

Cellulose, chitin, and hyaluronan are likely synthesized by a conserved mechanism involving a membrane-embedded glycosyltransferase that couples polymer synthesis with its translocation across the cell membrane. This unique mechanism stands in contrast to the translocation of most other biological polymers where *polymer synthases* function independently of or alongside dedicated translocation machineries. The structure presented here provides a basis for unraveling the details of this process.

### 3.2.2.4.2 Polymerization of Cellulose in Plants

Although the *A. xylinum* gene for cellulose synthase was cloned in 1990, it is not before 1996 that the first cellulose synthase genes were identified from higher plants, notably cotton fibers (*Gossypium hirsutum*).[56] The two cotton genes identified (*GhCesA1* and *GhCesA2*) are homologs to the bacterial genes encoding the catalytic subunit of cellulose synthase. The reported cotton genes are both highly expressed during active secondary wall cellulose synthesis.[57] The full-length ORF of *GhCesA1* encodes a 974 amino acid polypeptide of ~110 kDa that, like the bacterial

CesA proteins, is presumed to be a membrane-bound protein with eight transmembrane helices.[41] However, GhCesA1 and GhCesA2 differ from the bacterial CesAs in that they contain two large plant-specific insertions within the central domain: a *conserved plant-specific region* (CR-P) and a *hypervariable region* (HVR). Furthermore, the plant CesA proteins have an extended N-terminal region and a shorter C-terminal region compared to bacterial CesAs.

Since 1996, a large number of processive β-glycosyltransferase genes have been identified in plants. They all code for proteins that contain the D, D, D, QxxRW motif. Based on predicted protein sequences, these genes have been grouped into seven families:

- The *CesA* family that encodes catalytic subunits of cellulose synthase.
- Six families of structurally related genes designated as the cellulose synthase-like genes (*CslA*, *CslB*, *CslC*, *CslD*, *CslE*, and *CslG*).[57]

The Csl proteins are likely to catalyze the biosynthesis of noncellulosic polysaccharides. The seven families form the *CesA* superfamily.[58]

The correct localization of CesA proteins to sites of secondary wall deposition requires the presence of cortical microtubules (Figure 3.10).[59] In young xylem vessels,

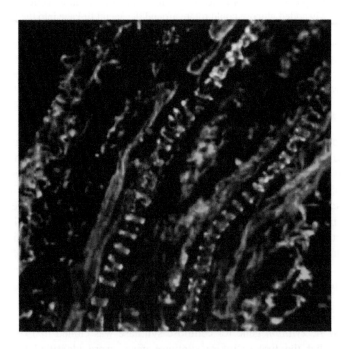

**FIGURE 3.10**  Localization of microtubules and the AtCesA7 (IRX3) protein in developing xylem vessels. Confocal image showing the localization of α-tubulin (green), AtCesA7 (red), and their overlap (orange), in a banded pattern, at late stages of vessel development. Band width ~3 μm. (Reproduced from Eckardt, N.A., *Plant Cell*, 15, 1685, 2003. With permission of American Society of Plant Biologists.)

the three CesA proteins involved in secondary wall synthesis localize within the cell and do not migrate to the microtubules, whereas in older vessels all three proteins colocalize with bands of microtubules.

### 3.2.3 CHAIN ASSEMBLY INTO MICROFIBRILS

The unique structure and properties of cellulose result not only from its linear glucan chain but also from the assembly of the chains into a microfibril. The terminal complex organization appears most important to determine the size and crystal structure of the microfibril formed.

#### 3.2.3.1 Chain Assembly in Plants

Single, hexagonal rosette structures were found to be exclusive to all higher plant cellulose assembly. The size and hexameric structure of the rosette fit with a model in which each subunit of the rosette synthesizes six glucan chains, leading to a 36-chain microfibril (Figure 3.11).[41,60] This is in agreement with the hypothesis that truly crystalline cellulose I requires an aggregate of at least 35 chains.[61]

The simultaneous presence of three different CesA proteins to produce a functional cellulose synthase complex is possibly explained by the geometrical constraints associated with assembling 36 proteins into the rosette.[41,62]

The mechanisms underlying the biosynthesis of cellulose in plants are complex and still poorly understood.[63] A key question is the mechanism of microfibril structure and how this is linked with the polymerization action of cellulose synthase. Furthermore, it remains unclear whether modification of microfibril structure can be achieved genetically, which could reduce its recalcitrance to enzymatic hydrolysis.

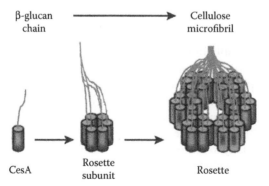

β-glucan              Cellulose
chain                microfibril

CesA      Rosette        Rosette
        subunit

**FIGURE 3.11** A model for the cellulose synthase rosette and rosette subunits. A rosette consists of six subunits, each of which possibly contains six CesA proteins. Each CesA protein polymerizes one glucan chain, and rosette would synthesize 36 chains that coalesce into the microfibril. (From Klemm, D. et al.: *Biopolymers–Cellulose*. 2001. Copyright Wiley-VCH Verlag GmbH & Co. KGaA. Reproduced with permission.)

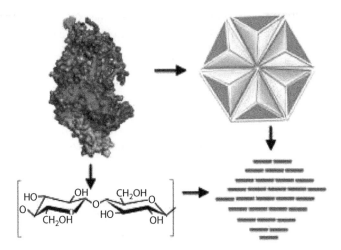

**FIGURE 3.12** Top left: GhCESA1 (catalytic subunit); top right: cellulose synthase complex: below left: two glucose monomers; bottom right: cellulose microfibril. (From Hirai, A. et al., *Cellulose*, 9, 105, 2002.)

In 2015, Kumar and Tuner[64] presented a model structure of a plant CESA protein suggesting considerable similarity between the bacterial and plant cellulose synthesis (Figure 3.12). The authors provided the following highlights:

- Modeling of plant CESA proteins suggests it is homologous to bacterial protein.
- A new structure of cellulose microfibril (MF) is suggested with less than 36 chains per MF as has been widely believed.
- These findings have major implications for the way plants synthesize cellulose. We may soon be able to relate the structure of the plant cellulose synthase complex with the microfibrils it synthesizes.

### 3.2.3.2 Chain Assembly in Bacteria

Under standard culture conditions, *A. xylinum* produces a 40–60-nm-wide ribbon, which is extruded parallel to the long axis of the cell through about one hundred pores in a linear row, and twisted in a right-handed manner.[65]

In contrast, *A. xylinum* produces band-like cellulose assemblies under some conditions.[66] The band-like assemblies consist of strand-like entities, each of which is extruded perpendicularly to the longitudinal axis of the cell. This raises the question of the nature and localization of the proteins in the cellulose-synthesizing particles within the linearly arranged terminal complexes.

## 3.3 ENZYMATIC HYDROLYSIS

Cellulose glycosidic bonds are strong and stable under various reaction conditions. However, cellulose can be degraded in nature by enzymes as an essential part of the carbon cycle.

Enzymatic degradation of cellulose is nevertheless hindered by the recalcitrance of plant cell walls, when cellulose is included in lignocellulosic structures. The major barrier for the development and cost reduction of biochemical cellulosic ethanol, and the implementation of related biorefineries, is to overcome this recalcitrance. In particular, the compact structure of crystalline cellulose and its embedding in a matrix of lignin and hemicelluloses impede enzymatic access. Moreover, enzymatic hydrolysis is more efficient in cellulose amorphous zones.

In a typical cellulose-degrading ecosystem, a variety of cellulolytic bacteria and fungi work in association with related microorganisms to convert insoluble cellulose to soluble sugars, primarily cellobiose and glucose, which are then assimilated by the cell.[66] For catalyzing this process, the cellulolytic microbes produce a wide diversity of enzymes, known as *cellulases*. Cellulases catalyze the hydrolysis of the β-1,4 glycosidic linkages in cellulose. They are members of the *glycoside hydrolase* (GH) families of enzymes that hydrolyze the polysaccharides in plant cell walls.[67] Due to the structural complexity and rigidity of cellulosic substrates, their efficient degradation generally requires multienzyme systems. Such systems include either a collection of free cellulases or/and multicomponent complexes called *cellulosomes*.

### 3.3.1 CELLULASES

Cellulases are produced not only by microorganisms such as bacteria, fungi, and protozoa but also by plants and some invertebrate animals.[2,68,69] Although it is widely accepted that most multicellular animals (Metazoa) do not have endogenous cellulases, relying instead on intestinal symbiotic microorganisms for cellulose digestion, cellulases have been found in the animal genomes of termites (insects), abalone (a mollusc), and sea squirts (tunicates).[70] Most likely, cellulases are not produced by vertebrate animals. Ruminants, for example, have highly specialized digestive tracts where mixtures of bacteria and protozoa degrade cellulose under anaerobic conditions.[70]

### 3.3.2 NON-COMPLEXED CELLULASE SYSTEMS

In order to degrade crystalline cellulose substrates, microorganisms produce multienzyme systems. Generally, two types of systems occur. One type consisting of a collection of independent extracellular cellulases that act synergistically is produced by aerobic bacterial and fungal microorganisms.[71,72] The second type consisting of an enzyme complex and produced by anaerobic bacteria will be reviewed in the next section.[73,74]

Non-complexed cellulases from the soft-rot fungus *Trichoderma reesei* and the soft-rot fungus *Humicola insolens* have been intensively studied. Cellulases from *T. reesei* are commercial especially under the trade name Celluclast from Novozymes. Also, *H. insolens* cellulases are commercial especially under the trade name Celluzyme from Novozymes.

#### 3.3.2.1 *Trichoderma reesei (Hypocrea jecorina)*

Fungal cellulases of the genus *Trichoderma* have been studied due especially to the high levels of cellulase produced by the species in this genus. The cellulases of *Trichoderma* are known to work at acid pH.[18]

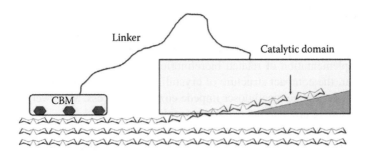

**FIGURE 3.13**  Schematic representation of a fungal cellulase consisting of a catalytic domain and a CBM (carbohydrate binding module) linked by a glycosylated peptide. The hexagons in the CBM are the three aromatic residues that interact with the glucose rings of the cellulose chain. The gray area represents the loops covering the active site. (From Divne, C. et al., *Science*, 265, 524, 1994.)

The *T. reesei* cellulolytic enzyme system consists of at least two cellobiohydrolases (Cel7A and Cel6A, formerly CBHI and CBHII, respectively), five endoglucanases, and two β-glucosidases.[75,76] Cel7A and Cel6A, are the main components of the enzymatic system, representing ~50% and ~20% of total cellulolytic enzyme.

The structure of Cel7A and Cel6A, like most fungal cellulases, displays a catalytic domain and a CBM connected by a glycosylated peptide linker (Figure 3.13).

The catalytic domain structures of Cel7A and Cel6A are very different but both display tunnel-shaped topologies formed by disulfide bridges.[74]

The crystal structure of Cel7A catalytic domain is shown in Figure 3.14.[77] The structure reveals a β sandwich of 12–14 β strands in two sheets with a 50 Å-long tunnel formed by the inner β strands and four long surface loops covering the active site.[77–80]

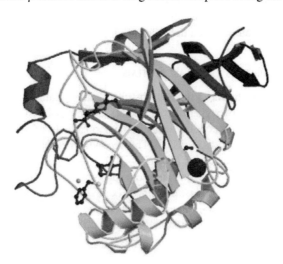

**FIGURE 3.14**  Three-dimensional crystal structure of the catalytic core of cellobiohydrolase Cel7A (formerly CBHI) from *Trichoderma reesei*. The structure displays a β sandwich fold of 12–14 β strands with a 50 Å-long tunnel. (From RCSB PDB Structure Summary—The structure was drawn from the pdb file found at 1CEL.)

Ten well-defined binding subsites for glucose units have been identified along the 50 Å-long tunnel.[79] Four tryptophan residues form a glucose-binding platform in the tunnel. Furthermore, it has been observed that the processivity of Cel7A action is toward non-reducing ends from reducing ends.[81]

Cel6A catalytic domain consists of a β/α barrel structure, similar to triose phosphate isomerase (TIM) but with seven instead of eight β strands.[77,82] A 20 Å-long substrate-binding tunnel adjacent to the α/β barrel is formed by two well-ordered loops covering the active site.[74] It contains four glucose unit-binding subsites.[77] Two additional subsites close to the tunnel entrance have been identified. Furthermore, it has also been observed that the processivity of Cel6A action is toward reducing ends from non-reducing ends.[83]

The tunnel-shaped topology of these enzymes allows for a structural explanation of the processivity of exoglucanases. The tunnel retains a single cellulose chain and prevents it from readhering to the cellulose crystal.[84,85] The catalytic sites of both CBHs are within this cellulose-binding tunnel near the outlet so that β-1,4 glycosidic bonds are cut with retaining (Cel7A) or inverting (Cel6A) mechanisms.[76] Both enzymes can cleave several bonds following a single adsorption event before dissociation of the enzyme substrate complex.

*T. reesei* Cel7A and Cel6A have a partial endo character.[76,86] It has been suggested that some exoglucanases are capable of both endo- and exo-actions due to temporary conformational changes of loops on the tunnel structure that encloses the active sites. This proposal has been supported by the observation that one of the loops of *T. reesei* Cel6A has substantial mobility and that the resulting tunnel could be either more tightly closed or more open.[86] The open conformation is likely to correspond to the endo-action. In addition, it should be noted that Cel6A contains fewer loops along the catalytic tunnel and exhibits greater endoglucanase activity than Cel7A.

The *T. reesei* CBMs, like all fungal CBMs, belong to family 1 (CBM1), characterized by a small wedge-shaped fold featuring a cellulose-binding surface with three exposed coplanar aromatic residues.[87–89] These aromatic residues are critical for the binding of a CBM1 onto crystalline cellulose, and the presence of a tryptophan (Trp or W) instead of a tyrosine (Tyr or Y) residue leads to an increased affinity on crystalline cellulose.[90,91] The spacing of the three aromatic residues coincides with the spacing of every second glucose ring on a glucan chain.[88]

### 3.3.2.2  *Humicola insolens*

Cellulases from the fungus *H. insolens*, developed by Novo, work at neutral or alkaline pHs, and are therefore adapted to laundry detergents.[75] *H. insolens* produces two cellobiohydrolases, Cel7A (formerly CBHI) and Cel6A (formerly CBHII) and six endoglucanases.[91–93]

*H. insolens* Cel6A displays a β/α barrel fold with a high degree of similarity with *T. reesei* Cel6A. The strong sequence similarity of *H. insolens* Cel7A with *T. reesei* Cel7A predicts an identical folding geometry.[86]

Cel6A can be described as an endo-processive cellobiohydrolase, whereas Cel7A is essentially a processive cellobiohydrolase.[86] Furthermore, the processivity of Cel6A action is from nonreducing ends, whereas that of Cel7A action is from reducing ends such as *T. reesei*.

### 3.3.2.3 Synergism

The enzymes involved in cellulose degradation appear to act synergistically. Several types of synergism in the degradation of cellulose by free enzymes have been described[76,86]:

- *Endo- and exo-acting enzymes*: This type of synergy is particularly important for degradation of crystalline cellulose. In such cooperation, the action of endoglucanases is to increase the number of chain ends which then become substrates for processive exoglucanases. In addition, β-glucosidases can work in synergy with cellulases by removing a cellobiose unit.
- *Exoglucanase and exoglucanase*: This synergy has been confirmed with fungal and bacterial cellobiohydrolases. Based on the observation that Cel6A and Cel7A from *T. reesei* and *H. insolens* act, respectively, from the nonreducing and reducing ends of the substrate, it was proposed that the differences in the chain end preference and in the directionality of action of the two cellobiohydrolases were responsible for the exo–exo synergy.[86,94] This synergy could also be explained by the partial endo activity of one or more cellobiohydrolases from a microorganism.

### 3.3.3 Multienzyme Complexes

Many cellulolytic microorganisms produce multienzyme complexes called cellulosomes that efficiently degrade cellulose and related plant cell wall polysaccharides.[95,96] The cellulosomes are composed of subunits, each of which comprises a set of interacting functional modules. A multifunctional integrating subunit called scaffoldin is responsible for organizing the catalytic subunits into the multienzyme complex. This is accomplished by the interaction of the *dockerin* domains present in the enzymatic subunits with the reiterated *cohesin* domains of the scaffoldin subunit. The high-affinity cohesin–dockerin interaction is responsible for the cellulosome architecture. The scaffoldin also bears a CBM that anchors the cellulosome to its substrate. In any given species, the interaction among the cohesins and dockerins usually proceeds in a nonspecific manner.[97,98]

The simplest cellulosomes, such as those produced by the mesophilic *Clostridium cellulolyticum*, include eight enzymes tightly bound to the scaffoldin.[99] The most elaborate cellulosomes, such as those produced by rumen bacteria, contain several interacting scaffoldins that can integrate dozens of enzymes. Two major types of cellulosome have so far been identified in relation with the cellulosomal gene cluster arrangement in the microorganisms.[98,99]

In this first type of cellulosome system, many of the cellulosomal genes are arranged in a cluster on the genome, consisting of a *primary* (enzyme-incorporating) *scaffoldin* gene and various dockerin-containing enzyme genes.[101] This enzyme-linked gene cluster arrangement is found in many mesophilic clostridia, such as *C. cellulolyticum*,[100] *C. cellulovorans*,[101] *C. josui*,[102] and *C. acetobutylicum*.[103] Hydrophilic domains present in the scaffoldin proteins from these bacteria have been proposed to fulfill a cell surface-anchoring role.[71,104]

In the other type of cellulosome system, a more elaborate arrangement allowing amplification of the number of enzymes in the complex comprises two or more genes that encode scaffoldins, at least one of which anchors the cellulosome onto the cell surface.[101,105] The genes for the enzymes are distributed elsewhere on the genome.[101] In addition to the cohesins, each anchoring protein usually bears an S-layer homology (SLH) module which is known to bind to the cell surface.[97] This multiple-scaffoldin gene cluster arrangement was first observed for *C. thermocellum*.[106] Similar multiple-scaffoldin gene clusters have been described for *Acetivibrio cellulolyticus*,[107] *Ruminococcus flavefaciens*,[105] and *Bacteroides cellulosolvens*.[107]

### 3.3.3.1 *Clostridium cellulolyticum*

*C. cellulolyticum* is an anaerobic, mesophilic, soil bacterium that is able to grow on cellulose as its sole carbon source.[108] Twelve genes encoding the key components of *C. cellulolyticum* cellulosomes are clustered.[109,110] Among them, the first upstream encodes the scaffoldin CipC (cellulosome integrating protein C), the second encodes the major processive endocellulase Cel48F and the fifth, the major cellobiohydrolase Cel9E which has retained some capacity for random attack mode.[111,112] The other cellulosomal enzymes encoded by the clustered genes include Cel8C, Cel9G, Cel9H, Cel9J, Man5K, Cel9M, and Cel9N.[114] CipC contains eight reiterated cohesins, a family 3a CBM and two domains of unknown function.[102,113] Three isolated genes encoding the cellulosomal cellulases CelA and CelD and non-cellulosomal cellulase CelI were found elsewhere on the chromosome.[112]

A model of a *C. cellulolyticum* cellulosome and its interaction with the cellulose substrate is shown in Figure 3.15.[99] Following the binding of the cellulosome

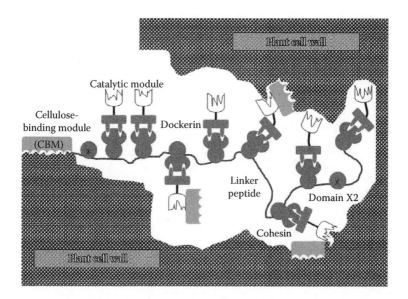

**FIGURE 3.15** Proposed schematic representation depicting a functional model of a cellulosome from *Clostridium cellulolyticum* and the interaction of its components with the cellulose substrate. The scaffoldin is bound to the substrate through the family 3a CBM. Some cellulosomal enzymes bear a CBM that mediates a relatively weak interaction with the substrate. (From Xu, Q. et al., *J. Bacteriol.*, 186, 968, 2004.)

to cellulose through the powerful family 3a CBM, the scaffoldin linkers between the cohesins would undergo large-scale rearrangement to adjust the positions of the enzyme subunits to the topography of the substrate. In this context, the relatively weak CBM borne by some cellulosomal enzymes would play a role in maintaining an extended conformation of the whole complex.

### 3.3.3.2 *Clostridium thermocellum*

*C. thermocellum* is a cellulolytic, thermophilic, anaerobic bacterium.[114] A schematic representation of the cellulosome from *C. thermocellum* is shown in Figure 3.16.[115]

The *C. thermocellum* cellulosome is characterized by a primary scaffoldin that contains nine copies of type I cohesins, each of which binds to a complementary type I dockerin, contained on each enzymatic subunit.[109,116,117] The primary scaffoldin also contains a *type II dockerin* that interacts selectively with complementary *type II cohesins*, contained on at least three different anchoring scaffoldins. In *C. thermocellum*, the nine cohesins of the primary scaffoldin CipA are unable to discriminate between the individual dockerins present in the various enzymes so that any cellulosome may comprise a different set of enzymatic subunits.[118] To date, at least 20 cellulosomal enzymes are known to be produced by *C. thermocellum*.[119] These catalytic subunits include mostly endoglucanases, exoglucanases, xylanases, and other hemicellulases.

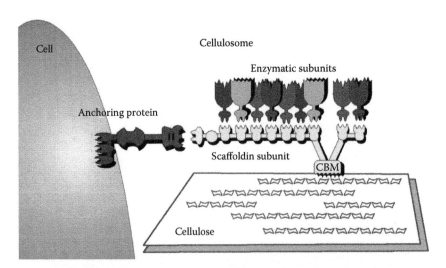

**FIGURE 3.16** Schematic diagram of the *Clostridium thermocellum* cellulosome.[117] The type I dockerins mediate attachment of the enzymatic subunits to the primary scaffoldin subunit, which is composed of nine cohesins, a CBM, a hydrophilic domain of unknown function and a type II dockerin. The scaffoldin binds through its type II dockerin to a type II cohesin-containing protein that anchors the cellulosome through an SLH module to the cell surface. (Courtesy of Edward A. Bayer.)

### 3.3.3.3  Designer Cellulosomes

The complexity and diversity of native cellulosomes prevent detailed analysis of the structural features responsible for their enhanced activity.[99] In addition, the lack of specificity of the cohesin–dockerin interaction within a species is an obstacle for the construction of homogeneous simplified cellulosomes containing more than one enzyme. There has been interest in constructing *designer cellulosomes* (or designer nanosomes), in which enzymes can be incorporated into defined positions, for both basic and applied purposes.

Such artificial cellulosomes are designed to comprise recombinant chimeric scaffoldin constructs and selected dockerin-containing enzyme hybrids, as a platform for promoting synergism among enzyme components.[97,119] In practice, a chimeric scaffoldin is produced that bears cohesins of different dockerin specificities. The designer cellulosome is constructed by mixing in solution the chimeric scaffoldin and complementary dockerin-containing components. Thus, in these cellulosomes, the composition and architecture of the complex can be controlled. The concept has potential applications as a molecular Lego in biotechnology and nanotechnology.[97]

## REFERENCES

1. A. Payen, *Compt. Rend. Acad. Sci.* 7, 1052, 1838.
2. T.P. Nevell and S.H. Zeronian, Eds., *Cellulose Chemistry and Its Applications*, Ellis Horwood, Chichester, 1985.
3. A.D. French, N.R. Bertoniere, R.M. Brown, H. Chanzy, D. Gray, K. Hattori and W. Glasser, *Encyclopedia of Polymer Science and Technology–Cellulose*, John Wiley & Sons, New York, 2003.
4. M. Jarvis, *Nature* 426, 611, 2003. http://scienceweek.com/2004/sb040130-5.htm.
5. R.H. Attala and D.L. Vanderhart, *Science* 223, 283, 1984.
6. D.L. Vanderhart and R.H. Attala, *Macromolecules* 17, 1465, 1984.
7. J. Sugiyama, R. Vuong and H. Chanzy, *Macromolecules* 24, 4168, 1991.
8. Y. Nishiyama, J. Sugiyama, H. Chanzy and P. Langan, *J. Am. Chem. Soc.* 125, 14300, 2003.
9. M. Wada, L. Heux and J. Sugiyama, *Biomacromolecules* 5, 1385, 2004. http://www.cermav.cnrs.fr/monos/publi-pdf/P04-48.pdf.
10. P. Langan, Y. Nishiyama and H. Chanzy, *J. Am. Chem. Soc.* 121, 9940, 1999.
11. Cellulose structure and hydrolysis challenges. https://public.ornl.gov/site/gallery/originals/Fig2_Cellulose_Structure_a.jpg.
12. J. Sugiyama and T. Imai, *Trends Glycosci. Glycotechnol.* 11, 23, 1999.
13. H. Kråssig, J. Schurz, R.G. Steadman, K. Schliefer, W. Albrecht, M. Mohring and H. Schlosser, *Ullmann's Encyclopedia of Industrial Chemistry–Cellulose*, Wiley-VCH, Hoboken, NJ, 2004.
14. H. Chanzy, *Cellulose Sources and Exploitation*, J.F. Kennedy, G.O. Phillips and P.A. Williams (Eds.), Ellis Horwood, Chichester, 1990.
15. H. Chanzy, Personal communication, April 15, 2005.
16. E.A. Bayer, H. Chanzy, R. Lamed and Y. Shoham, *Curr. Opin. Struct. Biol.* 8, 548, 1998.
17. D. Klemm, B. Philipp, T. Heinze, U. Heinze and W. Wagenknecht, *Comprehensive Cellulose Chemistry, Volume 1: Fundamentals and Analytical Methods*, Wiley-VCH, Weinheim, Germany, 1998.
18. H. Chanzy, Personal communication, June 23, 2008.
19. N.H. Kim, W. Herth, R. Vuong and H. Chanzy, *J. Struct. Biol.* 117, 195, 1996.
20. R.J. Vietor, R.H. Newman, M.A. Ha, D.C. Apperley and M.C. Jarvis, *Plant J.* 30, 721, 2002.
21. C. Somerville, S. Bauer, G. Brininstool, M. Facette, T. Hamann, J. Milne, E. Osborne et al., *Science* 306, 2206, 2004.

22. D.P. Delmer, *Annu. Rev. Plant Physiol. Plant Mol. Biol.* 50, 245, 1999.
23. M. Koyama, W. Helbert, T. Imai, J. Sugiyama and B. Henrissat, *Proc. Natl. Acad. Sci. USA* 94, 9091, 1997.
24. B. Alberts, A. Johnson, J. Lewis, M. Raff, K. Roberts and P. Walter, *Molecular Biology of the Cell*, 4th ed., Garland Publishing, New York, 2002. www.ncbi.nlm.nih.gov/books/bv.fcgi?call=bv.View..ShowTOC&rid=mboc4.TOC&depth=2.
25. A.I. Cano-Delgado, K. Metzlaff and M.W. Bevan, *Development* 127, 3395, 2000. http://dev.biologists.org/content/127/15/3395.full.pdf.
26. N.C. Carpita and M.C. McCann, *Trends Plant Sci.* 13, 415, 2008. http://www.science-direct.com/science/article/pii/S1360138508001817.
27. A. Endler and S. Persson, *Molecular Plant* 4, 199, 2011. http://mplant.oxfordjournals.org/content/4/2/199.full#ref-112.
28. S.G. Saupe, College of St. Benedict/St. John University, Plant Physiology (Biology 327), 2009. http://employees.csbsju.edu/SSAUPE/biol327/Lecture/cell-wall.htm.
29. U.P. Agarwal, *Planta* 224, 1141, 2006. www.fpl.fs.fed.us/documnts/pdf2006/fpl_2006_agarwal003.pdf.
30. E. Sjostrom, *Wood Chemistry. Fundamentals and Applications*, 2nd ed., Academic Press, San Diego, CA, 1993.
31. C. Plomion, G. Leprovost and A. Stokes, *Plant Physiol.* 127, 1513, 2001.
32. M.B. Sticklen, *Nat. Rev. Genet.* 9, 433, 2008. http://www.ncbi.nlm.nih.gov/pubmed/18487988.
33. B.A. O'Brien, W.T. Avigne, D.R. McCarty, A.M. Settles, L.C. Hannah, W. Vermerris, N.C. Carpita, M. McCann, S.P. Latshaw and K.E. Koch, *Annual Meeting of the American Society of Plant Biologists*, Abstract P17046, p. 106, Plant Biology 2007. http://hos.ufl.edu/sites/default/files/faculty/kekoch/publications/brent/PB2007.pdf.
34. K. Keegstra, *Plant Physiol.* 154, 483, 2010. http://www.plantphysiol.org/content/154/2/483.full.
35. G. Pogorelko, V. Lionetti, D. Bellincampi and O. Zabotina, *Plant Signal Behav.* 8(9), e25435, 2013. http://www.ncbi.nlm.nih.gov/pmc/articles/PMC4002593/.
36. R.M. Brown Jr. and D.L. Montezinos, *Proc. Natl. Acad. Sci. USA* 73, 143, 1976.
37. R.M. Brown Jr., *Pure Appl. Chem.* 71, 767, 1999.
38. http://www.cazy.org/GlycosylTransferases.html
39. W.D. Reiter and G.F. Vanzin, *Plant Mol. Biol.* 47, 95, 2001.
40. M.S. Doblin, I. Kurek, D. Jacob-Wilk and D.P. Delmer, *Plant Cell Physiol.* 43, 1407, 2002.
41. D. Klemm, H.P. Schmauder and T. Heinze, *Biopolymers–Cellulose*, Wiley-VCH, Hoboken, NJ, 2001.
42. S. Bielecki, A. Krystynowicz, M. Turkiewicz and H. Kalinowska, *Biopolymers–Bacterial Cellulose*, Wiley-VCH, 2001.
43. Bioenergy Science Center-Complex Carbohydrate Research Center. http://bioenergy.ccrc.uga.edu/People/people.htm.
44. R.M. Perrin, *Glycosyltransferases in Plant Cell Wall Synthesis*, 2008. http://www.els.net/WileyCDA/ElsArticle/refId-a0020102.html.
45. S.J. Charnock, B. Henrissat and G.J. Davies, *Plant Physiol.* 125, 527, 2001. http://www.plantphysiol.org/content/125/2/527.full.
46. J. Ross, Y. Li, E.K. Lim and D.J. Bowles, *Genome Biol.* 2, 3004, 2001. http://genomebiology.com/2001/2/2/reviews/3004.
47. UniProt, UniProtKB - Q9ZB73 (PBDGT_MYCGE). http://www.uniprot.org/uniprot/Q9ZB73.
48. I.M. Saxena and R.M. Brown Jr., *Cellulose* 4, 35, 1997.
49. I.M. Saxena, R.M. Brown Jr., M. Fevre, R.A. Geremia and B. Henrissat, *J. Bacteriol.* 177, 1419, 1995.
50. I.M. Saxena, F.C. Lin and R.M. Brown Jr., *Plant Mol. Biol.* 15, 673, 1990.

51. H.C. Wong, A.L. Fear, R.D. Calhoon, G.H. Eichinger, R. Mayer, D. Amikam, M. Benziman et al., *Proc. Natl. Acad. Sci. USA.* 87, 8130, 1990.
52. L. Glaser, *J. Biol. Chem.* 232, 627, 1958.
53. T. Imai, T. Watanabe, T. Yui and J. Sugiyama, *Biochem. J.* 374, 755, 2003.
54. J.L.W. Morgan, J. Strumillo and J. Zimmer, *Nature* 493, 181, 2013. http://www.ncbi.nlm.nih.gov/pmc/articles/PMC3542415/.
55. http://en.wikipedia.org/wiki/Gram-negative_bacteria
56. J.R. Pear, Y. Kawagoe, W.E. Schreckengost, D.P. Delmer and D.M. Stalker, *Proc. Natl. Acad. Sci. USA* 93, 12637, 1996.
57. T.A. Richmond and C.R. Somerville, *Plant Physiol.* 124, 495, 2000.
58. Carnegie Institution for Sciences, Plant Biology. http://www-ciwdpb.stanford.edu or http://carnegiedpb.stanford.edu.
59. J.C. Gardiner, N.G. Taylor and S.R. Turner, *Plant Cell* 15, 17470, 2003.
60. N.A. Eckardt, *Plant Cell* 15, 1685, 2003.
61. H. Chanzy, K. Imada and R. Vuong, *Protoplasma* 94, 299, 1978.
62. I. Kurek, Y. Kawagoe, D. Jacob-Wilk, M. Doblin and D.P. Delmer, *Proc. Natl. Acad. Sci. USA* 99, 11109, 2002.
63. D.M. Harris, K. Corbin, T. Wang, R. Gutierrez, A.L. Bertolo, C. Petti, D.M. Smilgies et al., *Proc. Natl. Acad. Sci. USA* 109, 4098, 2012. http://www.pnas.org/content/109/11/4098.abstract.
64. M. Kumar and S. Tuner, *Phytochemistry* 112, 91, 2015. http://www.sciencedirect.com/science/article/pii/S0031942214002799.
65. A. Hirai, M. Tsuji and F. Horii, *Cellulose* 9, 105, 2002.
66. E.A. Bayer, H. Chanzy, R. Lamed and Y. Shoham, *Curr. Opin. Struct. Biol.* 8, 548, 1998.
67. B. Henrissat, T.T. Teeri and R.A.J. Warren, *FEBS Lett.* 425, 352, 1998.
68. N. Lo, H. Watanabe and M. Sugimura, *Proc. R. Soc. Lond. B (Suppl.)* 270, S69, 2003.
69. G. Smant, J.P.W.G. Stokkermans, Y. Yan, J.M. de Boer, T.J. Baum, X. Wang, R.S. Hussey et al., *Proc. Natl. Acad. Sci. USA* 95, 4906, 1998.
70. A. Davison and M. Blaxter, *Mol. Biol. Evol.* 22, 1273, 2005.
71. R.H. Doi, A. Kosugi, K. Murashima, Y. Tamura, and S.O. Han, *J. Bacteriol.* 185, 5907, 2003.
72. C. Boisset, H. Chanzy, B. Henrissat, R. Lamed, Y. Shoham and E.A. Bayer, *Biochem. J.* 340, 829, 1999.
73. E.A. Bayer, E. Setter and R. Lamed, *J. Bacteriol.* 163, 552, 1985.
74. W.H. Schwarz, *Appl. Microbiol. Biotechnol.* 56, 634, 2001.
75. Y.H.P. Zhang and L.R. Lynd, *Biotech. Bioeng.* 88, 797, 2004.
76. A. Nutt, Thesis, Hydrolytic and oxidative mechanisms involved in cellulose degradation, Acta Universitatis Upsaliensis Uppsala, 2006. www.diva-portal.org/diva/getDocument?urn_nbn_se_uu_diva-6888-1__fulltext.pdf.
77. C. Divne, J. Stahlberg, T. Reinikainen, L. Ruohonen, G. Pettersson, J.K. Knowles, T.T. Teeri and T.A. Jones, *Science* 265, 524, 1994.
78. C. Divne, J. Stahlberg, T.T. Teeri and T.A. Jones, *J. Mol. Biol.* 275, 309, 1998.
79. M. Ike, Y. Ko, K. Yokoyama, J. Sumitani, T. Kawaguchi, W. Ogasawara, H. Okada and Y. Morikawa, *J. Mol. Cat. B: Enzymatic* 47, 159, 2007.
80. RCSB (Research Collaboratory for Structural Bioinformatics) PDB (Protein Data Bank), The Three-Dimensional Crystal Structure of the Catalytic Core of Cellobiohydrolase I from Trichoderma reesei. www.rcsb.org/pdb/cgi/explore.cgi?pdbId=1CEL.
81. T. Imai, C. Boisset, M. Samejima, K. Igarashi and J. Sugiyama, *FEBS Lett.* 432, 113, 1998.
82. J. Rouvinen, T. Bergfors, T. Teeri, J.K. Knowles and T.A. Jones, *Science* 249, 380, 1990.
83. H. Chanzy and H. Henrissat, *FEBS Lett.* 184, 285, 1985.
84. M. Schulein, *Biochim. Biophys. Acta* 1543, 239, 2000.
85. C. Boisset, C. Fraschini, M. Schülein, B. Henrissat and H. Chanzy, *Appl. Environ. Microbiol.* 66, 1444, 2000.

86. J. Zou, G.J. Kleywegt, J. Stahlberg, H. Driguez, W. Nerinckx, M. Claeyssens, A. Koivula, T.T. Teeri and T.A. Jones, *Structure* 7, 1035, 1999.

87. J. Kraulis, G.M. Clore, M. Nilges, T.A. Jones, G. Petterson, J. Knowles and A.M. Gronenborn, *Biochemistry* 28, 7241, 1989.

88. M.L. Mattinen, M. Linder, T. Drakenberg and A. Annila, *Eur. J. Biochem.* 256, 279, 1998.

89. J. Lehtiö, J. Sugiyama, M. Gustavsson, L. Fransson, M. Linder and T.T. Teeri, *Proc. Natl. Acad. Sci. USA* 100, 484, 2003.

90. T. Reinikainen, L. Ruohonen, T. Nevanen, L. Laaksonen, P. Kraulis, T.A. Jones, J.K. Knowles and T.T. Teeri, *Proteins* 14, 475, 1992.

91. M. Schulein, *J. Biotechnol.* 57, 71, 1997.

92. CAZy Glycoside Hydrolase Family 6. www.cazy.org/fam/GH6.html.

93. K. Igarashi, T. Ishida, C. Hori and M. Samejima, *Appl. Environ. Microbiol.* 74, 5628, 2008.

94. B.K. Barr, Y.L. Hsieh, B. Ganem and D.B. Wilson, *Biochemistry* 35, 586, 1996.

95. E.A. Bayer, Weizmann Institute of Science, The Cellulosome Complex. www.weizmann.ac.il/Biological_Chemistry/scientist/Bayer.

96. E.A. Bayer, J.P. Belaich, Y. Shoham and R. Lamed, *Annu. Rev. Microbiol.* 58, 521, 2004.

97. M. Hammel, H.P. Fierobe, M. Czjzek, V. Kurkall, J.C. Smith, E.A. Bayer, S. Finet and V. Receveur-Bréchot, *J. Biol. Chem.* 280, 38562, 2005.

98. A. Mechaly, H.P. Fierobe, A. Belaich, J.P. Belaich, R. Lamed, Y. Shoham and E.A. Bayer, *J. Biol. Chem.* 276, 9883, 2001.

99. Q. Xu, E.A. Bayer, M. Goldman, R. Kenig, Y. Shoham and R. Lamed, *J. Bacteriol.* 186, 968, 2004.

100. A. Belaich, G. Parsiegla, L. Gal, C. Villard, R. Haser and J.P. Belaich, *J. Bacteriol.* 184, 1378, 2002.

101. Y. Tamaru, S. Karita, A. Ibrahim, H. Chan and R.H. Doi, *J. Bacteriol.* 182, 5906, 2000.

102. M. Kakiuchi, A. Isui, K. Suzuki, T. Fujino, E. Fujino, T. Kimura, S. Karita, K. Sakka and K. Ohmiya, *J. Bacteriol.* 180, 4303, 1998.

103. F. Sabathe, A. Belaich and P. Soucaille, *FEMS Microbiol. Lett.* 217, 15, 2002.

104. M.T. Rincon, T. Cepeljnik, J.C. Martin, R. Lamed, Y. Barak, E.A. Bayer and H.J. Flint, *J. Bacteriol.* 187, 7569, 2005.

105. Q. Xu, Y. Barak, R. Kenig, Y. Shoham, E.A. Bayer and R. Lamed, *J. Bacteriol.* 186, 5782, 2004.

106. T. Fujino, P. Beguin and J.P. Aubert, *J. Bacteriol.* 175, 1891, 1993.

107. S.Y. Ding, E.A. Bayer, D. Steiner, Y. Shoham and R. Lamed, *J. Bacteriol.* 182, 4915, 2000.

108. D. Mandelman, A. Belaich, J.P. Belaich, N. Aghajari, H. Driguez and R. Haser, *J. Bacteriol.* 185, 4127, 2003.

109. H. Maamar, L. Abdou, C. Boileau, O. Valette and C. Tardif, *J. Bacteriol.* 188, 2614, 2006.

110. S. Pages, O. Valette, L. Abdou, A. Belaich and J. P Belaich, *J. Bacteriol.* 185, 4727, 2003.

111. C. Reverbel-Leroy, S. Pages, A. Belaich, J.P. Belaich and C. Tardif, *J. Bacteriol.* 179, 46, 1997.

112. C. Gaudin, A. Belaich, S. Champ and J.P. Belaich, *J. Bacteriol.* 182, 1910, 2000.

113. S. Pages, A. Belaich, H.P. Fierobe, C. Tardif, C. Gaudin and J.P. Belaich, *J. Bacteriol.* 181, 1801, 1999.

114. S. Spinelli, H.P. Fierobe, J.P. Belaich, B. Henrissat and C. Cambillau, *J. Mol. Biol.* 304, 189, 2000.

115. R. Lamed, The George S. Wise Faculty of Life Science, Tel Aviv University, Molecular Microbiology and Biotechnology. www.tau.ac.il/lifesci/departments/biotech/members/lamed/lamed.html.

116. U.T. Gerngross, M.P. Romaniec, T. Kobayashi, N.S. Huskisson and A.L. Demain, *Mol. Microbiol.* 8, 325, 1993.

117. T.W. Dror, A. Rolider, E.A. Bayer, R. Lamed and Y. Shoham, *J. Bacteriol.* 185, 5109, 2003.

118. A.L. Carvalho, F.M. V. Dias, J.A.M. Prates, T. Nagy, H.J. Gilbert, G.J. Davies, L.M.A. Ferreira, M.J. Romão and C.M.G.A. Fontes, *Proc. Natl. Acad. Sci. USA* 100, 13809, 2003.

119. E.A. Bayer, E. Morag and R. Lamed, *Trends Biotechnol.* 12, 378, 1994.

# 4 Structure and Biosynthesis of Hemicelluloses

## 4.1 INTRODUCTION

Hemicelluloses are a common but archaic term for all substances extractable from plant cell walls with molar concentrations of alkali.[1] These are a diverse group of carbohydrate polymers that vary among different plant groups and between primary and secondary walls.

Structurally, hemicelluloses are plant cell wall polysaccharides that have a backbone of β-1,4-linked-D-*pyranosyl* residues, such as glucose, mannose, and xylose, in which O4 is in the equatorial configuration.[2,3] To the backbone short side chains are attached. The structural similarity between hemicellulosic backbones and cellulose most likely gives rise to a conformational homology that can lead to a strong noncovalent association of hemicelluloses and cellulose microfibrils.

In the cell walls, hemicelluloses are matrix polymers of relatively low molecular weight that are associated with cellulose and other polymers. They are known to bind tightly to cellulose microfibrils via most hydrogen bonding.[4]

## 4.2 GENERAL STRUCTURE OF HEMICELLULOSES

In contrast with cellulose containing only D-glucose, hemicelluloses are composed of several different sugars. The hemicellulosic sugars include (Figure 4.1)

The five-carbon sugars include

- D-xylose, an analog of D-glucose that lacks a 6-hydroxymethyl group; xylose is the second most abundant sugar in the biosphere.
- L-arabinose, an analog of L-galactose that lacks a 6-hydroxymethyl group.

The six-carbon sugars include

- D-glucose, most abundant sugar in the biosphere.
- D-mannose, a C2 *epimer* of D-glucose.
- D-galactose, a C4 epimer of D-glucose.
- L-rhamnose (6-deoxy-L-mannose).
- L-fucose (6-deoxy-L-galactose).
- D-galacturonic acid, an oxidized form of D-galactose having a carboxylic group at C6; it is the main component of pectins.
- D-glucuronic acid, an oxidized form of glucose having a carboxylic group at C6, and its 4-*O*-methyl derivative. A certain proportion of these D-galacturonic and D-glucuronic acids are in the methyl ester form.

Hemicelluloses can be divided into four major groups. Their schematic structures are shown in Figures 4.2. and 4.3

- *Xyloglucans* have a glucose backbone and xylose-containing branches; they are the predominant hemicelluloses in many primary cell walls.
- *Xylans*, which are the most abundant hemicelluloses, include
  - *Glucuronoxylans* with a xylose backbone and glucuronic acid branches; they are found in the secondary cell walls of dicots.
  - *Arabinoxylans* with a xylose backbone and arabinose branches; they are found in endosperm walls of cereals.
  - *Glucuronoarabinoxylans* (GAXs) with a xylose backbone and branches of arabinose and glucuronic acid; they are found in primary walls of commelinid monocots.
- *Mannans* include:
  - Pure mannans with a mannose backbone.
  - *Glucomannans* with a mannose/glucose backbone.
  - *Galactomannans* with a mannose backbone and galactose branches.
  - *Galactoglucomannans* with a mannose/glucose backbone and galactose branches.
  In addition to being well-defined stored reserve material, mannan-based polysaccharides are also found as ubiquitous components of plant cell walls.[5–7]
- β-*1,3;1,4-glucans*, also called mixed-linkage glucans, have interspaced single β-1,3-linkages; they are highly prevalent in grasses.

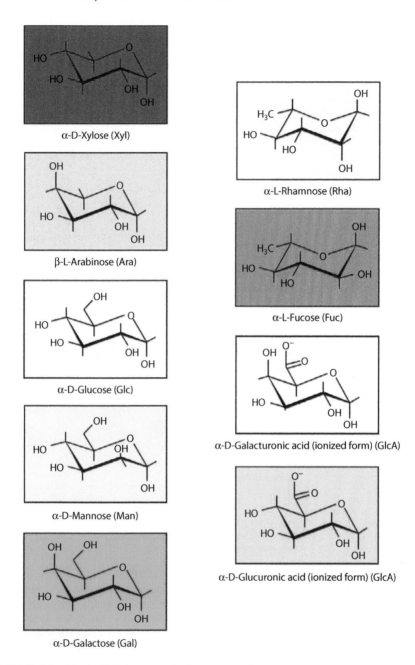

**FIGURE 4.1** Hemicellulosic sugars in the pyranose form (*p*).

Hemicelluloses and Lignin in Biorefineries

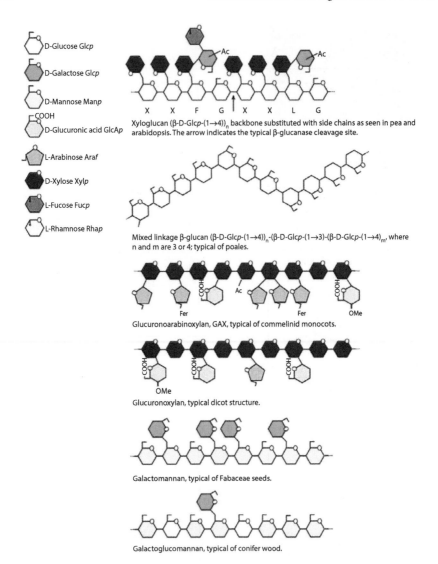

D-Glucose Glcp

D-Galactose Glcp

D-Mannose Manp

D-Glucuronic acid GlcAp

L-Arabinose Araf

D-Xylose Xylp

L-Fucose Fucp

L-Rhamnose Rhap

Xyloglucan (β-D-Glcp-(1→4))ₙ backbone substituted with side chains as seen in pea and arabidopsis. The arrow indicates the typical β-glucanase cleavage site.

Mixed linkage β-glucan (β-D-Glcp-(1→4))ₙ-(β-D-Glcp-(1→3)-(β-D-Glcp-(1→4))ₘ, where n and m are 3 or 4; typical of poales.

Glucuronoarabinoxylan, GAX, typical of commelinid monocots.

Glucuronoxylan, typical dicot structure.

Galactomannan, typical of Fabaceae seeds.

Galactoglucomannan, typical of conifer wood.

**FIGURE 4.2** Schematic structures of hemicelluloses divided into four major groups: (1) xyloglucans, (2) β-1,3;1,4-glucans, (3) glucuronoarabinoxylans, glucuronoxylans, (4) galactomannans and galactoglucomannans (same colors of sugars as in Figure 4.1). The letters under the xyloglucan structure represent the symbols used for the most common side chains. *Fer* represents esterification with ferulic acid, which is characteristic of xylans in commelinid monocots. Fabaceae or Leguminosae are dicot plants, commonly known as the legume family, bean family or pea family. (Reproduced from Scheller, H.V. and Ulvskov, P., *Annu. Rev. Plant Biol.*, 61, 263–289, 2010. With permission.)

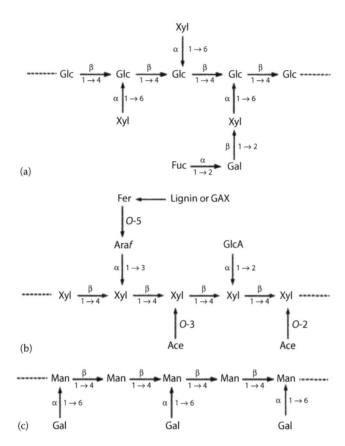

**FIGURE 4.3** Structures of xyloglucans (a), glucuronoarabinoxylans (b), and galactomannans (c) with a focus on linkages. Ace, acetate; Fer, ferulate; Fuc, fucose; Gal, galactose; Glc, glucose; GlcA, glucuronate; Man, mannose. (Reprinted from *Mol. Plant*, 2, Sandhu, A.P.S. et al., Plant cell wall matrix polysaccharide biosynthesis, 840–850, Copyright 2009, with permission from Elsevier.)

## 4.3 DETAILED STRUCTURE AND OCCURRENCE OF HEMICELLULOSES

### 4.3.1 XYLOGLUCANS

Xyloglucans (XyGs) are present in the cell walls of all land plants (Embryophyta) including mosses, but they are not present in green algae charophytes.[3,9]

XyGs are the predominant hemicellulosic polysaccharides in the primary walls of dicots and nongraminaceous monocots in which they may account for ~20%–25% of the dry weight of the primary wall (Table 4.1).[10–12]

XyGs consist of a β-1,4-glucan backbone with up to 75% of the glucosyl residues substituted at O6 with mono-, di-, or triglycosyl side chains.[2] Their backbone is ~0.15–1.5 μm long with 300–3,000 residues.[10] XyGs are made of repetitive units, and a one-letter code is used to denote the different XyG side chains.[13,14] For example, an

**TABLE 4.1**

**Occurrence of Hemicelluloses in Plant Cell Walls**

| Amount of Hemicelluloses (% w/w) | Primary Cell Wall | | | Secondary Cell Wall | | |
|---|---|---|---|---|---|---|
| | **Grasses** | **Conifers** | **Dicots** | **Grass** | **Conifers** | **Dicots** |
| Xyloglucans | 1–5 | 10 | 20–25 | Minor | ~0[a] | Minor |
| Glucuronoxylans | ~0 | ~0 | ~0 | ~0 | ~0 | 20–30 |
| Glucuronoarabinoxylans | 20–40 | 2 | 5 | 40–50 | 5–15 | ~0 |
| (Gluco)mannans | 2 | ~0 | 3–5 | 0–5 | ~0 | 2–5 |
| Galactoglucomannans | ~0 | Present | ~0 | ~0 | 10–30 | 0–3 |
| β-1,3;1,4-Glucans | 2–15 | 0 | 0 | Minor | 0 | 0 |

*Source:* Scheller, H.V. and Ulvskov, P., *Annu. Rev. Plant Biol.*, 61, 263–289, 2010.

[a] ~0, absent or minor.

unbranched glucosyl residue is designated **G**, whereas an α-xylose-1,6-β-glucose segment is designated **X**. The xylosyl residues can be substituted at O2 with β-galactose (**L** side chain) or α-arabinose (**S** side chain). The galactosyl residues substituted at O2 with α-fucose is designated **F** (Figure 4.4).

The branching pattern of XyGs is both of functional and taxonomic importance.[3] Less branched XyGs are less soluble, and this may correlate with functional aspects in those families with the lowly substituted XyGs. An important difference in XyG

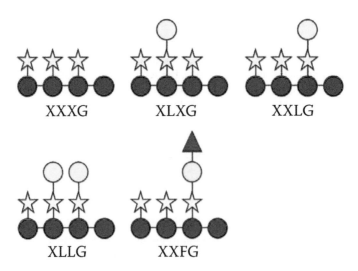

**FIGURE 4.4**   Structures of segments that build up XyGs in many land plants. The building blocks are composed of *glucosyl units* (*blue circle*) joined together by β-1,4-linkages, to which α-1,6-linked *xylosyl units* (*star*) may be added. *Galactose* (*yellow circle*) and *fucose* (*red triangle*) are incorporated into these structures via β-1,2- and α-1,2-linkages, respectively. (Open access figure reproduced by permission of BioMed Central LTD; From von Schantz, L. et al., *BMC Biotechnol.*, 9, 92, http://www.biomedcentral.com/1472-6750/9/92, 2009.)

structure is charged versus uncharged side chains. For example, moss and liverwort XyGs contain galacturonic acid and are structurally distinct from the XyGs synthesized by hornworts and vascular plants.[16]

Most XyGs are composed of either XXXG-type or XXGG-type building blocks.[14] XXXG-type XyGs have three consecutive backbone residues that are substituted with xylose and a fourth unbranched backbone residue. XXGG-type XyGs have two consecutive branched backbone residues and two unbranched backbone residues. The primary walls of a wide range of dicots, nongraminaceous monocots, and gymnosperms contain fucosylated xyloglucans with a XXXG-type structure.[2] The typical subunits in these XyGs are XXFG, XXXG, and XLFG. The less substituted XXGG structure predominates in cell walls of commelinid monocots (e.g., grasses), which also distinguished by the low content of XyGs in the primary cell walls, typically 1%–5% compared with 20%–25% in dicots.[3] Oligosaccharides of the XXGG-type are also found in *solanaceous* XyGs, which are further characterized by lacking fucosyl residues and featuring the arabinose-containing S chain. S chain is also found in XXXG-type subunits in olive.

The role of fucosyl residues in XyGs is a subject of debate.[2] Fucosyl residues are typically absent from seed XyGs but present on the XyGs in the vegetative portions of the same plant.[10] XLFG subunits are more abundant in pea leaf XyGs than in pea stem XyGs.

### 4.3.2 XYLANS

Xylans are a complex group of hemicellulosic polysaccharides with the common feature of a backbone of β-1,4-linked xylose residues (Figures 4.5 and 4.6).[3,17]

A common modification of xylans is substitution with glucuronosyl and 4-*O*-methyl glucuronosyl residues α-1,2-linked to the xylosyl residues of the

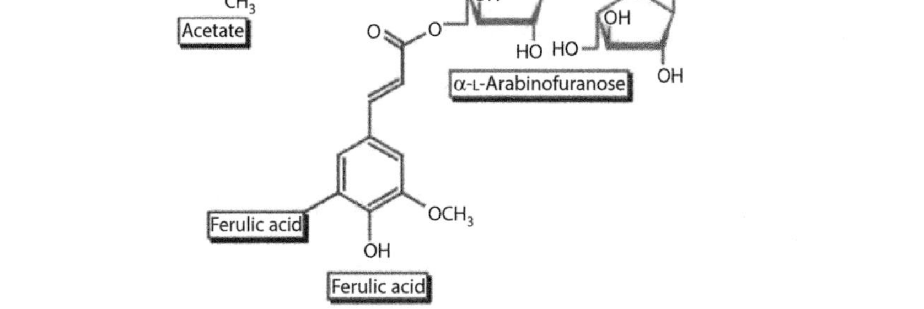

**FIGURE 4.5** General structure showing the various linkages found in a variety of xylans isolated from plant cell walls. (From Dodd, D. and Cann, I.K.O.: Enzymatic deconstruction of xylan for biofuel production. *Glob. Change Biol. Bioenergy.* 2009. 1. 2–17. Copyright Wiley-VCH Verlag GmbH & Co. KGaA. Reproduced with permission.)

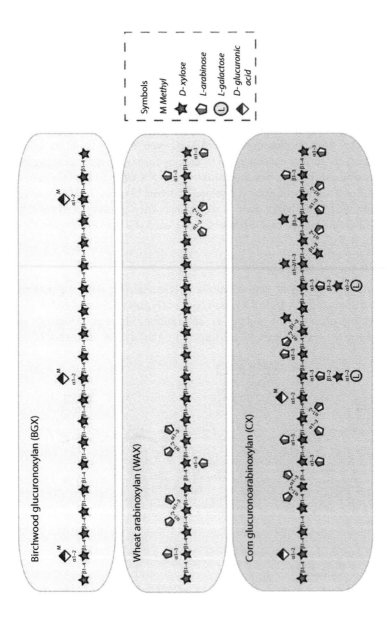

**FIGURE 4.6**  Schematic of the structures of the main classes of xylans. The xylans shown are from birchwood (birch glucuronoxylan; BGX), wheat flour (wheat arabinoxylan; WAX) and corn bran (corn glucuronoarabinoxylan; CX). (Reproduced from Nature/Creative Commons CC-BY; Reprinted by permission from Macmillan Publishers Ltd. *Nature Communications*, Rogowski. A. et al., 2015, copyright 2015.)

backbone.[3] Such xylans are generally known as glucuronoxylans (*GXs*) and are the main hemicellulosic component in the secondary walls of dicots. In commelinid monocots such as grasses, xylans are the major hemicellulosic component of the primary walls, constituting 20%–40% of the wall. These xylans usually contain many arabinose residues (typically α-1,3-linked in grasses) linked to the backbone and are known as arabinoxylans (*AXs*) and glucuronoarabinoxylans (*GAXs*) (Figures 4.7 and 4.8). Grass GAXs make up 35% of cell walls of vegetative tissues, whereas AXs are found mainly in the endosperm cell walls in cereals.[20]

Branching patterns in xylans, like those in XyGs, depend on species and tissues.[17] Arabinofuranose (Ara*f*) substitutions in grass walls are mostly from O3 of the backbone residues, and xylose residues doubly substituted with arabinofuranose at both O2 and O3 are common in grass *endosperm*.[3] Gymnosperm walls also contain arabinoxylans in relatively high amounts. Arabinose substitutions are less frequent in dicot xylans. The arabinose substitutions on dicot xylans are normally at O2 rather than O3 as in grasses.

Unlike XyGs, xylans do not have a repeated structure, and the substitution pattern of the backbone is not well known.[3,20] Most xylans are acetylated to various degrees.[22] Acetyl groups are attached to O3 of xylose residues and to a lesser extent to O2.

In addition to the xylosyl backbone, the reducing end of xylans from dicots and conifers contains the unique conserved oligosaccharide sequence β-xylose-1, 4-β-xylose-1,3-α-rhamnose-1,2-α-galacturonic acid-1,4-xylose (or, in an abbreviated form, β-Xyl-1,4-β-Xyl-1,3-α-Rha-1,2-α-GalA-1,4-Xyl) (Figure 4.8).[3,23]

An important feature of grass xylans is the presence of *ferulic acid* esters attached to O5 of some of the arabinofuranosyl residues.[3] Ferulic acid, also known as coniferic acid or 3-(4-hydroxy-3-methoxyphenyl)-2-propenoic acid, is a hydroxycinnamic acid derived from *phenylpropanoid* metabolism.

In contrast to monocots, ferulic acid in dicots is associated with pectic polysaccharides via ester linkages to O2 of arabinose residues or O6 of galactose residues. Ferulate esters bound to cell walls have been also described in gymnosperms, but it is not clear to which polysaccharide they are bound.[3]

Ferulic acid esterified to cell wall polysaccharides can form dimers crosslinking AXs. Two mechanisms have been proposed for the formation of ferulate crosslinking of hemicelluloses to lignin: (1) oxidative phenol coupling mediated by hydrogen peroxide and peroxidases and (2) dimerization and photoisomerism by UV light. Crosslinking through ferulate esters is assumed to render the cell wall recalcitrant to digestion by microorganisms and herbivores. However, the ferulate esters make grass cell walls also recalcitrant to enzymatic hydrolysis that greatly reduces the efficiency of biofuel and chemicals production.

### 4.3.3 Mannans

Mannans are widespread among land plants and are also present in many algal species, some of which completely lack cellulose in their cell walls.[24,25] Several varieties of mannan polysaccharides have been characterized. Structurally, each of these polysaccharides contains a β-1,4-linked backbone composed of mannose residues as in mannans or a combination of glucose and mannose residues in a nonrepeating

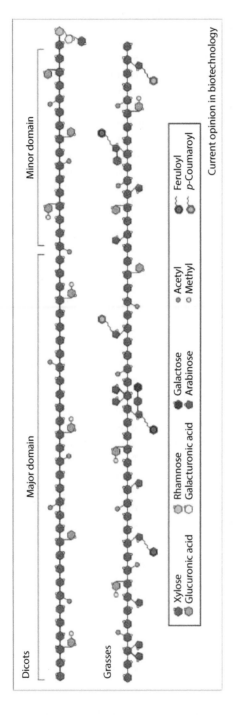

**FIGURE 4.7** Generalized structures of xylans. Dicot xylans are substituted with GlcA, Me-GlcA, and acetate. The major domain has Me-GlcA on evenly spaced about eight xylose units apart, whereas the minor domain has Me-GlcA more closely spaced. In grasses, xylan may also be substituted with arabinose, xylose, galactose, and ferulic and coumaric acid. (Reprinted from *Curr. Opin. Biotechnol.*, 26, Rennie, E.A. and Scheller, H.V., Xylan biosynthesis, 100–107, Copyright 2014, with permission from Elsevier.)

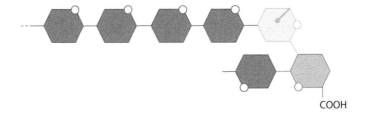

COOH

**FIGURE 4.8**   Oligosaccharide found in the reducing end of xylans from several different species. (Adapted from Scheller, H.V. and Ulvskov, P., *Annu. Rev. Plant Biol.*, 61, 263–289, 2010.)

pattern as in glucomannans; both mannans and glucomannans may be substituted with α-1,6-linked galactose side chains as in galactomannans and galactoglucomannans (Figure 4.9). Mannans and glucomannans are often acetylated. They have been much studied in their role as seed storage compounds, but they are found in variable amounts in all cell walls.[3]

Linear mannans fulfill primarily structural functions particularly in the seeds of many plants, such as ivory nut, green coffee, coconut kernel, and the cell walls of some algae.[26] A low degree of α-1,6-linked galactose substitution results in (1) linear mannan with polymerization degree (DP) of ~15 compacted into dense granular and crystalline structures (mannan I), or in (2) microfibrils (similar to cellulose microfibrils) that are less crystalline and with a higher DP of ~80 (mannan II). Both forms are insoluble and provide rigidity and protection to the endosperm. Once germination takes place, these mannans can be mobilized as nonstarch storage polysaccharides.

The α-1,6-linked galactose side groups in galactomannans prevent close association between adjacent polymers, resulting in a more amorphous structure that retains water and contributes to its water solubility.[26] They are commonly found in the endosperm of guar and other seeds of plants belonging to the legume family. Water retention by galactomannans is particularly important to the long-term survival of legume seeds in arid areas, whereas these storage polysaccharides are mobilized during germination.

Glucomannans (substituted or not) form the major hemicellulose fraction of softwoods and can represent up to 50% of the hemicelluloses in coniferous woods.[26] They typically contain linear chains of β-1,4-linked mannose and glucose in a 3:1 ratio with a DP greater than 200. The glucomannans in hardwoods have a glucose:mannose ratio of 1:1.5-2 with a DP of ~60–70. Glucomannans are closely associated with cellulose and xylans as cell wall components. They are organized in paracrystalline arrays and are adsorbed on the cellulose microfibrils. Glucomannans may also be acetylated up to a level of 18%.

Galactoglucomannans may contain α-1,6-linked galactose residues on glucose or mannose residues with a mannose:glucose:galactose ratio of 3:1:1.[26] They may also contain *O*-acetyl groups. Acetylated galactoglucomannans form the major hemicelluloses of softwoods and have a DP of 100–150, thus, fulfilling structural functions similar to that of xylans in hardwoods.

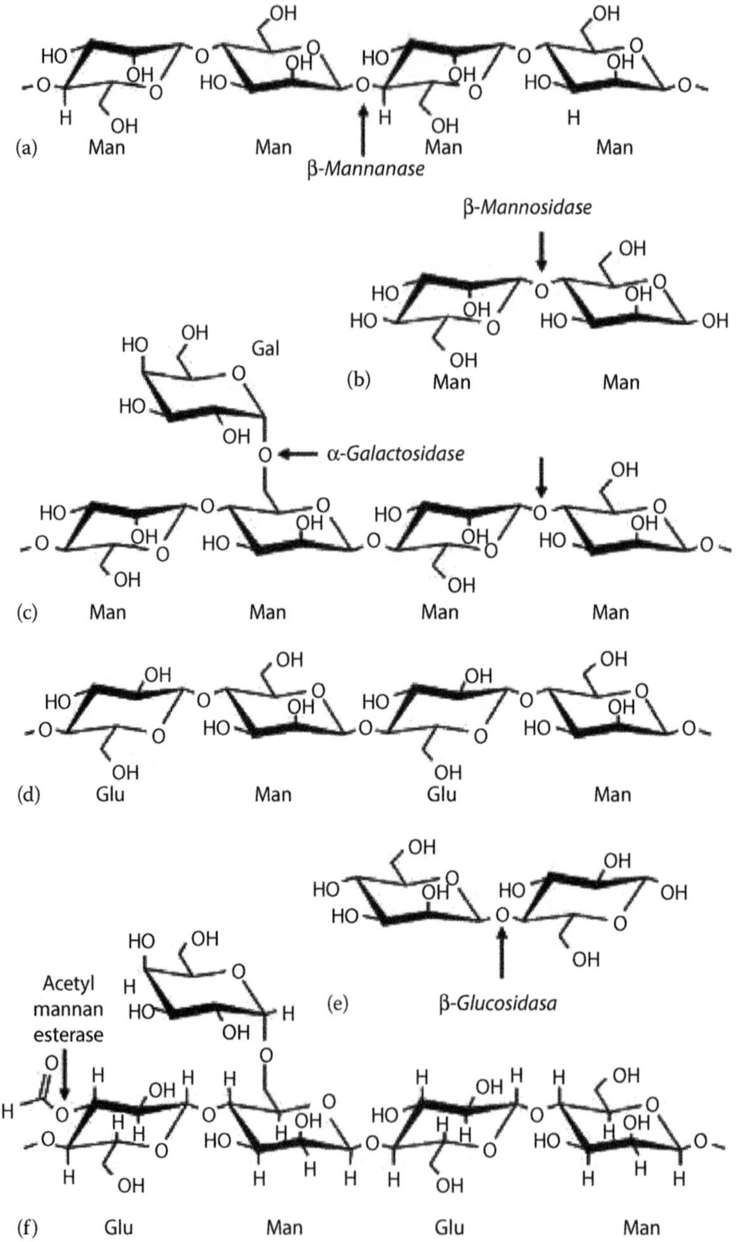

**FIGURE 4.9** Structures of different forms of mannans/glucomannans and the enzymes required for their hydrolysis. Typical structures of (a) linear mannan, (b) mannose dimer, (c) branched galactomannan, (d) linear glucomannan, (e) glucomannose, and (f) branched galactoglucomannan are shown. (Reprinted from *Process Biochem.*, 45, van Zyl, W.H. et al., Fungal β-mannanases: Mannan hydrolysis, heterologous production and biotechnological applications, 1203–1213, Copyright 2010, with permission from Elsevier.)

Mannans appear to have been very abundant in early land plants and are still abundant in mosses and lycophytes. In seed plants, mannans and glucomannans are generally much less abundant and it appears that other hemicelluloses have largely replaced them.[3]

### 4.3.4 MIXED-LINKAGE GLUCANS

(1,3;1,4)-β-Glucans, also known as mixed-linkage glucans, consist of unbranched and unsubstituted chains of (1,3)- and (1,4)-β-glucosyl residues. They are dominated by cellotriosyl units (trisaccharides) and cellotetrasyl units (tetrasaccharides) by β-1,3-linkages, but longer β-1,4-linked segments also occur (Figure 4.10).[3] The two types of linkage are arranged neither at random nor with predictable regularity.[27]

The ratio of (1,4)-β-glucosyl residues to (1,3)-β-glucosyl residues in these mixed-linkage glucans appears to influence not only the physicochemical properties of the polysaccharide and therefore its functional properties in cell walls but also its adoption by different plant species during evolution.[27] The (1,3;1,4)-β-glucans are widely distributed as noncellulosic matrix phase polysaccharides in cell walls of the Poaceae (also known as the Gramineae), which evolved relatively recently and consist of the grasses and commercially important cereal species, but they are less commonly found in lower vascular plants, such as the horsetails, in algae and in fungi.[29] The (1,3;1,4)-β-glucans have often been considered to be components mainly of primary cell walls, but recent observations indicate that they can also be located in secondary walls of certain tissues.[27]

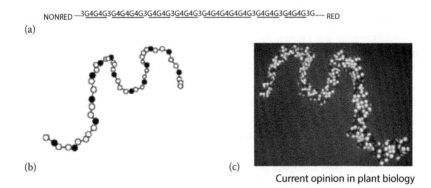

(a)

NONRED ····3G4G4G3G4G4G4G3G4G4G3G4G4G3G4G4G4G4G4G3G4G4G3G4G4G3G······ RED

(b)                          (c)

Current opinion in plant biology

**FIGURE 4.10** Structural features of cereal (1,3;1,4)-β-glucans (a) Representation of the structure of a typical cereal (1,3;1,4)-β-glucan showing the chain of glucosyl residues (G) linked by β-1,3-linkages (3) and β-1,4-linkages (4). The nonreducing and reducing ends of the chain are indicated. (b) Representation of the three-dimensional conformation of the barley mixed-linkage, showing (1,3)-β-glucosyl residues as black circles and (1,4)-β-glucosyl residues as white circles. (c) Space-filling model of a portion of the barley (1,3;1,4)-β-glucan. (Reprinted from *Curr. Opin. Plant Biol.*, 12, Fincher, G.B., Exploring the evolution of (1,3;1,4)-β-d-glucans in plant cell walls: Comparative genomics can help! 140–147, Copyright 2009, with permission from Elsevier.)

### 4.3.5  RELATED MOLECULES: PECTINS

Pectins, which are also found in plant cell walls especially as gel matrix, are heterogeneous and highly hydrated polysaccharides that contain α-1,4-linked-D-galacturonic acid residues.[30]

Three pectic polysaccharides have been structurally characterized: *homogalacturonans*, substituted galacturonans (especially *xylogalacturonans*), and *rhamnogalacturonans*. Homogalacturonans (HG) are linear chains of α-1,4-linked-D-galacturonic acid residues in which some of the carboxyl groups are methyl esterified or partially *O*-acetylated. Xylogalacturonans (XG) have β-D-xylose residues attached to C3 of the galacturonan backbone. Rhamnogalacturonans I (RG-I) are a group of pectic polysaccharides that contain a backbone of the repeating rhamnose-galacturonic acid disaccharide. From rhamnose, side chains of various neutral sugars branch off. The neutral sugars are mainly D-galactose and L-arabinose. Rhamnogalacturonans II (RG-II) are highly branched pectins that contain at least eleven different glycosyl residues.[31] Their backbone contains at least eight α-1,4-linked galacturonic acid residues.

## 4.4  BIOSYNTHESIS

### 4.4.1  GENERAL

In contrast to cellulose that is synthesized at the cell surface in the plasma membrane, hemicelluloses, like most noncellulosic polysaccharides, are synthesized in the Golgi apparatus and delivered to the wall by secretory vesicles (Figure 4.11).[32–35]

Hemicelluloses are synthesized from *sugar nucleotides* by *glycosyltransferases* (*GTs*) located in the Golgi membranes.[3,37] Many GTs needed for biosynthesis of XyGs and mannans are known. In contrast, the biosynthesis of xylans and β-1,3;1, 4-glucans remains elusive.

The majority of the enzymes involved in noncellulosic polysaccharide synthesis are *integral membrane proteins*.[32] Several Golgi proteins that are involved in glucomannan and XyG biosynthesis have been identified, including some *glycan synthases* that show sequence similarities to the *cellulose synthases*, and several GTs that add side chains to the polysaccharide backbones.[35] Recent progress in identifying the proteins needed for polysaccharide biosynthesis should lead to an improved understanding of these complex processes and possibly to an ability to manipulate them to generate plants that have improved properties for human uses.

### 4.4.2  XYLOGLUCANS

#### 4.4.2.1  Glycosyltransferases Involved in Xyloglucan Synthesis

Considerable progress has been made in the identification of GTs involved in the biosynthesis of XyGs.[3,38] The XyG fucosyltransferase (named FUT1; family GT37[39]) was one of the first cell wall biosynthetic enzymes to be identified.[40,41] The XyG α-1,2-fucosyltransferase directs addition of fucose residues to terminal galactose residues on XyG side chains. A XyG galactosyltransferase (named MUR3; family

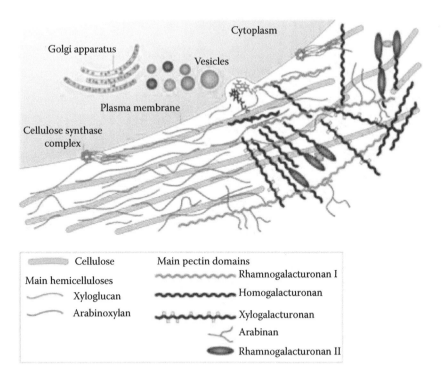

Cellulose

Main hemicelluloses
Xyloglucan
Arabinoxylan

Main pectin domains
Rhamnogalacturonan I
Homogalacturonan
Xylogalacturonan
Arabinan
Rhamnogalacturonan II

**FIGURE 4.11**   Structure of the primary cell wall. (Reprinted by permission from Macmillan Publishers Ltd. *Nature Reviews Molecular Cell Biology*, Cosgrove, D.J. et al., 2005, copyright 2005.)

GT47) that attaches a β-galactose residue specifically to the third xylose residue within the XXXG core structure has also been identified.[42] The galactosyltransferase has been shown to be highly specific for the third galactose in the repeating unit. Therefore, the second galactose in the XXXG must be incorporated by a different galactosyltransferase. A GT in the GT47 family has been shown to be responsible for the activity, but the final evidence has not yet been reported.[3,43]

The xylose residues in XyGs are transferred from UDP-xylose onto β-1,4-glucan chains by α-1,6-xylosyltransferases (XT), which are retaining enzymes from family GT34.[3,44] Two enzymes, XXT1 and XXT2, formerly named XT1 and XT2, have been identified in *Arabidopsis* and shown to be involved in the synthesis of XyGs.[45]

The backbone of XyGs is apparently synthesized by members of *cellulose synthase-like* (CSL) proteins belonging to the CSLC family (Figure 4.12). CSL proteins are a group of plant GTs that are predicted to synthesize β-1,4-linked polysaccharide backbones.[46]

### 4.4.2.2   Hydrolases Involved in Xyloglucan Synthesis

The GTs mentioned earlier are all localized in the Golgi and work together to produce a XyG precursor that is transported to the wall.[3] However, important changes to the XyG molecules take place after the initial synthesis in the Golgi. It has been

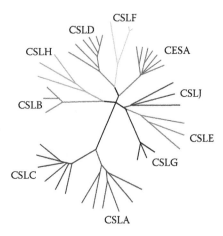

**FIGURE 4.12** Schematic illustration of the phylogeny of the cellulose synthase (*CESA*) and cellulose synthase-like (*CSL*) superfamily of GTs. Not all families are present in the same species. CSLH and CSLF are only known from grasses, and CSLB and CSLG are only known from dicots. CSLE is known from all angiosperms but not outside these groups. CESA, CSLA, CSLC, and CSLD are known from all plant genomes analyzed so far. CSLJ is present in some angiosperms, both dicots and grasses, but not in Arabidopsis and rice. (Adapted from Scheller, H.V. and Ulvskov, P., *Annu. Rev. Plant Biol.*, 61, 263–289, 2010.)

shown that specific *apoplastic* (extracellular or cell wall)[47] *glycosidases* are responsible for the trimming of nascent XyG chains. In general, hydrolases are likely to play an important role in determining hemicellulose structures in the wall, and it may be noted that many hydrolases are coexpressed with polysaccharide biosynthetic enzymes. It is interesting to note that the *Arabidopsis* genome contains ~300 membrane-bound GTs and more than 500 glycoside hydrolases and *lyases*, many of which are involved in modification of wall polysaccharides.

Another important modification of XyG is carried out by xyloglucan endotransglycosylase (XET) enzymes, which can cut and rejoin XyG chains (see Section 4.4.1). XET catalyzes molecular grafting between XyG molecules by endolytically cleaving the XyG backbone. This is followed by the formation of a covalent enzyme substrate intermediate and a deglycosylation step involving a XyG acceptor, which releases the enzyme and leads to the formation of a new β-1,4-glycosidic bond.[48]

### 4.4.2.3 Acetylation

XyGs, like xylans, mannans, and pectins are usually acetylated.[3] Acetylation of XyG is on the galactose residues, mostly on O6. Acetylation of cell wall polysaccharides takes place in the Golgi by means of transferases using acetyl-CoA.

### 4.4.3 XYLANS

In grasses and in dicot secondary cell walls, the major hemicellulose is xylan. Unlike cellulose, xylan is synthesized by enzymes in the Golgi apparatus. Xylan synthesis, thus, requires the coordinated action and regulation of these synthetic enzymes

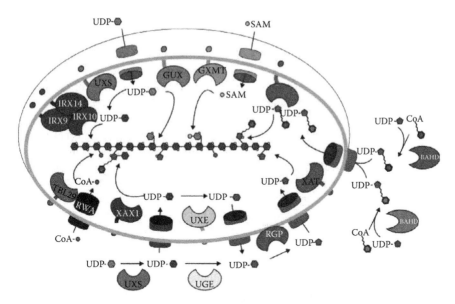

**FIGURE 4.13** Xylans are synthesized in the Golgi apparatus by Type II membrane proteins anchored by a single N-terminal transmembrane domain and with their catalytic domains in the Golgi lumen. Some proteins, such as IRX 10, are predicted to lack a transmembrane domain. Substrates are synthesized in both the cytosol and the lumen. UDP-GlcA is transported into the Golgi and converted into UDP-Xyl Synthase (UXS). UDP-Xyl is converted to UDP-Ara*p* inside the Golgi by UDP-Xyl Epimerase (UXE), and the UDP-Ara*p is* converted to UDP-Ara*f*. (From Curr. Opin. Biotechnol., 100, Rennie, E.A. and Scheller, H.V., Xylan biosynthesis, 26, Copyright 2014, with permission from Elsevier.)

as well as others that synthesize and transport substrates into the Golgi.[21] Recent research has identified several genes involved in xylan synthesis, some of which have already been used to create plants that are better suited for biofuel production.[21,49]

Several enzymes have been implicated in xylan synthesis, many in the last few years.[21] Figure 4.13. shows a schematic overview of xylan biosynthesis. Two members of glycosyltransferase family 43 (GT43) and one member of GT47 encode putative xylosyltransferases required for synthesizing the xylan backbone.

The biosynthesis of xylans will be presented below three subtitles: backbone, reducing end, and side chains.[22,23,50–53]

#### 4.4.3.1 Backbone

None of the CSL proteins is apparently involved in xylan biosynthesis.[3] Instead, characterization of xylan-deficient mutants *irx9* (*IRREGULAR XYLEM* 9), *irx14*, *irx10*, and *irx10-like* has indicated that the corresponding GTs belonging to families GT43 and GT47 are responsible for elongation of xylan backbone.[3] Unlike the CSL proteins with multiple transmembrane segments, the members of GT43 and GT47 are predicted to be type II membrane proteins with a single N-terminal membrane anchor.

In 2010, Lee et al.[23] studied the roles of the four members of the *Arabidopsis* family GT43 (IRX9, IRX14, I9H [IRX9 *homolog* or IRX9-L], and I14H [IRX14 homolog or IRX14-L]) in glucuronoxylan biosynthesis. Their results have shown that all four GT43 members are involved in glucuronoxylan biosynthesis and have suggested that they form two functionally nonredundant groups essential for the normal elongation of glucuronoxylan backbone.

In 2010 also, Wu et al.[54] investigated the role of eight genes from *Arabidopsis* in glucuronoxylan synthesis. Their results have revealed a set of four genes (*IRX9, IRX10, IRX14,* and *FRA8*) that perform the main role in glucuronoxylan synthesis during vegetative development, and a second set of four genes (*IRX9-L, IRX10-L, IRX14-L,* and *F8H*), homologs of the first set, which are able to partially substitute for their respective homologs and normally perform a minor function.

In 2012, Lee et al.[55] provided the first biochemical evidence that *Arabidopsis* IRX9 and IRX14, which have been shown to function nonredundantly, are xylosyltransferases (XylTs) that function cooperatively in the elongation of the xylan backbone. They further extended this finding in a work demonstrating that two poplar GT43 glycosyltransferases are xylan XylTs that act cooperatively in catalyzing the successive transfer of xylosyl residues during xylan backbone synthesis.[56]

### 4.4.3.2 Reducing End Oligosaccharide

Treatment of xylans with xylanases allows the isolation of the reducing end oligosaccharide with the structure β-Xyl-1,4-β-Xyl-1,3-α-Rha-1,2-α-GalA-1,4-Xyl. It has been shown that IRX7, IRX8, and PARVUS[57] proteins are specifically involved in its biosynthesis.[3]

It is unclear how many GTs are required to make the oligosaccharide.[3] There must be at least:

- A xylosyltransferase specific for Rha-GalA-Xyl
- A rhamnosyltransferase specific for GalA-Xyl
- A galacturonic acid transferase specific for the terminal xylose

The rhamnose-specific xylosyltransferase and the galacturonic acid-specific rhamnosyltransferase are expected to be *inverting enzymes*, whereas the xylose-specific galacturonic acid transferase would be a *retaining enzyme*. IRX8, a member of GT8, is the most obvious candidate for the galacturonic acid transferase. PARVUS has been reported to be located in the endoplasmic reticulum, indicating an earlier biosynthetic step than the subsequent Golgi-localized steps. Hence, PARVUS, also a member of GT8, is presumably an α-xylosyltransferase transferring the reducing end xylose to a primer, which may be a lipid. IRX7, an inverting enzyme belonging to GT47, is a candidate for the rhamnose-specific xylosyltransferase.

The function of the reducing end oligosaccharide is not yet clearly established.[3] At first sight, the most obvious function would be a primer for elongation of the xylan backbone via addition of xylose residues at the nonreducing end, because polymerization of cell wall polysaccharides is generally assumed to take place at the nonreducing ends of the growing chains.[58] However, it has been proposed that xylan backbone could be elongated from its reducing end, with the oligosaccharide functioning as a terminator sequence.[59] This could explain why *irx7* and *irx8* mutants have unusually long xylan chains.

### 4.4.3.3  Side Chains

The most important side chains of xylans should be formed by α-glucuronosyltransferases and α-arabinofuranosyltransferases.[3] Both activities have been detected *in vitro*, but the GTs responsible for the transfer have not been identified. In 2010, Zeng et al.[60] revealed that a GAX synthase complex from wheat contains members of the families GT43, GT47, and GT75 and functions cooperatively. Their hypothesis was that the GT43 protein is the XylT responsible for the synthesis of the xylan backbone of GAX polymers, and the most logical function of the GT47 protein is GAX-AraT.

GAX in primary walls has been shown to be synthesized in a more highly arabinosylated form and subsequently trimmed by arabinofuranosidases, resulting in much fewer substituted polymers in mature cells.[3] Highly substituted arabinoxylans are more soluble and does not interact with cellulose. Presumably, the soluble form is ideal for initial integration into an expanding wall, whereas the less substituted polymer functions in the mature wall by interaction between insoluble xylan molecules and cellulose.

Ferulic acid esters attached to O5 of some of the arabinofuranosyl residues are important components of grass arabinoxylans.[3] The feruloyl transferases are not known, but the transfer takes place intracellularly in the Golgi. Feruloyl-CoA is the most likely substrate, although feruloyl-glucoside was suggested as a possible substrate. Nevertheless, it is interesting to note that Mitchell et al.[61] identified candidate genes for the synthesis and feruloylation of arabinoxylan in 2007. They suggested that genes in the family GT43 encode β-1,4-xylan synthases, genes in the family GT47 encode xylan α-1,2- or α-1,3-arabinosyl transferases, and genes in the family GT61 encode the xylosyl transferases responsible for adding β-1,2-xylosyl residues to the feruloyl arabinose residues, and identified putative acyl-transferases as candidate feruloyl transferases.

### 4.4.4  Mannans

A mannan synthase involved in making the backbone of galactomannans in guar was shown to be a member of the CSLA family.[3] Later, several CSLA members have been shown to have mannan and glucomannan synthase activity, apparently being able to utilize both GDP-mannose and GDP-glucose as substrates.

In 2011, Verhertbruggen et al.[46] gathered evidence that three CSLD proteins are also mannan synthases. Furthermore, their work revealed a complex interaction among the three GTs and brought new evidence regarding the formation of noncellulosic polysaccharides through multimeric complexes.

A galactosyltransferase involved in making galactomannans was identified in fenugreek (a plant in the family Fabaceae or Leguminosae) and was the first GT involved in synthesis of plant cell walls for which the activity of the pure enzyme was shown.[3,62] Although the fenugreek enzyme, which belongs to GT34, is involved in making seed galactomannans, it appears to have *orthologs* in dicots that have not been reported to have seed galactomannans, such as *Arabidopsis*.[3] Scheller et al.[3] proposed that the apparent orthologs in *Arabidopsis* signify that the seeds do contain galactomannan, in addition to oil. Seed galactomannans are known outside legumes, for example, in tobacco and coffee, and these species have galactosyltransferases similar to the fenugreek enzyme. Mannans are likely to be synthesized as more highly substituted polymers that are subsequently trimmed by α-galactosidases.

### 4.4.5  MIXED-LINKAGE GLUCANS

The biosynthesis of β-1,3;1,4-glucans has been shown to involve CSLF and CSLH proteins.[3] The corresponding gene families are absent in *Arabidopsis* and poplar and present in rice and *Brachypodium* (a wild annual grass) consistent with a grass-specific occurrence. It appears that the CSLF and CSLH proteins do not need to be present simultaneously for β-1,3;1,4-glucan synthase activity to occur. It has been shown that *CSLF6* is the major *CSLF* gene expressed in wheat and barley seedlings, with very low level of expression of other *CSLF* isoforms and *CSLH*, whereas *Bachypodium* seedlings show a high expression level of *CSLH*. Isoforms of *CSLH* and *CSLF* have been localized in the Golgi, and *in vitro* activity studies have been consistent with synthesis of β-1,3;1,4-glucan in the Golgi. However, unlike other matrix polysaccharides, β-1,3;1,4-glucan has not been detected in the Golgi.

### 4.4.6  ROLE OF CESA-CSL SUPERFAMILY

The cellulose synthase and cellulose synthase-like superfamily of GTs are involved in synthesis of cellulose (CESA), XyG (CSLC), glucomannans (CSLA and CSLD), and β-1,3;1,4-glucans (CSLF and CSLH).[3] It seems highly likely that the other members of the CESA-CSL superfamily have a similar function.[63]

## 4.5  FUNCTION IN PLANT CELL WALLS

The structural similarity between the hemicelluloses and cellulose facilitates a strong noncovalent association of the hemicelluloses with cellulose microfibrils.[64]

### 4.5.1  XYLOGLUCANS

The cellulose-XyG network is believed to be the major load-bearing structure in the primary wall.[2] XyGs coat the surface of the cellulose microfibrils, limiting their aggregation and connecting them via tethers that directly or indirectly regulate the mechanical properties of the wall.

#### 4.5.1.1  Xyloglucan Endo-Transglycosylase/Hydrolases

The breaking and reformation of the glucosidic bonds in the XyG backbone by *xyloglucan endotransglycosylase/hydrolases* (*XTHs*) have been proposed as a mechanism whereby the wall can expand and grow under turgor pressure without losing its mechanical strength.[65–71] The *XTH* genes encode proteins that can potentially have two distinct catalytic activities, with radically different effects on XyG:XyG endotransglycosylase (XET) activity (formally, xyloglucan:xyloglucosyl transferase)

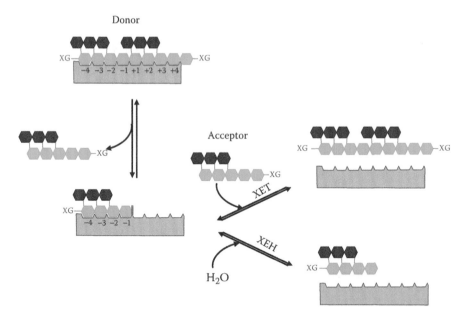

**FIGURE 4.14** A schematic representation of the mechanism used by XETs and XEHs. Left, XyG (glucosyl units in green and xylosyl units in purple) binds to XETs and XEHs in both negative and positive subsites (subsite nomenclature of Davies et al.[73] as per Figure 2.13; nonreducing end of the substrate drawn on the left; *reducing end* of the substrate drawn on the *right*). After binding, the substrate is cleaved, resulting in a covalent glycosyl-enzyme intermediate (bond indicated in red). Right, in the last step, the glycosyl-enzyme intermediate is broken down by an incoming acceptor, either water (XEH activity) or the nonreducing end of a XyG molecule (XET activity). (Adapted from Eklof, J.M. and Brumer, H., *Plant Physiol.*, 153, 456, 2010.)

results in the nonhydrolytic cleavage and ligation of XyG chains, whereas xyloglucan endohydrolase (XEH) activity (formally, xyloglucan-specific endo-β-1,4-glucanase) yields irreversible chain shortening.[71]

XET acts similarly to the glycoside hydrolases in the first reaction step but prefers a carbohydrate acceptor over a water molecule in the second reaction step (Figure 4.14).[72] This strong preference for transglycolyzation instead of hydrolysis allows XET to perform its natural function (i.e., religation of the nascent donor end of one XyG molecule to the nonreducing, acceptor, end of another XyG molecule).

Since the major review on the XTH family by Rose et al.[69] in 2002, there have been significant advances in the protein–function relationships encoded by the *XTH* gene family, including the first three-dimensional structure of a plant (poplar) XET solved by Johansson et al.[72] in 2004, and the first three-dimensional structure of a plant (nasturtium, cabbage family) XEH solved by Baumann et al.[74] in 2007.[71,75]

Poplar XET (PttXET16A) exhibits a curved β-sandwich (β-jellyroll) that is typical for enzymes in the glycoside hydrolase family GH16 except for the long C-terminal linker that forms the only α-helix in the molecule and a short additional β-strand (Figure 4.15).[72] The extension of the acceptor binding site, together with variations

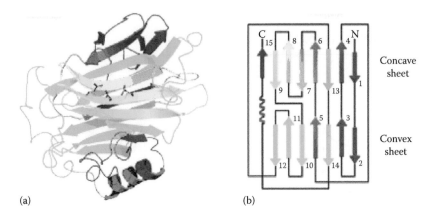

(a)                                                           (b)

**FIGURE 4.15** Features of the poplar XET molecular structure. (a) Poplar XET consists of two large β-sheets arranged in a sandwich-like manner. The active site residues E85, D87, and E89 are included in the figure. (b) XET topology. N- to C-terminals are color-coded red to blue. The nucleophile and the general acid/base (catalytic acid/base) reside on β7. (Reproduced from Johansson, P. et al., *Plant Cell*, 16, 874, http://www.plantcell.org/cgi/content/full/16/4/874, 2004. With permission of American Society of Plant Biologists.)

in length and conformation of loops connecting the strands in the β-sandwich, produces an active site that is unique to the XET family of enzymes.

The structure of the archetypal XEH from nasturtium, TmNXG1 (NXG for nasturtium xyloglucanase), was solved three years after that of the poplar XET (Figure 4.16).[74]

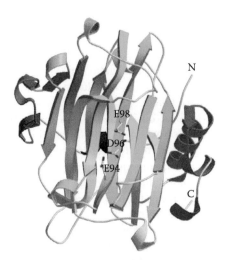

**FIGURE 4.16** Three-dimensional structure of TmNXG1. The polypeptide chain is colored from blue (N terminus) to red (C terminus). The three strictly conserved amino acids forming the catalytic machinery, E94, D96, and E98, are labeled. (Reproduced from Baumann, M.J. et al., *Plant Cell*, 19, 1947, http://www.plantcell.org/cgi/content/full/19/6/1947?ijkey=eb13d280656f78 b1561142d8f8f7350e65429869, 2007. With permission of American Society of Plant Biologists.)

Both the PttXET16 and TmNXG1 structures display the β-jellyroll fold common to all members of GH16, but with notable differences that reflect the specialization of these enzymes toward their highly branched substrate.[71] In addition to three loop variants (called loop 1, loop 2, and loop 3), an N-terminal extension, a C-terminal truncation, three small loop truncations, and a shortening of the C-terminal α-helix were observed in the TmNXG1 enzymes relative to PttXET16.[74]

### 4.5.1.2 Action of XTHs

The structure of the cellulose-xyloglucan network can be seen as including three XyG domains[2]:

- Regions of XyGs that are bound directly to the cellulose surface.
- Regions of XyGs that are not in direct contact with the cellulose and act as tethers between two neighboring microfibrils (these regions are the cross-linking tethers).
- Regions of XyGs that are entrapped within the cellulose microfibrils.

It is likely that XTHs act on the tether domains rather than the bound or entrapped XyG domains.[2,69]

Two main roles have been proposed for XTHs in growing cells[69]:

1. Wall restructuring (Figure 4.17). XTH may reversibly (XET) or irreversibly (XEH) loosen existing cell wall material, enabling cell expansion. In favor of this hypothesis, XET activity is often correlated with growth rate. Moreover, XyG turnover is associated with rapid cell expansion.
2. Integration into the cell wall (Figure 4.18). XTHs may also catalyze the integration of newly synthesized XyGs into the cell wall through XET activity. Such integration is necessary for wall synthesis in meristems.

The binding of cellulose with XyGs is likely to be a complex topological process.[2] Conformational energy calculations suggest that for binding to cellulose to occur the XyG backbone must adopt a *flat ribbon* conformation of which surface is complementary to that of cellulose. In solution, XyGs are likely to adopt a *twisted* conformation that is not complementary to cellulose. It has been presumed that binding of xyloglucan to cellulose untwists the XyG backbone.

### 4.5.2   Xylans, Mannans, and Mixed-Linkage Glucans

Xylans are the main hemicelluloses that crosslink with cellulose in secondary cell walls of dicot plants.[23] The precise role of xylans in the secondary cell walls is not clear.[22] However, it has been suggested that xylans coat the cellulose microfibrils[76] and may influence the helicoidal orientation of the microfibrils, while being a host structure for lignin precursors during lignification of the wall.[77] Xylans have been shown to be a major load-bearing structure in grasses.[3] The important role of xylans in strengthening secondary walls is very clear from analysis of xylan-deficient mutants. All these contain collapsed xylem vessels and have severely impacted

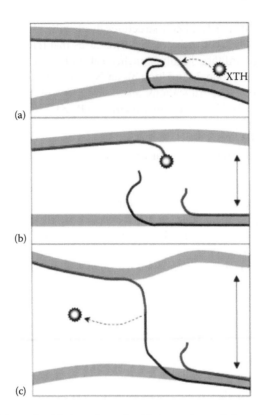

**FIGURE 4.17** Polysaccharide-to-polysaccharide transglucosylation of xyloglucan in the cell wall showing restructuring-type transglycosylation. (*Step 1; a and b*) XTH cleaves a xyloglucan chain (red line), which is acting as a tether between two neighboring microfibrils. This xyloglucan chain is broken, and a xyloglucan-XTH complex is formed. If the cell is turgid, the microfibrils can now move further apart, as indicated by the unbroken arrows. (*Step 2; b and c*) The xyloglucan-XTH complex is now out of reach of the new nonreducing end but within reach of the *nonreducing* end of an adjacent xyloglucan chain (blue line). The latter acts as an *acceptor* substrate, and a tether is thereby reformed between the two microfibrils. The reducing ends of the XyG chains are to the right, using the common nomenclature. (From Rose, J.K.C. et al., *Plant Cell Physiol.*, 2002. By permission of Oxford University Press.)

growth and fertility. Xylans contribute to the recalcitrance of secondary walls to enzymatic degradation.[78] They become a barrier to efficient cellulase function and must be removed for efficient cell wall saccharification.[79] Thus, reducing the xylan content of secondary walls and altering xylan structure, molecular weight, ease of extractability, and susceptibility to enzymatic degradation are key targets for the genetic improvement of plants.

Mannans are widespread in *Arabidopsis* tissues and may be of particular significance in both lignified and nonlignified thickened cell walls.[7] It was shown that *Arabidopsis* can be used as a model plant in studies of synthesis and functions of

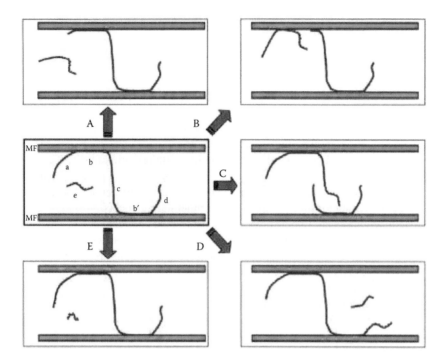

**FIGURE 4.18** Polysaccharide-to-polysaccharide transglucosylation of XyG in the cell wall showing integrational transglucosylation. There are five possible modes of integrational endo-transglucosylation between a newly secreted XyG (*blue line*) and a previously wall-bound XyG chain (*red line*). The *reducing end is the right-hand end* of each chain. In the initial state (left center panel), four different segments of the wall-bound chain are distinguished: (a) a nonreducing loose end, (b, b′) regions anchored to cellulose microfibrils by hydrogen bonding, (c) a tether between the microfibrils, and (d) a reducing loose end. Arrows (A to D) indicate what happens when XTH cleaves the wall-bound chain at sites a to d respectively, and the newly secreted chain acts as acceptor. Arrow E shows what happens if the XTH attacks the newly secreted chain at site e and the previously wall-bound chain acts as an acceptor (nonreducing). (From Rose, J.K.C. et al., *Plant Cell Physiol.*, 2002. By permission of Oxford University Press.)

mannans. Apart for the role of galactomannans as seed storage compounds, it is unclear what specific roles mannans have.[3] Mannans and glucomannans are highly conserved through plant evolution. Preliminary evidence suggests that mannan biosynthesis is affected in the mutants.

The exact role of (1,3;1,4)-β-glucans in the cell wall architecture of the Poaceae and other families with wall containing (1,3;1,4)-β-glucans remains unclear.[80] Models of the architecture of these walls have been proposed in which there are two coextensive but independent polymer networks. Initially, (1,3;1,4)-β-glucans and GAX were considered to form crosslinks, but the hypothesis was abandoned. From a new study, it was concluded that much of the (1,3;1,4)-β-glucans are tightly associated with microfibrils, rather than forming a gel-like matrix.

## 4.6 SUMMARY

The key points on the structure and biosynthesis of hemicelluloses include the following:

- Hemicelluloses are cell-wall polysaccharides and constitute a major part of lignocellulosic biomass.
- All the hemicelluloses show structural differences between different species and in different cell types within plants.
- The main role of hemicelluloses is to tether cellulose microfibrils, thereby strengthening the cell wall.
- Xyloglucans dominate in primary walls of dicots and conifers, whereas glucuronoarabinoxylans dominate in commelinid monocots.
- Hemicelluloses are important components in food and feed.[3]
- Hemicelluloses are synthesized by glycosyltransferases located in the Golgi membranes. The backbones of XyG, mannan, and mixed-linkage glucan are synthesized by members of the CSL family, which are multimembrane-spanning proteins.
- Xylan backbones are apparently synthesized by GTs that are type II membrane proteins with a single membrane-spanning segment.

## REFERENCES

1. D. Hildebrand, *Plant Biochemistry*, BCH/PPA/PLS 609, 2010. www.uky.edu/~dhild/biochem/11B/lect11B.html.
2. Complex Carbohydrate Research Center, The University of Georgia, *Plant Cell Walls*, Hemicelluloses, 2007. http://www.ccrc.uga.edu/~mao/xyloglc/Xtext.htm.
3. H.V. Scheller and P. Ulvskov, *Ann. Rev. Plant Biol.* 61, 263, 2010. http://arjournals.annualreviews.org/doi/full/10.1146/annurev-arplant-042809-112315?amp;searchHistoryKey=%24%7BsearchHistoryKey%7D.
4. K. Keegstra, *Plant Physiol.* 154, 483, 2010. http://www.plantphysiol.org/content/154/2/483.full.
5. P.M. Dey and J.B. Harbone, Eds., *Plant Biochemistry*, Academic Press, San Diego, CA, 1997.
6. R.H. Atalla, J.M. Hackney, I. Uhlin and N.S. Thompson, *Int. J. Biol. Macromol.* 15, 109, 1993. http://www.fpl.fs.fed.us/documnts/pdf1993/atall93a.pdf.
7. M.G. Handford, T.C. Baldwin, F. Goubet, T.A. Prime, J. Miles, Y. Xiaolan and P. Dupree, *Planta* 218, 27, 2003. http://cat.inist.fr/?aModele=afficheN&cpsidt=15433283.
8. A.P.S. Sandhu, G.S. Randhawa and K.S. Dhugga, *Mol. Plant* 2, 840, 2009. http://mplant.oxfordjournals.org/content/2/5/840.full and in http://www.cell.com/molecular-plant/abstract/S1674-2052(14)60700-0.
9. Z.A. Popper and M.G. Tuohy, *Plant Physiol.* 153, 373, 2010. http://www.plantphysiol.org/cgi/content/full/153/2/373.
10. S.C. Fry, *J. Exp. Bot.* 40, 1, 1989. http://jxb.oxfordjournals.org/cgi/content/abstract/40/1/1.
11. N. Raikhel, Center for Plant Cell Biology at UC Riverside. www.cepceb.ucr.edu/members/raikhel.htm.
12. Complex Carbohydrate Research Center, The University of Georgia. http://www.ccrc.uga.edu/~mao/intro/ouline.htm.

13. S.E. Marcus, Y. Verhertbruggen, C. Herve, J.J. Ordaz-Ortis, V. Farkas, H.L. Pedersen, W.G.T. Willats and J.P. Knox, *BMC Plant Biol.* 8, 60, 2008. http://www.biomedcentral.com/1471-2229/8/60.

14. J.P. Vincken, W.S. York, G. Beldman and A.G.J. Voragen, *Plant Physiol.* 114, 9, 1997. http://www.plantphysiol.org/cgi/reprint/114/1/9.pdf.

15. L. von Schantz, F. Gullfot, S. Scheer, L. Filonova, L. Cicortas Gunnarsson, J.E. Flint, G. Daniel, E. Nordberg Karlsson, H. Brumer and M. Ohlin, *BMC Biotechnol.* 9, 92, 2009. http://www.biomedcentral.com/1472-6750/9/92.

16. M.J. Pena, A.G. Darvill, S. Eberhard, W.S. York and M.A. O'Neill, *Glycobiology* 18, 891, 2008. http://glycob.oxfordjournals.org/content/18/11/891.full.

17. W. Zeng, M. Chatterjee and A. Faik, *Plant Physiol.* 147, 78, 2008. http://www.ncbi.nlm.nih.gov/pmc/articles/PMC2330321/.

18. D. Dodd and I.K.O. Cann, *Glob. Change Biol. Bioenergy* 1, 2, 2009. http://www.ncbi.nlm.nih.gov/pmc/articles/PMC2860967.

19. A. Rogowski, J.A. Briggs, J.C. Mortimer, T. Tryfona, N. Terrapon, E.C. Lowe, A. Baslé et al., *Nat. Commun.* 6, 7481, 2015. doi:10.1038/ncomms8481.

20. A. Faik, *Plant Physiol.* 153, 396, 2010. http://www.plantphysiol.org/content/153/2/396.full.

21. E.A. Rennie and H.V. Scheller, *Curr. Opin. Biotechnol.* 26, 100, 2014. http://www.sciencedirect.com/science/article/pii/S0958166913007143 and http://www.debiq.eel.usp.br/aferraz/T%C3%B3p.%20Esp.:%20Forma%C3%A7%C3%A3o%20e%20degrada%C3%A7%C3%A3o%20da%20parede%20celular/aula%203a%20hemicellulose%20biosynthesis%20review%202014%20Curr%20Opn%20Biotechnol%20Scheller.pdf.

22. S. Persson, K.H. Caffall, G. Freshour, M.T. Hilley, S. Bauer, P. Poindexter, M.G. Hahn, D. Mohnen and C Somerville, *Plant Cell* 19, 237, 2007. http://www.ncbi.nlm.nih.gov/pmc/articles/PMC1820957.

23. C. Lee, Q. Teng, W. Huang, R. Zhong and Z.H. Ye, *Plant Physiol.* 153, 526, 2010. http://www.plantphysiol.org/cgi/content/full/153/2/526.

24. A.H. Liepman, C.J. Nairn, W.G.T. Willats, I. Sorensen, A.W. Roberts and K. Keegstra, *Plant Physiol.* 143, 1881, 2007. http://www.ncbi.nlm.nih.gov/pmc/articles/PMC1851810/.

25. T.C. Baldwin, M.G. Handford, M.I. Yuseff, A. Orellana and P. Dupree, *Plant Cell* 13, 2283, 2001. http://www.plantcell.org/content/13/10/2283.full.

26. W.H. van Zyl, S.H. Rose, K. Trollope and J.F. Gorgens, *Process Biochem.* 45, 1203, 2010. http://www.sciencedirect.com/science/article/pii/S1359511310001923.

27. R.A. Burton and G.B. Fincher, *Mol. Plant* 2, 873, 2009. http://mplant.oxfordjournals.org/content/2/5/873.full.

28. G.B. Fincher, *Curr. Opin. Plant Biol.* 12, 140, 2009. http://www.sciencedirect.com/science/article/pii/S1369526609000077.

29. G.B. Fincher, *Plant Physiol.* 149, 27, 2009. http://www.plantphysiol.org/content/149/1/27.full.

30. www.ccrc.uga.edu/~mao/galact/gala.htm

31. www.ccrc.uga.edu/~mao/rg2/intro.htm

32. C. Somerville, S. Bauer, G. Brininstool, M. Facette, T. Hamann, J. Milne, E. Osborne et al., *Science* 306, 2206, 2004.

33. E.M. Kerr and S.C. Fry, *Planta* 217, 327, 2003.

34. G.M. Cooper, *The Cell: A Molecular Approach*, 2nd ed., Boston University, Sinauer Associates, 2000. http://www.ncbi.nlm.nih.gov/bookshelf/br.fcgi?book=cooper&part=A1498.

35. O. Lerouxel, D.M. Cavalier, A.H. Liepman and K. Keegstra, *Curr. Opin. Plant Biol.* 9, 621, 2006. http://www.ncbi.nlm.nih.gov/pubmed/17011813.

36. D.J. Cosgrove, *Nat. Rev. Mol. Cell Biol.* 6, 850, 2005. http://www.nature.com/nrm/journal/v6/n11/fig_tab/nrm1746_F2.html.

37. G.J. Seifert, *Curr. Opin. Plant Biol.* 7, 277, 2004. http://www.ncbi.nlm.nih.gov/pubmed/15134748.

38. A.S. Karthikeyan, The Arabidopsis Information Resource (TAIR), MetaCyc Pathway: Xyloglucan biosynthesis, 2008. http://metacyc.org/META/new-image?type=PATHWAY&object=PWY-5936.

39. http://www.cazy.org/GlycosylTransferases.html

40. R.M. Perrin, A.E. Derocher, M. Bar-Peled, W. Zeng, L. Norambuena, L. Orellana, N.V. Raikhel and K. Keegstra, *Science* 284, 1976, 1999. http://www.ncbi.nlm.nih.gov/pubmed/10373113.

41. G.F. Vanzin, M. Madson, N.C. Carpita, N.V. Raikhel, K. Keegstra and W.D. Reiter, *Proc. Natl. Acad. Sci. USA* 99, 3340, 2002. http://www.ncbi.nlm.nih.gov/pmc/articles/PMC122520.

42. M. Madson, C. Dunand, X. Li, R. Verma, G.F. Vanzin, J. Caplan, D.A. Shoue, N.C. Carpita and W.D. Reiter, *Plant Cell* 15, 1662, 2003. http://www.plantcell.org/cgi/content/full/15/7/1662.

43. X.M. Li, I. Cordero, J. Caplan. M. Molhoj and W.D. Reiter, *Plant Physiol.* 134, 940, 2004. http://www.plantphysiol.org/cgi/content/full/134/3/940?ijkey=c2c19c81525f47696cc2006861f28875124e8d.

44. A. Faik, N.J. Price, N.V. Raikhel and K. Keegstra, *Proc. Natl. Acad. Sci. USA* 99, 7797, 2002. http://www.pnas.org/content/99/11/7797.long.

45. D.M. Cavalier and K. Keegstra, *J. Biol. Chem.* 281, 34197, 2006. http://www.ncbi.nlm.nih.gov/pubmed/16982611.

46. Y. Verhertbruggen, L. Yin, A. Oikawa and H.V. Scheller, *Plant Signal. Behav.* 6, 1620, 2011. http://www.ncbi.nlm.nih.gov/pmc/articles/PMC3256401.

47. M.C. Giraldo and B. Valent, *Nature Rev. Microbiol.* 11, 8000, 2013. http://www.nature.com/nrmicro/journal/v11/n11/box/nrmicro3119_BX1.html.

48. V. Bourquin, N. Nishikubo, H. Abe, H. Brumer, S. Denman, M. Eklund, M. Christiernin, T.T. Teeri, B. Sundberg and E.J. Mellerowicz, *Plant Cell* 14, 3073, 2002. http://www.plantcell.org/content/14/12/3073.full.

49. D. Loque, H.V. Scheller and M. Pauly, *Curr. Opin. Plant Biol.* 25, 151, 2015. http://www.sciencedirect.com/science/article/pii/S1369526615000692.

50. R. Zhong, M.J. Pena, G.K. Zhou, C.J. Nairn, A. Wood-Jones, E.A. Richardson, W.H. Morrison, A.G. Darvill, W.S. York, Z.H. Ye, *Plant Cell* 17, 3390, 2005. http://www.ncbi.nlm.nih.gov/pubmed/16272433.

51. M.J. Pena, R. Zhong, GK. Zhou, E.A. Richardson, M.A. O'Neill, A.G. Darvill, W.S. York and Z.H. He, *Plant Cell* 19, 549, 2007. http://www.plantcell.org/cgi/content/short/19/2/549.

52. A. Winzell, Investigation of genes and proteins involved in xylan biosynthesis, KTH, Stockholm, 2010. http://kth.diva-portal.org/smash/record.jsf?pid=diva2:288482.

53. M. Busse-Wicher et al., *Plant J.* 79, 492, 2014. http://www.ncbi.nlm.nih.gov/pubmed/24889696.

54. A.M. Wu, E. Hornblad, A. Voxeur, L. Gerber, C. Rihouey, P. Lerouge and A. Marchant, *Plant Physiol.* 153, 542, 2010. http://www.plantphysiol.org/cgi/content/full/153/2/542.

55. C. Lee, R. Zhong and Z.H. Ye, *Plant Cell Physiol.* 53, 135, 2012. http://pcp.oxfordjournals.org/content/53/1/135.abstract.

56. C. Lee, R. Zhong and Z.H. Ye, *Plant Signal. Behav.* 7, 332, 2012. http://www.landesbioscience.com/journals/psb/article/19269/?nocache=1087295047.

57. N.T. Lao, D. Long, S. Kiang, G. Coupland, D.A. Shoue, N.C. Carpita and T.A. Kavanagh, *Plant Mol. Biol.* 53, 647, 2003. http://www.ncbi.nlm.nih.gov/pubmed/15010604.

58. M. Koyama, W. Helbert, T. Imai, J. Sugiyama and B. Henrissat, *Proc. Natl. Acad. Sci. USA* 94, 9091, 1997. http://www.pnas.org/content/94/17/9091.full.

59. W.S. York and M.A. O'Neill, *Curr. Opin. Plant Biol.* 11, 258, 2008. http://www.ncbi. nlm.nih.gov/pubmed/18374624.

60. W. Zeng, N. Jiang, R. Nadella, T.L. Killen, V. Nadella and A. Faik, *Plant Physiol.* 154, 78, 2010. http://www.plantphysiol.org/content/154/1/78.full.

61. R.A.C. Mitchell, P. Dupree and P.R. Shewry, *Plant Physiol.* 144, 43, 2007. http://www. ncbi.nlm.nih.gov/pmc/articles/PMC1913792/.

62. M.E. Edwards, C.A. Dickson, S. Chengappa, C. Sidebottom and M.J. Gidley, *Plant J.* 19, 691, 1999. http://www.ncbi.nlm.nih.gov/pubmed/10571854.

63. L. Yin, Y. Verhertbruggen, A. Oikawa, C. Manisseri, B. Knierim, L. Prak, J.K. Jensen et al., *Mol. Plant* 4, 1024, 2011. http://mplant.oxfordjournals.org/content/early/ 2011/04/06/mp.ssr026.full.

64. O.O. Obembe, E. Jacobsen, R.G.F. Visser and J.P. Vincken, *Biotechnol. Mol. Biol. Rev.* 1, 76, 2006. http://www.academicjournals.org/bmbr/PDF/Pdf2006/Sep/ Obembe%20et%20al.pdf.

65. CAZY, Carbohydrare Active enzymes, 2010. http://www.cazy.org/Glycoside-Hydrolases.html.

66. S.C Fry, R.C Smith, K.F. Renwick, D.J. Martin, S.K. Hodge and K.J. Matthews, *Biochem J.* 282, 821, 1992. http://www.ncbi.nlm.nih.gov/pmc/articles/PMC1130861.

67. P. Campbell and J. Braam, *Trends Plant Sci.* 4, 361, 1999. http://www.ncbi.nlm.nih.gov/ pubmed/10462769.

68. J.K.C. Rose, Cornell University, 2006. http://labs.plantbio.cornell.edu/XTH/overview.htm.

69. J.K.C. Rose, J. Braam, S.C. Fry and K. Nishitani, *Plant Cell Physiol.* 43, 1421, 2002. http://pcp.oxfordjournals.org/cgi/content/full/43/12/1421.

70. Y.R. Soro, Purification et caractérisation de l'alpha-glucosidase du suc digestif de Archachatina ventricosa (Achatinidae) ; Application à la synthèse de polyglucosyl-fructosides, Thèse, Toulouse, 2007. http://eprint.insa-toulouse.fr/archive/00000198/01/ Soro.pdf.

71. J.M. Eklof and H. Brumer, *Plant Physiol.* 153, 456, 2010. http://www.plantphysiol.org/ cgi/content/full/153/2/456.

72. P. Johansson, H. Brumer, III, M.J. Baumann, A.M. Kallas, H. Henriksson, S.E. Denman, T.T. Teeri and T.A. Jones, *Plant Cell* 16, 874, 2004. http://www.plantcell.org/cgi/ content/full/16/4/874.

73. G.J. Davies, K.S. Wilson and B. Henrissat, *Biochem. J.* 321, 557, 1997. http://www.ncbi. nlm.nih.gov/pmc/articles/PMC1218105/pdf/9020895.pdf.

74. M.J. Baumann, J.M. Eklof, G. Michel, A.M. Kallas, T.T. Teeri, M. Czjzek and H. Brumer, III, *Plant Cell* 19, 1947, 2007. http://www.plantcell.org/cgi/content/full/19/ 6/1947?ijkey=eb13d280656f78b1561142d8f8f7350e65429869.

75. N.A. Eckardt, *Plant Cell* 16, 792, 2004. http://www.plantcell.org/cgi/content/ full/16/4/792.

76. T. Awano, K. Takabe and M. Fujita, *Protoplasma* 219, 106, 2002. http://www.ncbi.nlm. nih.gov/pubmed/11926061.

77. D. Reis and B. Vian, *C. R. Biol.* 327, 785, 2004. http://www.ncbi.nlm.nih.gov/ pubmed/15587069.

78. http://www.ccrc.uga.edu/~mao/intro/ouline.htm

79. R. Brunecky, T.B. Vinzant, S.E. Porter, B.S. Donohoe, D.K. Johnson and M.E Himmel, *Biotechnol. Bioeng.* 102, 1537, 2009. http://www3.interscience.wiley.com/cgi-bin/ fulltext/121542373/PDFSTART.

80. A. Bacic, G.B. Fincher and B.A. Stone, Eds., *Chemistry, Biochemistry and Biology of (1,3)-β-Glucans and Related Polysaccharides*, Academic Press, Amsterdam, the Netherlands, 2009.

# 5 Biodegradation of Hemicelluloses

## 5.1 INTRODUCTION

The degradation of cellulose and hemicelluloses is carried out by microorganisms that can be found either free in nature or as part of the digestive tract of higher animal.[1] Microorganisms have evolved sophisticated mechanisms to degrade plant cell wall polysaccharides and consequently exploit this rich energy and carbon source.[2] The variable structure and organization of hemicelluloses require the concerted action of many enzymes for its complete degradation.

*Hemicellulases* are a group of diverse enzymes that hydrolyze hemicelluloses.[1,3] They are vital for maintaining the carbon cycle in nature, allowing the degradation of plant biomass to saccharides, which in turn can be utilized as carbon and energy sources for microorganisms and higher animals.[4] These enzymes have many biotechnological applications, and their structure/function relationships are the subject of intensive research. Hemicellulases are typically modular proteins with a catalytic module and other functional modules including a carbohydrate-binding module. Recently, new structures of catalytic and noncatalytic domains of hemicellulases have been elucidated and reveal, together with biochemical studies, the principles of catalysis and specificity for these enzymes.

As substrates, hemicelluloses are readily available for the production of value-added products such as bioethanol, xylitol, lactic acid, and 2,3-butanediol.[5,6] Xylitol is commonly used as a sweetener; lactic acid is used as a monomer for PLA (polylactic acid); and 2,3-butanediol as a precursor for the manufacture of a range of chemical products such as methyl ethyl ketone, gamma-butyrolactone, and 1,3-butadiene.[7] Xylitol can be produced either by chemical synthesis or by fermentation. Industrially, it is produced by catalytic hydrogenation of xylose.[8] Intensive research efforts in the last 25 years have led the way for the successful conversion of hemicelluloses into fermentable constituents by pretreatment technologies and engineered hemicellulases. A major challenge is the isolation of microbes with the ability to ferment a broad range of sugars and withstand fermentative inhibitors that are usually present in hemicellulosic sugar syrup.

A review of the applications of hemicelluloses and products derived from hemicellulose degradation is given in Chapter 7.

## 5.2  HEMICELLULASES: OVERVIEW

The classic model of enzymatic hemicellulose degradation is shown in Figure 5.1 and compared to that of enzymatic cellulose degradation. Cellulases include cellobiohydrolases acting either on the nonreducing end or on the reducing end, endoglucanases; in addition, β-glucosidases hydrolyze cellobiose. Xylans, as predominant hemicelluloses, include endoxylanases acting on the backbone, and glucuronidases, arabinofuranosidases, feruloyl esterases, and galactosidases acting on the branches; in addition, β-xylosidases hydrolyze short xylooligomers.

The catalytic modules of hemicellulases are either glycoside hydrolases (GHs) that hydrolyze glycosidic bonds, or carbohydrate esterases (CEs) that hydrolyze ester linkages of acetate or ferulic acid side groups.[1] Based on the homology of their primary sequence, these catalytic modules can be classified into families.[9] The main substrate of typical hemicellulases and their classification into GH and CE families are shown in Table 5.1.

The understanding of the structure/function relationship of hemicellulases has progressed considerably from the combination of high-resolution crystal structures and in-depth catalytic analysis.[1]

## 5.3  BIODEGRADATION OF XYLOGLUCANS

### 5.3.1  Overview

Presently, a wide variety of Biodegradation of Xyloglucan (XyG)-specific hydrolytic activities have been demonstrated in members of GH families 5, 7, 9, 12, 16, 44, and 74 (Table 5.1)[15]:

- XyG-specific endo-β-1,4-glucanase (XEG or *endoxyloglucanase*) activity (EC 3.2.1.151; EC number for Enzyme Commission number).[12,16]
- XyG-specific exo-β-1,4-glucanase (*exoxyloglucanase*) activity (EC 3.2.1.155), so as to successively remove XyG oligosaccharides from the chain end.[17]

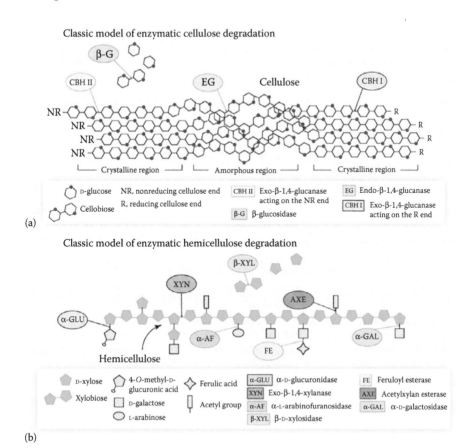

**FIGURE 5.1** Overview of enzymatic polysaccharide degradation. (a) Cellulose. (b) Hemicelluloses (xylans). Enzymatic breakdown of biomass: enzyme-active sites, immobilization, and biofuel production. (From Duttaa, S. and Wu, K.C.-W., *Green Chem.*, 16, 4615–4626, 2014.)

- Oligoxyloglucan (OXG) reducing-end-specific cellobiohydrolase (OXG-RCBH) activity (EC 3.2.1.150), so as to hydrolyze cellobiose from the reducing end of XyGs.[18]

Many of these families are classic *cellulase* families and it is not unreasonable to expect that xyloglucanase activity may be found in other known β-1,4-glucanase-containing GH families, as well as in novel, emerging families.[15]

The initial attack in XyG degradation is made by specific endoglucanases (i.e., endoxyloglucanases or xyloglucan endohydrolases) that depolymerize high molecular mass XyG into XyG oligosaccharides.[10,19,20] While most of these endoxyloglucanases act by cleaving XyG after an unsubstituted glucosyl residue (Figure 5.2), exceptions have been found in GH44 and GH74.[19] Two GH74 enzymes have been shown to be reducing-end-specific cellobiohydrolases releasing XG, LG, or FG from XyG.

**TABLE 5.1**

**Typical Hemicellulolytic Enzymes and Their Classification into Major GH and CE Families**

| Enzymes | Main Substrates | GH Family[a] |
|---|---|---|
| Xyloglucan-specific endo-β-1,4-glucanases | β-1,4-Xyloglucans | GH5, 7, 9, 12, 16, 44, 74[10] |
| Xyloglucan-specific exo-β-1,4-glucanases | β-1,4-Xyloglucans | GH74[11] |
| Oligoxyloglucan reducing-end-specific cellobiohydrolases | Oligoxyloglucans | GH74[12] |
| Endo-β-1,4-xylanases | β-1,4-Xylans | GH5, 8, 10, 11, 30, 43 |
| Exo-β-1,4-xylosidases | β-1,4-Xylans and xylooligomers | GH3, 39, 43, 52, 54 |
| α-Arabinofuranosidases | α-Arabinofuranosyl-containing hemicelluloses | GH3, 43, 51, 54, 62 |
| α-Glucuronidases | 4-O-Methyl-glucuronoxylans | GH67 |
| Acetyl xylan esterases | O-Acetyl xylans | CE1, 2, 3, 4, 5, 6, 7 |
| Feruloyl esterases | Feruloyl-polysaccharides | CE1 |
| Endo-β-1,4-mannanases | β-1,4-Mannans | GH5, 26 |
| Exo-β-1,4-mannosidases | β-1,4-Mannooligomers | GH1, 2, 5 |
| α-Galactosidases | Galacto(gluco)mannans | GH4, 27, 36, 57, 97, 110[13] |
| Endo-α-1,5-arabinanases | α-1,5-Arabinans | GH43 |
| Endo-1,4-β-galactanases | β-1,4-Arabinogalactans | GH53[14] |
| β-Galactosidases | β-Galactosides | GH1, 2, 3, 35, 42 |

*Source:* Shallom, D. and Shoham, Y., *Curr. Opin. Microbiol.* 6, 219, http://biotech.technion.ac.il/Media/Uploads/Documents/Yuval/Publications/Yuval_67.pdf, 2003; CAZY, Carbohydrate Active enzymes, http://www.cazy.org/Glycoside-Hydrolases.html, 2010.

[a] GH, Glycoside hydrolase; CE, carbohydrate esterase.

After conversion of XyG into XyG oligosaccharides, exo-acting enzymes trim the branches of the oligosaccharides, so as to typically form the xylosylated oligosaccharides that are then degraded to monosaccharides by α-xylosidases and β-glucosidases (Figure 5.2).[19]

### 5.3.2 XYLOGLUCANASES AND OLIGOXYLOGLUCAN-SPECIFIC GLYCOSIDASES

XG-degrading enzymes have attracted the attention of researchers especially since the beginning of the twenty-first century.[21]

In 2002, Yaoi et al.[22] isolated a novel exo-β-1,4-glycosidase, OXG-RCBH, from the fungus *Geotrichum*. OXG-RCBH had exoglucanase activity. It recognizes the reducing end of oligoxyloglucan and releases two glucosyl residues from the reducing end of the main chain (XG, LG, or FG).[22] The protein shows ~35% identity to members of GH family 74, indicating that OXG-RCBH can be classified in this

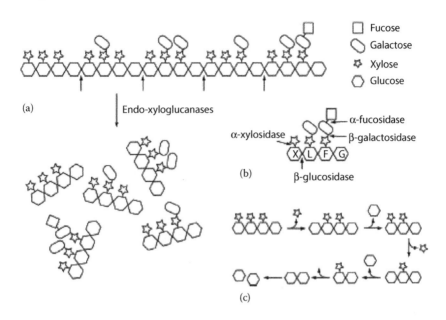

**FIGURE 5.2** The degradation of high molecular mass XyG to monosaccharides. The first step in XyG degradation is to depolymerize XyG into XyG oligosaccharides (XXXG) by endoxyloglucanases (a). In (b), some of the exo-acting side chain-trimming enzymes are shown. In (c), the degradation of XXXG into monosaccharides by the sequential action of α-xylosidases and β-glucosidases as observed in *Arabidopsis* and bacteria. (From Eklov, J., Plant and microbial xyloglucanases: Function, structure and phylogeny, Doctoral thesis, KTK, Stockholm, http://kth.diva-portal.org/smash/record.jsf?pid=diva2:405550, 2011.)

family. It was suggested that OXG-RCBH has at least four subsites (−2 to +2 according to the nomenclature of subsites proposed by Davies et al.[23]; see Section 2.3.7). It is very important that the reducing end (subsite +2) is unbranched. It is also important that the third glucosyl residue from the reducing end (subsite −1) is branched with an unmodified xylose residue (Figure 5.3).[24]

In 2004, Yaoi et al.[25] determined the crystal structure of OXG-RCBH. The GH74 enzyme consists of a tandem repeat of two seven-bladed β-*propeller* domains. There is a cleft that can accommodate the substrate oligosaccharide between the two domains, which is a putative substrate-binding subsite. A unique loop encloses one

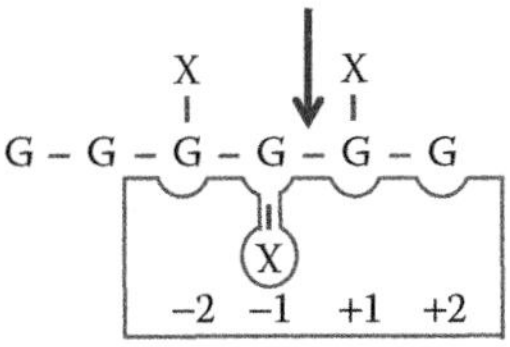

**FIGURE 5.3** Subsite image of OXG-RCBH. The arrow indicates the cleavage point. Subsite +2: reducing end. (From Yaoi, K. and Mitsuishi, Y., *J. Biol. Chem.*, 277, 48276, http://www.jbc.org/content/277/50/48276.full, 2002.)

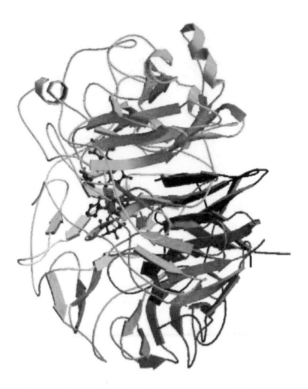

**FIGURE 5.4** Crystal structure of oligoxyloglucan reducing-end-specific cellobiohydro-lase (OXG-RCBH) complexed with a XyG heptasaccharide.[27,26] OXG-RCBH consists of two seven-bladed β-propeller domains. There is a large cleft between the two domains, and a unique loop encloses one side of the active site cleft. The substrate binds to the cleft, and its reducing end is arranged near the loop region that is believed to impart OXG-RCBH with its activity. (From RCSB PDB Structure Summary—2EBS.)

side of the cleft. Asp35 and Asp465, located in the binding cleft, have crucial roles in hydrolytic activity. In 2007, Yaoi et al.[26] reported the X-ray crystal structure of the OXG-RCBH-substrate complex determined to a resolution of 2.4 Å (Figure 5.4). It is suggested that the loop closing off the positive aglycone subsites imparts unique substrate specificity with exo-mode hydrolysis in OXG-RCBH.

In 2004, Grishutin et al.[28] reported that three specific xyloglucanases were iso-lated from *Aspergillus japonicus*, *Chrysosporium lucknowense*, and *Trichoderma reesei*. The *T. reesei* xyloglucanase represents Cel74A, whose gene has been discov-ered. The three enzymes had high specific activity toward tamarind XyG, whereas the activity toward carboxymethylcellulose and barley β-glucan was absent or very low. All enzymes produced XXXG, XXLG/XLXG, and XLLG oligosaccharides as the end products of XyG hydrolysis. *A. japonicus* xyloglucanases displayed an endo-type of attack on the polymeric substrate, while the mode of action of the two other xyloglucanases was similar to the exo-type, when oligosaccharides con-taining four glucose residues in the main chain were split off the ends of XyG molecules. These results, together with the former literature data, show that specific xyloglucanases may represent a new class of GHs, which are different from regular endo-1,4-β-glucanases.

In 2005, Zverlov et al.[29] showed that the xyloglucanase Xgh74A (XyG hydrolase) in the cellulosome of the bacterium *Clostridium thermocellum* hydrolyzes every fourth β-1,4-glucan bond in the XyG backbone, thus producing decorated cellote-traose units. Its low activity on CMC and lack of activity on amorphous cellulose indicates recognition of the xylosidic side chains present in XyG. Xgh74A is the first xyloglucanase identified in *C. thermocellum* and the only enzyme in the cellulosome that hydrolyzes tamarind XyG.

In 2005, Bauer et al.[30] identified an oligoxyloglucan reducing-end-specific xylo-glucanobiohydrolase (OREX[31]) from the fungus *Aspergillus nidulans*. The GH74 enzyme acts on XyG oligomers and releases the first two glycosyl residue segments from the reducing end, provided that neither the first glucose nor the xylose attached to the third glucose residue from the reducing end is further substituted.[19]

In 2005, Yaoi et al.[32] isolated two xyloglucanases from the gram-positive bac-terium *Paenibacillus* sp. Strain KM21, belonging to families GH5 and GH74. The GH5 enzyme is a typical endo-type enzyme that randomly cleaves the xyloglucan main chain, while the GH74 enzyme has dual endo- and exo-mode activities or pro-cessive endo-mode activity. The GH5 enzyme digested the XyG oligosaccharide XXXGXXXG to produce XXXG, whereas the GH74 enzyme digested XXXGXXXG to produce XXX, XXXG, and GXXXG, suggesting that this enzyme cleaves the gly-cosidic bond of unbranched glucosyl residues.

In 2006, Martinez-Fleites et al.[2] determined the crystal structures of *C. thermocel-lum* endoxyloglucanase Xgh74A in both apo- and ligand-complexed forms, thereby revealing the structural basis for XyG recognition and degradation. Xgh74A consists of a tandem repeat of two seven-bladed β-propeller domains. The two domains form a substrate-binding cleft at the interface.[33] The catalytic residues are located in the middle of this cleft. Its overall topology is very similar to the only member of family GH74 previously published, the OXG-RCBH from *Geotrichum*.

The Xgh74A structure reveals a complex-binding architecture in which sub-sites accommodate 17 distinct sugar moieties accounting for the pattern of XyG recognition. In both Xgh74A and OXG-RCBH, the substrate-binding cavities are open grooves well exposed to solvent. One side of the binding cleft of OXG-RCBH is blocked by a so-called exo-loop which is found only in exo-acting enzymes of family 74.[33]

Xgh74A processes the XyG chain in an endo fashion, releasing four glucosyl resi-due segments. In contrast, OXG-RCBH is an exoglucanase that releases two glucosyl residues from the reducing end of the polymer.

In 2007, Desmet et al.[34] investigated the substrate specificity of the xyloglucanase Cel74A from *Hypocrea jecorina* (*T. reesei*) using several polysaccharides and oligo-saccharides. XyG chains are hydrolyzed at substituted glucosyl residues, in contrast to the action of all known xyloglucan endoglucanases. The building block of XyG, XXXG, was rapidly degraded to XX and XG, which has only been observed before with OXG-RCBH.

### 5.3.3 Xyloglucan Oligosaccharide-Degrading Enzymes

The end result of endoglucanases activity on XyG is the release of XyG oligosaccharides with an unsubstituted glucosyl residue at the reducing end.[35] Endoglucanase treatment of XXXG-type XyGs typically generates a series of well-defined oligosaccharides that have a tetrasaccharide backbone.[36] In contrast, the types of oligosaccharide generated from XXGG-type XyGs depend on the substrate specificity of the enzyme and on the presence or absence of *O*-acetyl groups on the unbranched glucosyl residues. Once the oligosaccharides are formed, specific exoglycosidases (Table 5.2) are necessary to release each type of residue.[37,38]

α-Xylosidase activities in both pea (*Pisum sativum*)[45] and nasturtium (*Tropaeolum majus*)[46] can only remove unsubstituted xylosyl residues from the nonreducing end of the molecule.[35,47] A β-glucosidase[48,49] is then required to remove the unsubstituted glucosyl residue before α-xylosidase can act again.[50] β-Galactosidase and α-fucosidase are also required for the complete disassembly of the *Arabidopsis* XyG oligosaccharides.

The *apoplastic* degradation of XyG oligosaccharides in *Arabidopsis* was studied by Iglesias et al.[51] in 2006. All four glucanases necessary for the degradation of xyloglucan oligosaccharides were found in the apoplastic fluid. These enzymes acted cooperatively on XyG oligosaccharides (XLFG), leading to the sequential formation of XXFG, XXLG, XXXG, GXXG, and XXG. Thus, side chains are likely removed by α-1,2-L-fucosidases[52] (GH 95) and β-galactosidases (GH1, 2, 3, 35, and 42; EC 3.2.1.23)[53] to form xylosylated XyG oligosaccharides.[19]

The latter are then hydrolyzed into monosaccharides starting from the nonreducing end by the sequential and concerted actions of α-xylosidases[54] (GH31) and

---

## TABLE 5.2
### XyG Oligosaccharide-Degrading Enzymes

| Enzymes | Functions | EC Code | GH Family |
|---|---|---|---|
| β-Galactosidase | Hydrolysis of terminal nonreducing β-galactose residues in β-galactosides[39] | EC 3.2.1.23 | GH1, 2, 3, 35, and 42 |
| α-Xylosidase | Hydrolysis of terminal, nonreducing α-xylose residues with release of α-xylose[40,41] | EC 3.2.1.177 | GH31 |
| β-Glucosidase | Hydrolysis of terminal, nonreducing β-glucosyl residues with release of β-glucose[42] | EC 3.2.1.21 | GH1, 3, 5, 9, 30, and 116 |
| 1,2-α-Fucosidase | Hydrolysis of nonreducing terminal fucose residues linked to galactose residues by a 1,2-α-linkage[43] | EC 3.2.1.63 | GH95 |
| α-Arabinofuranosidases | Hydrolysis of terminal nonreducing α-arabinofuranoside residues in α-arabinosides[44] | EC 3.2.1.55 | GH3, 10, 43, 51, 54, and 62 |

β-glucosidases[55] (GH1, 3, 5, 9, 30, and 116 and in particular GH1 and 3 in XyG degradation).[56] In other species, other debranching, exo-acting enzymes are most likely present, such as α-arabinofuranosidases (GH3, 10, 43, 51, 54, and 62).[19]

### 5.3.3.1 β-Galactosidases

β-Galactosidases (EC 3.2.1.23) hydrolyze the terminal nonreducing β-D-galactose residues in β-D-galactosides, such as lactose, proteoglycans, glycolipids, oligosaccharides, and polysaccharides.[53,57]

Distributed across kingdoms, β-galactosidases are represented in bacteria, fungi, plants, and animals.[53] Based on the sequence and structural similarity, EC 3.2.1.23 β-galactosidases can be placed in five of the current GH families: GH1, GH2, GH3, GH35, and GH42. Plant β-galactosidases have been found only in GH35; β-galactosidases from the other four families have been observed solely in bacteria and archaea.

GH2 and GH35-β-galactosidases have been studied intensively.

### 5.3.3.2 Microbial GH2 β-Galactosidase

β-Galactosidase is specified by the first structural gene (*lac Z*) of the *lac* operon in *Escherichia coli.* The amino acids of *E. coli* GH2 β-galactosidase were first tentatively sequenced in 1970.[58] Its amino acid sequence was determined in 1977.[59] The protein contains 1021 amino acids in a polypeptide single chain. The protein is a tetramer of four identical, unusually long, polypeptide chains.

The crystal structure of the *E. coli* GH2 enzyme was first determined in 1994 by Jacobson et al.[60] (Figure 5.5). The authors showed that the protein is a tetramer with a

**FIGURE 5.5** Tetrameric structure of β-galactosidase. (From RCSB PDB Structure Summary—4V40; http://www.rcsb.org/pdb/explore.do?structureId=1bgm.)

222-point symmetry. The 1023-amino acid polypeptide chain folds into five sequential domains, with an extended segment at the amino terminus.

The third domain is a $(\beta/\alpha)_8$ barrel that comprises much of the active site.[62] This site does, however, include elements from other domains and other subunits. The N-terminal region of the polypeptide chains helps form one of the subunit interfaces. Taken together these features provide a structural basis for the well-known property of *alpha-complementation*.[63] Catalytic activity proceeds via the formation of a covalent galactosyl intermediate with Glu537 as catalytic nucleophile, and includes *shallow* and *deep* modes of substrate binding. Monovalent potassium ions and divalent magnesium ions are required for the enzyme's optimal activity.

### 5.3.3.3  Plant and Microbial GH35 β-Galactosidase

The majority of GH35 enzymes are β-galactosidases.[64,65] GH35 enzymes have been isolated from microorganisms such as fungi, bacteria, and yeasts, as well as higher organisms such as plants, animals, and human cells. As with many other CAZy families, GH35 members tend to be represented by multi-gene families in plants. Moreover, plant GH35 β-galactosidases have been divided into two classes: members of the first are capable of hydrolyzing pectic β-1,4-galactans, while those of the second can specifically cleave β-1,3- and β-1,6-galactosyl linkages of arabinogalactan proteins.

GH35 β-galactosidases catalyze the hydrolysis of terminal β-galactosyl residues via a retaining mechanism.[64] Glu268 was first identified as the catalytic nucleophile and Glu200 was inferred to be the general acid/base.

Crystal structures of β-galactosidase from *Penicillium* sp. (Psp-β-gal) and its complex with galactose were reported in 2004 by Rojas et al. (Figure 5.6).[66–68]

This model was the first 3D structure for a member of GH35. The enzyme is a 120 kDa monomer composed of five distinct structural domains. The overall structure is built around the first, TIM barrel, domain. Domain 2 is an all β-sheet domain containing an immunoglobulin-like subdomain; domain 3 is based on a Greek-key β-sandwich; and domains 4 and 5 are jelly rolls. Glu200 (by convention, peptide sequences are written *N*-terminus [start] to *C*-terminal) was identified as the proton donor and residue Glu299 as the nucleophile involved in catalysis. *Penicillium* sp. β-galactosidase is a glycoprotein containing seven N-linked oligosaccharide chains and is the first structure of a glycosylated β-galactosidase. Domains 1–5 are colored in cyan, red, yellow, green, and magenta, respectively.

The structures of β-galactosidases from *Bacteriodes thetaiotamicron* (Btm-β-gal) and *T. reesei* (Tr-β-gal) were subsequently reported.[64] The comparison of the native structures of Psp-β-gal, Tr-β-gal, and Btm-β-gal reveals a difference in the number of domains: three domains for Btm-β-gal, five domains for Psp-β-gal, and six domains for Tr-β-gal (Figure 5.7a). The comparison of the active sites of the β-galactosidases shows a remarkable similarity (Figure 5.7b). In addition to the catalytic residues, the active sites of the GH35 β-galactosidases contain many identical residues.

**FIGURE 5.6** Native structure of β-galactosidase from *Penicillium* sp. Protein chains are colored from the N-terminal to the C-terminal using a rainbow (spectral) color gradient. The monomeric enzyme is composed of five domains. Domain 1 is an $(\alpha/\beta)_8$ TIM barrel, domain 2 is an all β-sheet domain, domain 3 is based on a Greek-key β-sandwich, and domains 4 and 5 are jelly rolls. Domains 1–5 are colored in cyan, red, yellow, green, and magenta, respectively. The long linker peptide connecting domains 3 and 4 is depicted in blue. (Reprinted from *J. Mol. Biol.*, 343, Rojas, A.L. et al., Crystal structures of β-galactosidase from *penicillium* sp. and its complex with galactose, 1281–1292, Copyright 2016, with permission from Elsevier.)

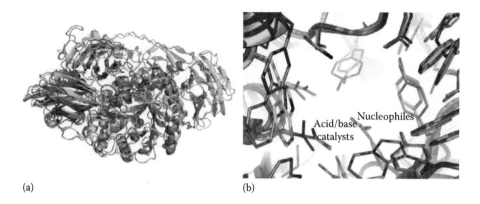

(a)  (b)

**FIGURE 5.7** (a) Comparison of the native structures of GH35 β-galactosidases. Psp-β-gal, Tr-β-gal, and Btm-β-gal are colored in green, brown, and blue, respectively. (b) Comparison of the active sites of GH35 β-galactosidases. Psp-β-gal, Tr-β-gal, and Btm-β-gal are colored as in (a). (From Mirko Maksimainen, http://www.cazypedia.org/index.php/Glycoside_Hydrolase_Family_35. With permission.)

#### 5.3.3.4  α-Xylosidases

The first crystal structure of a GH 31 enzyme was that of the α-xylosidase YicI from
*E. coli*, published in 2005 by Lovering et al.[69] The structure shows an intimately
associated hexamer. The catalytic domain of each monomer forms a $(\beta/\alpha)_8$ barrel.[70]
YicI is most likely responsible for xyloglucan degradation. It is a retaining enzyme
that uses a double displacement mechanism in which Asp482 acts as an acid/base
catalyst to assist the formation of a covalent β-xylosyl-enzyme that involves Asp416
as the catalytic nucleophile.

The structure of a YicI monomer can be divided into five distinct domains: the
N-terminal domain (residues 1–245), the catalytic domain (residues 245–349 and
387–588), an insert in the catalytic domain (residues 349–387), a proximal C-terminal
domain (residues 588–665), and a distal C-terminal domain (residues 665–772).

#### 5.3.3.5  β-Glucosidase

The most common known enzymatic activities for GH1 enzymes are β-glucosidases
and β-galactosidases.[71] GH1 β-glycosidases are retaining enzymes and follow a double displacement mechanism. The catalytic nucleophile was first identified in the
*Agrobacterium* sp. β-glucosidase as Glu358. The general acid/base catalyst was first
identified as Glu170 in the same enzyme.

Three-dimensional structures are available for a large number of GH1 enzymes,
the first solved being that of the white clover cyanogenic β-glucosidase. As members
of clan GH-A, they have a classical $(\alpha/\beta)_8$ TIM barrel fold with the two key active
site glutamic acids being approximately 200 residues apart in sequence and located
at the C-terminal ends of β-strands 4 (acid/base) and 7 (nucleophile).

The GH3 family currently groups together exo-acting β-glucosidases, α-L-
arabinofuranosidases, β-xylopyranosidases, and *N*-acetyl-β-glucosaminidases.[72]
GH3 β-glucosidases from barley, which are more precisely referred to as β-glucan
glucohydrolases, are also broad specificity exo-hydrolases that remove single glucosyl residues from the nonreducing ends of a range of β-glucans, β-oligoglucosides,
and aryl β-glucosides, including some β-oligoxyloglucosides.

The three-dimensional structure of a barley β-glucan exohydrolase was determined
by Varghese et al.[73] This was the first reported structure of a GH3 (Figure 5.8).[74]

The enzyme is a two-domain, globular protein of 605 amino acid residues and is
N-glycosylated at three sites. The first 357 residues constitute an $(\alpha/\beta)_8$ TIM-barrel
domain. The second domain consists of residues 374–559 arranged in a six-stranded
β-sandwich, which contains a β-sheet of five parallel β-strands and one antiparallel β-strand, with three α helices on either side of the sheet. A glucose moiety is
observed in a pocket at the interface of the two domains, where Asp285 and Glu491
are believed to be involved in catalysis.

### 5.4  BIODEGRADATION OF XYLANS

Interest in xylan-degrading enzymes stems from their potential applications in the
paper and pulp industry for bio-bleaching and for bioconversion of lignocellulosic
materials to fermentative products.[4] These enzymes, as well as other glycosidases,

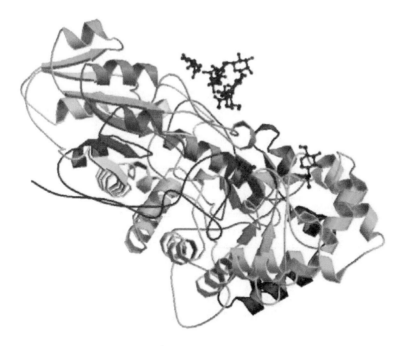

**FIGURE 5.8**  Structure of barley β-glucan exohydrolase. (From RCSB PDB Structure Summary—1EX1; http://www.rcsb.org/pdb/explore.do?structureId=1ex1.)

were also demonstrated as potential glycosynthases for oligosaccharides and thio-glycoside synthesis.[76]

The enzymatic degradation of plant cell wall xylans requires the concerted action of a diverse enzymatic syndicate.[77] Among these enzymes, there are (Figure 5.1 for the various chemical bonds hydrolyzed)[78]:

- *Endoxylanases*, such as endo-β-1,4-xylanases (EC 3.2.1.8) and glucurono-arabinoxylan endo-β-1,4-xylanases (EC 3.2.1.136), which both cleave the xylan backbone, yielding short xylooligomers. Most known xylanases belong to families GH 10 and 11.[79]
- *Oligosaccharide reducing-end xylanases* (EC 3.2.1.156), which release the xylose unit from the reducing end of oligosaccharides.[80]
- *Xylan β-1,4-xylosidases* (EC 3.2.1.37), which hydrolyze xylans to remove successive xylose residues from the nonreducing end. They also hydrolyze xylobiose and xylo-oligosaccharides.[81] Xylosidases are found in families GH 3, 39, 43, 52, and 54.[82,83]
- *α-Arabinofuranosidases* (EC 3.2.1.55), which release the terminal nonre-ducing α-arabinofuranoside residues in α-arabinosides, thereby cleaving the arabinofuranoside side chains.[84] They act on α-L-arabinofuranosides, α-arabinans containing (1,3)- and/or (1,5)-linkages, arabinoxylans and arabinogalactans. α-Arabinofuranosidases belong to families GH 3, 43, 51, 54, and 62.[85]

**TABLE 5.3**

**Function and Classifications of Various Xylan-Active Enzymes**

| Enzymes | Function | EC Code | GH Family |
|---|---|---|---|
| Endoxylanases | Cleave xylan backbone | EC 3.2.1.8; EC 3.2.1.136 | GH 10 and 11 |
| Exoxylanases | Remove xylose residues | EC 3.2.1.37 | GH 3, 39, 43, 52, and 54 |
| Glucuronosidases | Remove the glucuronic acid side chain | EC 3.2.1.139 | GH 67 |
| α-Arabino-furanosidases | Cleave the arabinofuranoside side chains of different hemicelluloses | EC 3.2.1.55 | GH 3, 43, 51, 54, and 62 |
| Acetyl xylan esterases | Remove acetyl substitutions | EC 3.1.1.72 | CE 1 to 7 |
| Feruloyl esterases | Remove ferulate groups | EC 3.1.1.73 | CE 1 |

- *Acetyl xylan esterases* (EC 3.1.1.72), which release the acetyl substitutions on xylans and xylo-oligosaccharides.[86] They are found in carbohydrate esterase (CE) families 1–7.[87]
- *Feruloyl esterases* (EC 3.1.1.73), which hydrolyze the feruloyl group from an esterified sugar, which is commonly arabinose in natural substrates.[88,89] This ester bond is usually involved in cross-linking xylan to lignin. Feruloyl esterases are found in CE family 1.[90]
- *α-Glucuronidases* (EC 3.2.1.139), which cleave the α-1,2-glycosidic bond of the 4-*O*-methyl-glucuronic acid side chain of xylans. They are found exclusively in family GH 67. The enzymes are active mainly on small xylooligomers and therefore are dependent on the action of endoxylanases.

Table 5.3 summarizes the functions of various xylan-degrading enzymes and their EC and GH classifications.

## 5.4.1 ENDOXYLANASES

Endoxylanases are industrially important enzymes which randomly cleave the β-1,4 backbone of xylans and produce xylooligosaccharides, xylobiose, and xylose.[91] They are mainly present in microbes and plants but not in animals. Xylanases from fungi and bacteria have been extensively studied and produced commercially. Their potential use in paper industries has been discussed in relation with reduction of pollution. They have been applied to bio-bleaching processes. Furthermore, they can be exploited for ethanol production and as an additive in animal feedstock to improve its nutritional value. Endoxylanases can also be exploited in baking and fruit juice industries.

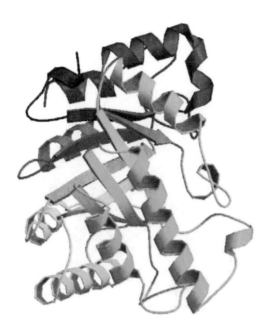

**FIGURE 5.9**   Crystal structure of the *Streptomyces lividans* xylanase A.[97] The 32-kDa catalytic domain of *S. lividans* xylanase A is folded into a complete $(\alpha/\beta)_8$ barrel. (From RCSB PDB Structure Summary—1XAS.)

Diverse forms of these enzymes exist, displaying varying folds, mechanisms of action, substrate specificities, hydrolytic activities (yields, rates, and products), and physicochemical characteristics.[92] Research has mainly focused on xylanases belonging to GH10 and 11 but xylanases have also been classified in GH5, 8, 30, and 43 in the CAZypedia database.[93]

The majority of GH10 enzymes are endo-$\beta$-1,4-xylanases, although a few show endo-$\beta$-1,3-xylanase activity.[94] GH10 xylanases are retaining enzymes. The catalytic nucleophile and the general acid/base residue were first identified in the *Cellulomonas fimi* endoxylanase as Glu-233 (earlier numbered as 274) and Glu-127.[95] Three-dimensional structures are available for a large number of GH10 enzymes, the first solved being that of the *Streptomyces lividans* xylanase A (Figure 5.9).[96] As members of Clan GH-A, they have a classical $(\alpha/\beta)_8$ TIM barrel fold with the two key active site glutamic acids located at the C-terminal ends of beta-strands 4 (acid/base) and 7 (nucleophile).

The GH11 enzymes are only endo-$\beta$-1,4-xylanases.[98] GH11 xylanases are retaining enzymes. The catalytic nucleophile and the acid/base residue were first identified in *Bacillus circulans* endoxylanase as Glu-78 and Glu-127.

Three-dimensional structures are available for several GH11 members, the first solved being that of the *B. circulans* xylanases (Figure 5.10).[99] As members of Clan GH-C, they have a jelly roll fold.

**FIGURE 5.10**  Crystal structure of the *Bacillus circulans* xylanases.[100] The 20-kDa catalytic domain of *B. circulans* xylanase is folded into a β-jellyroll. (From RCSB PDB Structure Summary—1BCX.)

### 5.4.2  β-XYLOSIDASES

β-Xylosidases are exo-type glycosidases that hydrolyze xylans and short xylo-oligomers into single xylose units and that are found in families GH3, 39, 43, 52, and 54.[1,101] The spatial similarity between D-xylopyranose and L-arabinofuranose leads to bifunctional xylosidase–arabinosidase enzymes, found mainly in GH3, 43, and 54. With the exception of β-xylosidases from GH43 that operate via an inverting mechanism, the hydrolytic reaction of all β-xylosidases from GH3, 39, 52, and 54 results in retention of the anomeric configuration.[4]

Family 39 enzymes belong to clan GH-A, whose members adopt a (β/α)$_8$ barrel. GH-A is also termed as the 4/7 superfamily, because the proton donor and the nucleophile are found on strands 4 and 7 of the (β/α)$_8$ barrel, respectively. The catalytic nucleophile of family 39 was first identified in 1998 in the β-xylosidase from *Thermoanaerobacterium saccharolyticum* as Glu-277.[102,103] The general acid/base residue was identified in *T. saccharolyticum* β-xylosidase as Glu-160.[104] The three-dimensional structure of the GH39 β-xylosidase from *T. saccharolyticum* was first solved in 2004 and represented the first crystal structure of any β-xylosidase of the previous five GH families (Figure 5.11).[101] Each monomer of *T. saccharolyticum* β-xylosidase comprises three distinct domains, such as a catalytic (β/α)$_8$ barrel domain, a β-sandwich domain of unknown function, and a small α-helical domain.[105]

In 2005, the three-dimensional structure for another GH39 β-xylosidase from *Geobacillus stearothermophilus* has also been solved (Figure 5.12).[4] The protein in solution consists of four identical subunits.

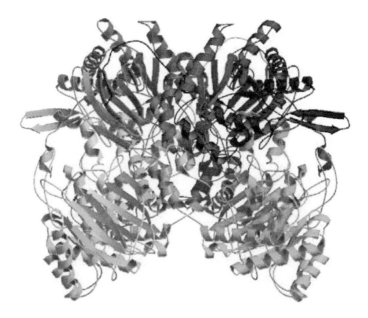

**FIGURE 5.11** Tetrameric crystal structure of β-D-xylosidase from *Thermoanaerobacterium saccharolyticum*, a family 39 GH. (From RCSB PDB Structure Summary—1UHV; From RSCB Protein Data Bank, http://www.rcsb.org/pdb/explore.do?structureId=1UHV.)

**FIGURE 5.12** Homotetrameric assembly of the GH39 β-xylosidase from *Geobacillus stearothermophilus*; this assembly consists of 4 molecules. (From RCSB PDB Structure Summary—2BS9; From RCSB, PDB, http://www.rcsb.org/pdb/explore.do?structureId=2bs9; EMBL-EBI, Protein Databank in Europe, http://www.ebi.ac.uk/pdbesrv/view/entry/2bs9/summary.html.)

All GH39 β-xylosidases structurally characterized up to 2012 displayed a modular multidomain organization that assembles a tetrameric quaternary structure.[109] In 2012, the crystal structure and the SAXS (Small Angle X-ray Scattering) molecular envelope of a new GH39 β-xylosidase from *Caulobacter crescentus* (CcXynB2) have been determined. Interestingly, CcXynB2 is a monomer in solution and comparative structural analyses suggest that the shortened C-terminus prevents the formation of a stable tetramer.

### 5.4.3  FERULOYL ESTERASES

Feruloyl esterases (Faes) represent a subclass of carboxyl esterases (or esterases; EC 3.1.1.x) that can release phenolic acids, such as ferulic acids or other cinnamic acids, from esterified polysaccharides, especially xylans and pectins.[88] Therefore, Faes are regarded as the key enzymes to loosen the internal cross-linking of plant cell walls by acting as important accessory enzymes in synergy with (hemi)cellulases in wall hydrolysis. Ferulic acids are the main phenolic acids to covalently link to the polysaccharides through ester bonds in the plant cell wall. In xylans, ferulic acids are attached to arabinose residues, which are linked to the xylan backbone, while in pectins, ferulic acids are ester-linked to pectin polysaccharides mainly through arabinose or galactose subunits. These cross-links formed via ferulic acids dramatically reduce the biodegradability of plant cell walls by microorganisms.[88]

Since the 1990s, microbial Faes have been studied due to their potential application in biotechnological processes.[78] Faes are not only relevant in biofuel production but also used in the medical industry for synthesis of nutraceutics. A wide range of microorganisms, including fungi and bacteria, have been reported to secrete Faes. Despite having various amino acid sequences, Faes share similar three-dimensional structures, with an α/β-*hydrolase* fold having a serine, histidine, and aspartic acid catalytic triad (Figure 5.13).[88]

The mechanism for ester hydrolysis or formation is composed of four steps[110]: First, the substrate is bound to the active serine, yielding a tetrahedral intermediate stabilized by the catalytic His and Asp residues. Next, the alcohol is released and an

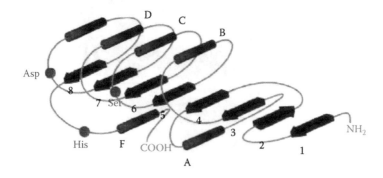

**FIGURE 5.13**  Schematic representation of the α/β-hydrolase fold. β-Sheets (1–8) are shown as blue arrows; α-helices (A–F) as red columns. The relative positions of the amino acids of the catalytic triad are indicated as orange circles. (Bornscheuer, U.T., *FEMS Microbiol. Rev.*, 2002. By permission of Oxford University Press.)

acyl-enzyme complex is formed. Attack of a nucleophile—water in hydrolysis, alcohol, or ester in (trans)esterification—forms again a tetrahedral intermediate which yields the product (an acid or an ester) and free enzyme.

In 2011, Li et al.[88] identified three different active feruloyl esterases in the bacterium *Cellulosilyticum ruminicola* H1 and proposed that the three esterases act synergistically to hydrolyze ferulate esters in natural hemicelluloses.

## 5.4.4 Acetyl Xylan Esterases

Many plant cell wall polysaccharides, including xylans, mannans, and pectins, are present in acetylated forms.[77] Acetylation not only modifies the physicochemical properties of polysaccharides but also means they are less readily attacked by phytopathogen-derived cell wall-degrading endoglycosidases. To overcome the steric problems provided by acetyl substituents, plant cell wall-degrading microorganisms have developed a wide range of acetyl esterases whose function is to deacetylate the polysaccharides prior to, or concomitant with, its complete hydrolysis by a consortium of exo- and endo-acting glycoside hydrolases. Such microbial esterases have found widespread applications in both biomass conversion and for the synthesis of various esters.

Carbohydrate esterases (CE), which catalyze the de-O- or de-N-acylation of saccharides, are classified into sequence-based families defined by Henrissat et al.[111,112] Until 2006, structures of ferulate and acetyl xylan esterases have revealed a $\alpha/\beta$ hydrolase fold with a Ser-His-Asp catalytic triad.[77] Enzymes in the largest sequence-based esterase family, CE4, do not, however, display the standard $\alpha/\beta$ hydrolase fold. CE4 members have been reported to be metal ion-dependent. Furthermore, family CE4 contains not only members with both de-O-acetylase activities, such as acetyl xylan esterases, but also de-N-acetylases involved in chitin and peptidoglycan degradation.

In 2006, Taylor et al.[77] reported the first structures for two distinct de-O-acetylases from family CE4, the *C. thermocellum* (CtCE4) and *S. lividans* (SlCE4) acetyl xylan esterases. The structure of both enzymes is best described as displaying greatly distorted $(\beta/\alpha)_8$ barrel folds. These enzymes were both metal-dependent and possess a single-metal center with a chemical preference for $Co^{2+}$. The *C. thermocellum* enzyme displays different ligand coordinations to the metal, utilizing an aspartate (Asp488), a histidine (His539) and four water molecules, as opposed to the more classical His-His-Asp family CE4 *consensus* displayed by the *S. lividans* enzyme (His 62, His 66, and Asp 13). Different metal ion preferences for the two enzymes may reflect the diversity with which the metal ion coordinates residues and ligands in the active center environment of the two enzymes.

About 60%–70% of xylose residues are esterified at the hydroxyl group by acetic acid; these acetylated xylans are abundant in hardwood.[113] The existence of acetyl xylan esterases (EC 3.1.1.72) was first reported in fungal cultures. It was also shown that acetyl xylan esterases and xylanases act synergistically in hydrolyzing acetyl xylans.[114]

Due to the complexity of lignocellulosic materials, a complete enzymatic hydrolysis into fermentable sugars requires a variety of cellulolytic and xylanolytic enzymes.[115] Hydrolysis of xylans can be enhanced by the removal of xylan side groups; it has been reported that the hydrolysis of isolated hardwood xylans by xylanases was restricted by increasing the degree of acetylation of the xylans.[115]

Chemical deacetylation of xylans of aspen wood and wheat straw increased the enzymatic solubilization of xylans and consequently enhanced cellulose accessibility. The improvement of xylan hydrolysis by the concerted action of endoxylanase and acetyl xylan esterase has been also observed.

In 2011, Zhang et al.[115] showed that the removal of acetyl groups in xylans by acetyl xylan esterase increased the accessibility of xylans to xylanase and improved the hydrolysis of xylans. Solubilization of xylans led to an increased accessibility of cellulose to cellulases and thereby increased the hydrolysis extent of cellulose.

### 5.4.5  α-L-Arabinanases and α-L-Arabinofuranosidases

L-arabinose is the most common pentose with the L-configuration present in nature.[116] It is found mainly in plant cell wall components such as pectic arabinans, arabinoxylans, and arabinogalactans.[117,118]

Among the enzymes participating to the process for obtaining monosaccharides useful in different industrial applications, are those releasing arabinose residues. These include endo-acting arabinanases, which hydrolyze α-1,5-arabinofuranosidic linkages in the arabinan backbone and exo-acting arabinofuranosidases (ABFs), which hydrolyze nonreducing arabinofuranose side chains (Figure 5.14).[119]

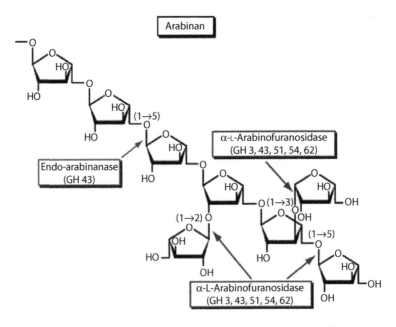

**FIGURE 5.14** Enzymatic activities associated with arabinan hydrolysis. Endoarabinanases cleave α-1,5 linkages between main chain arabinose residues. α-L-Arabinofuranosidases release arabinose monomers by cleaving α-1,2, α-1,3, or α-1,5 linked arabinose residues from the nonreducing end.[120] (Reprinted from *Adv. Appl. Microbiol.*, 70, Yeoman, C.J. et al., Chapter 1—Thermostable enzymes as biocatalysts in the biofuel industry, 1–55, Copyright 2016, with permission from Elsevier.)

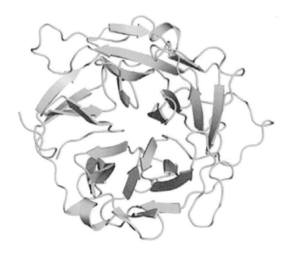

**FIGURE 5.15** Structure of *Cellvibrio japonicus* α-L-arabinanase Arb43A. (Courtesy of John Berrisford.)

### 5.4.5.1 α-L-Arabinanases

Endo-arabinanases (EC 3.2.1.99) hydrolyze 1,5-α-arabinofuranosidic linkages found in the backbone of 1,5-arabinans.[121] They are mainly found in GH 43. The three-dimensional structure of *Cellvibrio japonicus* arabinanase Arb43A revealed a novel five-bladed β-propeller catalytic domain and an inverting mechanism (Figure 5.15).[122] Members of GH family 43 share this structure and work in such a mechanism.[119]

The active site of enzymes within GH 43 possesses three essential catalytic acidic residues.[119] One carboxylate acts as a general acid, protonating the leaving aglycon and making it a better leaving group. A second acidic residue functions as a general base, activating a water molecule for a single-displacement attack on the anomeric carbon, resulting in inversion of the anomeric configuration. The roles of the third residue are still putative.[124]

### 5.4.5.2 α-L-Arabinofuranosidases

Because the structure of L-arabinose-containing polysaccharides is highly variable and complex, a wide variety of nonreducing end α-L-arabinofuranosidases (EC 3.2.1.55) that have various substrate specificities are necessary for the hydrolysis of such polysaccharides and for the production of L-arabinose.[117] α-L-Arabinofuranosidases are the key enzymes in the complete degradation of the plant cell wall.[84] They can be distinguished by their amino acid sequence. They belong to GH 3, 43, 51, 54, and 62.[125]

α-L-Arabinofuranosidases are hemicellulases that cleave the glycosidic bond between L-arabinofuranosides side chains and different oligosaccharides. These enzymes are part of an array of glycoside hydrolases responsible for

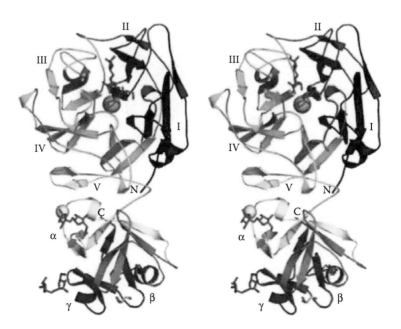

**FIGURE 5.16** Stereoview of exo-1,5-α-arabinofuranosidase from *Streptomyces avermitilis* (SaAraf43A). α-1,5-L-Arabinotriose complex structure is viewed along the pseudo-5-fold axis of the catalytic domain. The bound sugars are shown as stick models, and the bound calcium ion and sodium ion are shown as *pink* and light-blue spheres. Five blades in the catalytic domain and three subdomains in the binding module CBM42 are in different rainbow-ordered colors. (Courtesy of American Society for Biochemistry and Molecular Biology, Rockville, MD; From Fujimoto, Z. et al., *J. Biol. Chem.*, 285, 34134, http://www.ncbi.nlm.nih.gov/pmc/articles/PMC2962512/, 2010.)

the degradation of hemicelluloses such as arabinoxylan, arabinogalactan, and L-arabinan.

Exo-1,5-α-L-arabinofuranosidases belonging to GH 43 have been much studied. They have strict substrate specificity.[125] These enzymes hydrolyze only the α-1,5-linkages of linear arabinan and arabino-oligosaccharides in an exo-acting manner. The enzyme from *Streptomyces avermitilis* (SaAraf43A) contains a N-terminal catalytic domain belonging to GH family 43 and a C-terminal arabinan-binding module belonging to carbohydrate-binding module family 42 (Figure 5.16). The catalytic module of the enzyme is composed of a 5-bladed β-propeller (marked I to V in Figure 5.16). The arabinan-binding module had three similar subdomains assembled against one another around a pseudo-3-fold axis (referred to as subdomains α, β, and γ). A sugar complex structure with α-1,5-L-arabinofuranotriose revealed three subsites in the catalytic domain, and a sugar complex structure with α-L-arabinofuranosyl azide revealed three arabinose-binding sites in the carbohydrate-binding module. A mutagenesis study revealed that substrate specificity was regulated by residues Asn159, Tyr192, and Leu289.

### 5.4.6  α-Glucuronidases

α-D-Glucuronidases (EC 3.2.1.139) cleave the α-1,2-glycosidic bond of the 4-*O*-methyl-D-glucuronic acid side chain of xylans. They are found exclusively in GH67 (Equation 5.1).[1,126]

$$\alpha\text{-D-glucuronoside} + H_2O \rightarrow \text{an alcohol} + \text{D-glucuronate} \qquad (5.1)$$

In 2001, the first crystal structure of an α-glucuronidase was reported by Nurizzo et al. (Figure 5.17).[127] The enzyme, GlcA67A, from *C. japonicus* (previously known as *Pseudomonas cellulosa*) is a dimeric protein that comprises three domains. The N-terminal domain forms a two-layer β-sandwich, whereas the C-terminal domain consists mainly of long α-helices that form the dimer interface. The central domain is a classical $(\beta/\alpha)_8$ barrel housing the catalytic apparatus. The structure of GlcA67A in complex with its reaction products implicated Glu292 as the catalytic acid.[128]

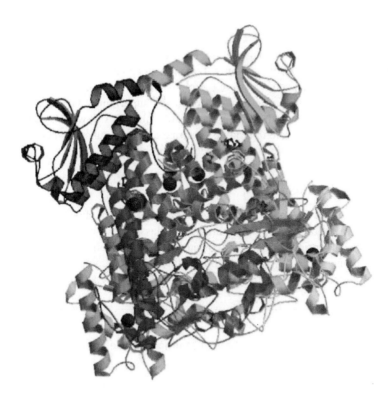

**FIGURE 5.17**  Structure of *Cellvibrio japonicus* α-D-glucuronidase. (From RCSB PDB Structure Summary—1GQI; From Nurizzo, D. et al., Structure, 10, 547, http://www.ncbi.nlm.nih.gov/pubmed/11937059?dopt=Abstract, 2002; http://www.rcsb.org/pdb/explore.do?structureId=1GQI.)

## 5.5   BIODEGRADATION OF MANNANS

The complete conversion of galactomannan into galactose and mannose requires the activity of three types of microbial enzymes, namely (Table 5.4)[130]:

- Mannanases or endo-1,4-β-mannanases (EC 3.2.1.78), which belong to GH 5, 26, and 113.[131]
- β-mannosidases (EC 3.2.1.25), which are found in GH 4, 27, 36, and 57.[132]
- α-galactosidases (EC 3.2.1.22), which are found in GH 1, 2, and 5.

Galactomannan hydrolysis results from the concerted action of microbial endo-mannanases, mannosidases, and α-galactosidases and is a mechanism of intrinsic biological importance.

Mannanases hydrolyze mannan-based hemicelluloses and liberate short β-1,4-manno-oligomers, which can be further hydrolyzed to mannose by β-mannosidases.[1,133,134] α-Galactosidases remove the galactose side branches from the mannose residues of the galacto(gluco)mannan backbone.

### 5.5.1   Mannanases

The crystal structure of mannanase 26A from *Pseudomonas cellulosa* was solved by Hogg et al.[135] The enzyme, PcMan26A, comprises $(\beta/\alpha)_8$-barrel architecture with two catalytic glutamates at the ends of β-strands 4 and 7 in precisely the same location as the corresponding glutamates in other 4/7-superfamily GH enzymes. The family 26 GH enzymes are therefore members of clan GH-A. It was also shown that Trp360 played a critical role in binding substrate at the −1 subsite, whereas Tyr285 was important to the function of the nucleophile catalyst.

### 5.5.2   β-Mannosidases

The structure of the GH2 *Bacteroides thetaiotaomicron* β-mannosidase, *Bt*Man2A, was solved by Tailford et al.[136–138] *Bt*Man2A is a classical exo-enzyme.

The three-dimensional structure indicates that the protein is a dimer in solution. Each monomer has a five-domain structure which encompasses the catalytic center in the central position. This organization is globally similar to the *E. coli*

---

## TABLE 5.4
### Function and Classifications of Typical Mannan-Active Enzymes

| Enzymes | Function | EC Code | GH Family |
|---|---|---|---|
| Mannanases | Cleave mannan (galactomannan, glucomannan) backbone | EC 3.2.1.78 | GH 5, 26, and 113 |
| β-Mannosidases | Remove mannosyl units from nonreducing ends of mannooligosaccharides | EC 3.2.1.25 | GH 1, 2, and 5 |
| α-Galactosidases | Remove the α-galactose side chain | EC 3.2.1.22 | GH 4, 27, 36, and 57 |

GH2 β-galactosidase.[60] The catalytic center is housed mainly within a $(\beta/\alpha)_8$ barrel, although the N-terminal domain that displays a CBM-fold also contributes to the active site topology. Domain 1 (N-terminal domain), consisting of residues 28–218, is a β-sheet domain containing a five-stranded antiparallel β-sheet, a four-stranded antiparallel β-sheet, and an α helical region, reminiscent of carbohydrate-binding modules. Domains 2 (cyan) and 4 (yellow), which comprise residues 219–331 and 676–780, respectively, are structurally very similar to domain 1, again displaying immunoglobulin-like folds (sandwich-like structure formed by two sheets of anti-parallel β-strands).[139,140] Both these domains consist of two antiparallel β-sheets: one containing four β-strands and the other three. Domain 3 (green) contains the catalytic center consisting of residues 332–675. Domain 5 (C-terminal domain; red) contains β-sheets and consists of residues 786–860. The structure of *Bt*Man2A differs from that of the *E. coli* β-galactosidase most significantly in the orientation of domain 5. The dimerization interface of *Bt*Man2A is formed by interactions between domain 5 of both monomers, whereas in the *E. coli* β-galactosidase the equivalent domain also forms the basis of the interactions which lead to tetrameric oligomerization.

The catalytic domain residues (332–675) are structurally similar to a number of clan GH-A glycoside hydrolases (Section 2.3.1). The catalytic residues, implicated as Glu-462 and Glu-555, lie on strands 4 and 7, respectively, of the $(\beta/\alpha)_8$ barrel as expected, and the active site comprises a deep pocket.

The nature of the substrate-binding residues is quite distinct from other GH2 enzymes of known structure, instead they are similar to other clan GH-A enzymes specific for manno-configured substrates.[136] A WDW (Trp198-Asp199-Trp200) motif in the N-terminal domain that contributes to the base of the active site pocket makes a significant contribution to catalytic activity. This motif is invariant in GH2 mannosidases.

### 5.5.3 α-Galactosidases

α-Galactosidases (α-D-galactopyranoside galactohydrolase, EC 3.2.1.22) hydrolyze terminal, nonreducing α-D-galactose residues in α-D-galactosides, including galactose oligosaccharides, galactomannans, and galactolipids commonly found in legumes and seeds.[13,141] These enzymes occur widely in microorganisms, plants, and animals. α-Galactosides are exclusively fermented by microbial enzymes. Based on their sequence similarities, α-galactosidases have been classified into families GH4, GH27, GH36, GH57, GH97, and GH110.[13]

Characterization of the complete gene sequence encoding the α-galactosidase from *Phanerochaete chrysosporium* confirmed that this enzyme is a member of the GH27 family.[142–144] This family, together with the GH31 and GH36 enzymes, forms the Clan GH-D, a superfamily of α-galactosidases, α-N-acetylgalactosaminidases, and isomaltodextranases which are likely to share a common catalytic mechanism and structural topology.[145,146] In 2002, Asp-130 was identified as the catalytic nucleophile in the GH27 α-galactosidase from *P. chrysosporium*.[143] The general acid/base residue in the GH27 family was first identified as Asp-201 in the chicken (*Gallus gallus*) N-acetylgalactosaminidase.[147]

The crystal structure of GH27 α-galactosidase from *T. reesei* and its complex with galactose has been determined by Golubev et al.[148]

The refined crystallographic model of the enzyme consists of two domains, an *N*-terminal catalytic domain of the $(\beta/\alpha)_8$ barrel topology and a C-terminal domain which is formed by an antiparallel $\beta$-structure. The protein contains four *N-glycosylation*[149] sites located in the catalytic domain. Some of the oligosaccharides were found to participate in inter-domain contacts. The galactose molecule binds to the active site pocket located in the center of the barrel of the catalytic domain. Analysis of the $\alpha$-galactosidase-galactose complex reveals the residues of the active site and offers a basis for identifying the mechanism of the enzymatic reaction. The structure of the $\alpha$-galactosidase closely resembles those of GH27 enzymes. The conservation of two catalytic Asp residues, identified for the GH27 family, is consistent with a retaining mechanism for the $\alpha$-galactosidase.

## 5.6   BIODEGRADATION OF ARABINANS, GALACTANS, AND ARABINOGALACTANS

Arabinans, galactans, and arabinogalactans (Figure 5.1) constitute together the so-called neutral pectic substances.[150] Pure arabinans and galactans occur in plant cell walls but are relatively rare. Pure arabinans are typically composed of $\alpha$-1,5-arabinosyl residues which are decorated with $\alpha$-1,2- and $\alpha$-1,3-arabinosyl units. Pure galactans consist mainly of $\beta$-1,4-linear polymers. Arabinogalactans are branched polysaccharides which fall into two groups, namely arabinogalactans type I and arabinogalactans type II. The important structural difference between these two types of arabinogalactans is the galactose units, which are 1,4-linked in type I and 1,3- and 1,6-linked in type II.[151] The backbone of arabinogalactans type I is composed of $\beta$-1,4-linked galactosyl residues, bearing 20%–40% of $\alpha$-arabinosyl residues 1,5-linked in short chains, in general at position 3.[152] The backbone of arabinogalactans type II is composed of $\beta$-1,3-linked galactosyl residues, which are substituted at position 6 with $\beta$-1,6-linked galactosyl residues and $\alpha$-1,3-linked-arabinosyl or arabinans sidechains.[153]

The enzymatic degradation of arabinans and arabinogalactans requires the action of several enzymes, among which:

- *Endo-1,5-$\alpha$-arabinanases* (EC 3.2.1.99), which cleave the arabinan backbone. They are found in family GH 43.[154]
- *Endo-1,4-$\beta$-galactanases* (EC 3.2.1.89), which cleave the galactan backbone in type I arabinogalactans. They belong to GH 53.[155]
- *Endo-1,3-$\beta$-galactanases* (EC 3.2.1.181), which cleave the galactan backbone in type II arabinogalactans. They belong to GH12.
- *Galactan 1,3-$\beta$-galactosidases* or *exo-$\beta$-1,3-galactanases* (EC 3.2.1.145), which hydrolyze terminal, nonreducing $\beta$-galactose residues in 1,3-$\beta$-galactan backbones of type II arabinogalactans. They belong to GH43.[156,157]
- *Galactan-endo-1,6-$\beta$-galactosidases* or *endo-$\beta$-1,6-galactanases* (EC 3.2.1.164), which cleave the 1,6-$\beta$-galactosidic linkages in arabinogalactans and 1,3;1,6-$\beta$-galactans.[158] They belong to GH5 (Clan GH-A).[159]
- *GH42 $\beta$-Galactosidases*, (EC 3.2.1.23) which hydrolyze arabinogalactans type I oligomers.[160]

**TABLE 5.5**
**Function and EC and GH Classifications of Enzymes That Degrade Neutral Pectins**

| Enzymes | Functions | EC Code | GH Family |
|---|---|---|---|
| Endo-1,5-α-arabinases | Cleave arabinan backbone | EC 3.2.1.99 | GH43 |
| Endo-1,4-β-galactanases | Cleave 1,4-β-galactosidic linkages in type I arabinogalactans. | EC 3.2.1.89 | GH53 |
| Endo-1,3-β-galactanases | Cleave 1,3-β-galactosidic linkages in type II arabinogalactans | EC 3.2.1.181 | GH16 |
| Galactan 1,3-β-galactosidases or exo-β-1,3-galactanases | Cleave terminal, nonreducing β-galactose residues in 1,3-β-galactans | EC 3.2.1.145 | GH43 |
| Galactan-endo-1,6-β-galactosidases or endo-β-1,6-galactanases | Cleave 1,6-β-galactosidic linkages in arabinogalactans and 1,3;1,6-β-galactans | EC 3.2.1.164 | GH5 |
| GH42 β-Galactosidases | Hydrolyze arabinogalactans type I oligomers | EC 3.2.1.23 | GH42 |

Table 5.5 summarizes the function of various enzymes that degrade neutral pectic polymers, and their EC and GH classifications.

## 5.7 PERSPECTIVES

The enzymatic degradation of plant polysaccharides is a multienzyme process of fundamental importance in nature.[161] Microorganisms produce the multiple enzymes necessary for plant degradation. These microorganisms play an important role in the carbon cycle.[162]

Degradation of plant polysaccharides is of major importance in the food and feed, beverage, textile, and paper and pulp industries, as well as in several other industrial production processes.[76] Enzymatic degradation of these polymeric carbohydrate structures is becoming a more and more attractive complement to chemical and mechanical processes.

The enzymatic degradation of plant polysaccharides is emerging as one of the key environmental goals of the early twenty-first century, impacting on many processes in the textile and detergent industries as well as biomass conversion to biofuels.[163] One of the issues with the use of nonedible substrates such as the plant cell wall is that the cellulose fibers are embedded in a network of diverse polysaccharides that render access difficult. The goal of pretreatments is to make the polysaccharides accessible to enzymatic hydrolysis prior to conversion into biofuels and biobased products.[164]

## REFERENCES

1. D. Shallom and Y. Shoham, *Curr. Opin. Microbiol.* 6, 219, 2003. http://biotech.technion. ac.il/Media/Uploads/Documents/Yuval/Publications/Yuval_67.pdf.
2. C. Martinez-Fleites, C.I.P.D. Guerreiro, M.J. Baumann, E.J. Taylor, J.A.M. Prates, L.M.A. Feirrera, C.M.G.A. Fontes, H. Brumer and G.J. Davies, *J. Biol. Chem.* 281, 24922, 2006. http://www.jbc.org/content/281/34/24922.full.
3. T.W. Jeffries and C. Ratledge, Eds., *Biochemistry of Microbial Degradation*, 233, Kluwer Academic Publishers, Dordrecht, the Netherlands, 1994. http://www.fpl.fs.fed. us/documnts/pdf1994/jeffr94b.pdf.
4. M. Czjzek, A. Ben David, T. Bravman, G. Shoham, B. Henrissat and Y. Shoham, *J. Mol. Biol.* 353, 838, 2005. http://www.ncbi.nlm.nih.gov/pubmed/16212978?dopt= Abstract.
5. A.K. Chandel, O.V. Singh and L.V. Rao, Biotechnological applications of hemicellulosic derived sugars: State-of-the-art, in *Sustainable Biotechnology*, 63, Springer Netherlands, 2010. http://link.springer.com/chapter/10.1007%2F978-90-481-3295-9_4#page-1.
6. J. Zhao, L. Xu, Y. Wang, X. Zhao, J. Wang, E. Garza, R. Manow and S. Zhou, *Microb. Cell Fact.* 12, 57, 2013. http://www.microbialcellfactories.com/content/pdf/1475-2859-12-57.pdf.
7. M. Kopke, C. Mihalcea, F.M. Liew, J.H. Tizard, M.S. Ali, J.J. Conolly, B. Al-Sinawi, and S.D. Simpson, *Appl. Environ. Microbiol.* 77, 15, 2011. http://www.ncbi.nlm.nih. gov/pmc/articles/PMC3147483/.
8. S.S. da Silva and A.K. Chandel, Eds., *D-Xylitol, Fermentative Production, Application and Commercialization*, Springer, 2012. http://www.springer.com/life+sciences/ microbiology/book/978-3-642-31886-3.
9. CAZY, Carbohydrate Active enzymes, 2010. http://www.cazy.org/Glycoside-Hydrolases.html.
10. H.J. Gilbert, *Plant Physiol.* 153, 444, 2010. http://www.ncbi.nlm.nih.gov/pmc/articles/ PMC2879781/.
11. http://solcyc.solgenomics.net//META/NEW-IMAGE?type=ENZYME-IN-RXN-DISPLAY&object=MONOMER-16608&detail-level=2
12. CAZY, Glycoside Hydrolase Family 74. http://www.cazypedia.org/index.php/Glycoside_ Hydrolase_Family_74.
13. M. Cervera-Tison, L. Tailford, C. Fuell, L. Bruel, G. Sulzenbacher, B. Henrissat, J.G. Berrin, M. Fons, T. Giardina and N. Juge, *Appl. Environ. Microbiol.* 78, 7720, 2012. http://aem.asm.org/content/78/21/7720.full.
14. http://biocyc.org/META/NEW-IMAGE?type=ENZYME-IN-RXN-DISPLAY& object=BSU34120-MONOMER
15. H.J. Gilbert, *Cellulases, Series: Methods in Enzymology*, vol. 510, Academic Press/ Elsevier, San Diego, CA, 2012.
16. EBI.    http://www.ebi.ac.uk/thornton-srv/databases/cgi-bin/enzymes/GetPage.pl?ec_ number=3.2.1.151.
17. http://biocyc.org/META/NEW-IMAGE?type=EC-NUMBER&object=EC-3.2.1.155
18. http://www.uniprot.org/uniprot/Q8J0D2
19. J. Eklov, Plant and microbial xyloglucanases: Function, structure and phylogeny, Doctoral thesis, KTK, Stockholm, 2011. http://kth.diva-portal.org/smash/record.jsf? pid=diva2:405550.
20. C.P. Kubicek, *Fungi and Lignocellulosic Biomass*, Wiley-Backwell, 2012. http:// onlinelibrary.wiley.com/doi/10.1002/9781118414514.ch1/summary.

21. B.Y. Wang, Ed., *Environmental Biodegradation Research Focus*, Nova Science Publishers, New York, 2007. http://books.google.be/books?id=WtHCXlVMH1EC&pg=PA15&lpg=PA15&dq=%22biodegradation+of+xyloglucan%22&source=bl&ots=RSZyJ71kfC&sig=N6OIWXYeAE965Y-HJWyeQ76SQNc&hl=fr#v=onepage&q=%22biodegradation%20of%20xyloglucan%22&f=false.

22. K. Yaoi and Y. Mitsuishi, *J. Biol. Chem.* 277, 48276, 2002. http://www.jbc.org/content/277/50/48276.full.

23. G.J. Davies, K.S. Wilson and B. Henrissat, *Biochem. J.* 321, 557, 1997. http://www.ncbi.nlm.nih.gov/pmc/articles/PMC1218105/pdf/9020895.pdf.

24. Ecocyc, SRI International, 2012. http://www.ecocyc.org/META/new-image?type=PATHWAY&object=PWY-6812&detail-level=2&ENZORG=NONE.

25. K. Yaoi, H. Kondo, N. Noro, M. Suzuki, S. Tsuda and Y. Mitsuishi, *Structure* 12, 1209, 2004. http://www.ncbi.nlm.nih.gov/pubmed/15242597?dopt=Abstract.

26. K. Yaoi, H. Kondo, A. Hiyoshi, N. Noro, H. Sujimoto, S. Tsuda, Y. Mitsuishi and K. Miyazaki, *J. Mol. Biol.* 370, 53, 2007. http://www.ncbi.nlm.nih.gov/pubmed/17498741.

27. RCSB, PDB, 2EBS. http://www.rcsb.org/pdb/explore.do?structureId=2EBS.

28. S.G. Grishutin, A.V. Gusakov, A.V. Markov, B.B. Ustinov, MV. Semenova and A.P. Sinitsyn, *Biochim. Biophys. Acta* 1674, 268, 2004. http://www.ncbi.nlm.nih.gov/pubmed/15541296.

29. V.V. Zverlov, N. Schantz, P. Schmitt-Kopplin and W.H. Schwarz, *Microbiology* 151, 3395, 2005. http://mic.sgmjournals.org/content/151/10/3395.full.pdf+html.

30. S. Bauer, P. Vasu, A.J. Mort and C.R. Somerville, *Carbohydr. Res.* 340, 2590, 2005. http://www.ncbi.nlm.nih.gov/pubmed/16214120.

31. http://www.uniprot.org/uniprot/Q5BD38

32. K. Yaoi, T. Nakai, Y. Kameda, A. Hiyoshi and Y. Mitsuishi, *Appl. Environ. Microbiol.* 71, 7670, 2005. http://www.ncbi.nlm.nih.gov/pmc/articles/PMC1317386.

33. CAZypedia, Glycoside Hydrolase Family 74, 2011. http://www.cazypedia.org/index.php/Glycoside_Hydrolase_Family_74.

34. T. Desmet, T. Cantaert, P. Gualfetti, W. Nerinckx, L. Gross, C. Mitchinson and K. Piens, *FEBS J.* 274, 356, 2007. http://www.ncbi.nlm.nih.gov/pubmed/17229143.

35. J. Sampedro, B. Pardo, C. Gianzo, E. Guitian, G. Revilla and I. Zarra, *Plant Physiol.* 154, 1105, 2010. http://www.plantphysiol.org/content/154/3/1105.full.

36. Complex Carbohydrate Research Center, University of Georgia, Hemicelluloses, Xyloglucans. http://www.ccrc.uga.edu/~mao/xyloglc/Xtext.htm.

37. http://biocyc.org/META/NEW-IMAGE?type=NIL&object=PWY-6791

38. http://biocyc.org/META/NEW-IMAGE?type=NIL&object=PWY-6807

39. http://www.chem.qmul.ac.uk/iubmb/enzyme/EC3/2/1/23.html

40. http://www.ebi.ac.uk/thornton-srv/databases/cgi-bin/enzymes/GetPage.pl?ec_number=3.2.1.177

41. http://www.chem.qmul.ac.uk/iubmb/enzyme/EC3/2/1/177.html

42. http://www.chem.qmul.ac.uk/iubmb/enzyme/EC3/2/1/21.html

43. http://www.chem.qmul.ac.uk/iubmb/enzyme/EC3/2/1/63.html

44. http://www.chem.qmul.ac.uk/iubmb/enzyme/EC3/2/1/55.html

45. R.A. O'Neill, P. Albersheim and A.G. Darvill, *J. Biol. Chem.* 264, 20430, 1989. http://www.jbc.org/content/264/34/20430.long.

46. C. Fanutti, M.J. Gidley and J.S.G. Reid, *Planta* 184, 137, 1991. http://cel.webofknowledge.com/InboundService.do?SID=V2ljjpfp3PpkEgonaBN&product=CEL&UT=A1991FF05400020&SrcApp=Highwire&Init=Yes&action=retrieve&Func=Frame&customersID=Highwire&SrcAuth=Highwire&IsProductCode=Yes&mode=FullRecord.

47. J. Sampredo, C. Sieiro, G. Revilla, T. Gonzales-Villa and I. Zarra, *Plant Physiol.* 126, 910, 2001. http://www.ncbi.nlm.nih.gov/pubmed/11402218.
48. http://www.brenda-enzymes.org/php/result_flat.php4?ecno=3.2.1.23
49. http://www.chem.qmul.ac.uk/iubmb/enzyme/EC3/2/1/23.html
50. H.J. Crombie, S. Chengappa, A. Hellyer and J.S. Reid, *Plant J.* 15, 27, 1998. http://www.ncbi.nlm.nih.gov/pubmed/9744092?dopt=Abstract.
51. N. Iglesias, J.A. Abelenda, M. Rodino, J. Sampedro, J. Revilla and I. Zarra, *Plant Cell Physiol.* 47, 55, 2006. http://pcp.oxfordjournals.org/content/47/1/55.full.
52. F. de la Torre, J. Sampedro, I. Zarra and G. Revilla, *Plant Physiol.* 128, 247, 2002. http://www.ncbi.nlm.nih.gov/pubmed/11788770.
53. N. Hobson and M.K. Deyholos, *BMC Genomics* 14, 344, 2013. http://www.ncbi.nlm.nih.gov/pmc/articles/PMC3673811/.
54. J. Sampedro, C. Sieiro, G. Revilla, T. Gonzales-Villa and I. Zarra, *Plant Physiol.* 126, 910, 2001. http://www.ncbi.nlm.nih.gov/pubmed/11402218.
55. J.R. Ketudat Cairns and A. Esen, *Cell. Mol. Life Sci.* 67, 3389, 2010. http://www.ncbi.nlm.nih.gov/pubmed/20490603.
56. M.D. Sweeney and F. Xu, *Catalysts* 2, 244, 2012. doi:10.3390/catal2020244.
57. Q. Husain, *Crit. Rev. Biotechnol.* 30, 41, 2010. http://bio.uqam.ca/upload/files/etudiants_menu-principal/ressources/articles/husain%202010.pdf.
58. A.V. Fowler and I. Zabin, *J. Biol. Chem.* 245, 5032, 1970. http://www.jbc.org/content/245/19/5032.
59. A.V. Fowler and I. Zabin, *Proc. Natl. Acad. Sci. USA* 74, 1507, 1977. http://www.pnas.org/content/74/4/1507.full.pdf.
60. R.H. Jacobson, X.J. Zhang, R.F. Dubose and B.W. Matthews, *Nature* 369, 761, 1994. http://www.ncbi.nlm.nih.gov/pubmed/8008071?dopt=Abstract.
61. http://www.rcsb.org/pdb/explore.do?structureId=1bgm
62. B.W. Matthews, *C. R. Biol.* 328, 549, 2005. http://www.ncbi.nlm.nih.gov/pubmed/15950161.
63. D.H. Juers, R.H. Jacobson, D. Wigley, X.J. Zhang, R.E. Huber, D.E. Tronrud and B.W. Matthews, *Protein Sci.* 9, 1685, 2000. http://www.ncbi.nlm.nih.gov/pmc/articles/PMC2144713/pdf/11045615.pdf.
64. http://www.cazypedia.org/index.php/Glycoside_Hydrolase_Family_35
65. D. Gantulga, Y. Turan, D.R. Bevan and A. Esen, *Phytochemistry* 69, 161, 2008. http://www.ncbi.nlm.nih.gov/pubmed/18359051.
66. A.L. Rojas, R.A.P. Nagem, K.N. Neustroev, M. Arand, M. Adamska, E.V. Eneyskaya, A.A. Kulminskaya, R.C. Garratt, A.M. Golubev and I. Polikarpov, *J. Mol. Biol.* 343, 1281, 2004. http://www.sciencedirect.com/science/article/pii/S0022283604011416.
67. http://www.proteopedia.org/wiki/index.php/1tg7
68. http://www.proteopedia.org/wiki/index.php/1xc6
69. A.L. Lovering, S.S. Lee, Y.M. Kim, S.G. Withers and N.C.J. Strynadka, *J. Biol. Chem.* 280, 2105, 2005. http://www.jbc.org/content/280/3/2105.long.
70. http://www.ebi.ac.uk/pdbe-srv/view/entry/1xsk/summary
71. http://www.cazypedia.org/index.php/Glycoside_Hydrolase_Family_1
72. http://www.cazypedia.org/index.php/Glycoside_Hydrolase_Family_3
73. J.N. Varghese, M. Hrmova and G.B. Fincher, *Structure* 7, 179, 1999. http://www.ncbi.nlm.nih.gov/pubmed/10368285?dopt=Abstract.
74. http://www.ebi.ac.uk/pdbe-srv/view/entry/1ex1/summary.html
75. http://www.rcsb.org/pdb/explore.do?structureId=1ex1
76. R.P. de Vries and J. Visser, *Microbiol. Mol. Biol. Rev.* 65, 497, 2001. http://mmbr.asm.org/content/65/4/497.full.
77. E.J. Taylor, T.M. Gloster, J.P. Turkenburg, F. Vincent, A.M. Brzozowski, C. Dupont, F. Shareck et al., *J. Biol. Chem.* 281, 10968, 2006. http://www.jbc.org/content/281/16/10968.full.pdf.

78. MetaCyc, MetaCyc Pathway: (1,4)-β-xylan degradation, 2013. http://ecocyc.org/META/new-image?type=PATHWAY&object=PWY-6717.
79. O. Gallardo, M. Fernandez-Fernandez, C. Valls, S.V. Valenzuela, M.B. Roncero, T. Vidal, P. Díaz and F.I.J. Pastor, *Appl. Environ. Microbiol.* 76, 6290, 2010. http://www.ncbi.nlm.nih.gov/pmc/articles/PMC2937475/.
80. IUBMB Biochemical Nomenclature, EC 3.2.1.156. http://www.chem.qmul.ac.uk/iubmb/enzyme/EC3/2/1/156.html.
81. EBI, EC 3.2.1.37 Xylan 1,4-β-xylosidase. http://www.ebi.ac.uk/thornton-srv/databases/cgi-bin/enzymes/GetPage.pl?ec_number=3.2.1.37.
82. A. Knob, C.R.F. Terrasan and E.C. Carmona, *World J. Microbial. Biotechnol.* 26, 389, 2010. http://link.springer.com/article/10.1007%2Fs11274-009-0190-4#page-1.
83. MetaCyc, MetaCyc Enzyme: xylan-1,4-β-xylosidase. http://biocyc.org/META/NEW-IMAGE?type=ENZYME-IN-RXN-DISPLAY&object=MONOMER-16539.
84. D. Shallom, V. Belakhov, D. Solomon, G. Shoham, T. Baasov and Y. Shoham, *J. Biol. Chem.* 277, 43667, 2002. http://www.jbc.org/content/277/46/43667.long.
85. IUBMB Biochemical Nomenclature, EC 3.2.1.55. http://www.chem.qmul.ac.uk/iubmb/enzyme/EC3/2/1/55.html.
86. http://www.chem.qmul.ac.uk/iubmb/enzyme/EC3/1/1/72.html
87. http://www.cazy.org/Carbohydrate-Esterases.html
88. J. Li, S. Cai, Y. Luo and X. Dong, *Appl. Environ. Microbiol.* 77, 6141, 2011. http://www.ncbi.nlm.nih.gov/pmc/articles/PMC3165382.
89. http://www.chem.qmul.ac.uk/iubmb/enzyme/EC3/1/1/73.html
90. CAZY, Carbohydrate Esterase Family 1. http://www.cazy.org/CE1.html.
91. M. Sharma and A. Kumar, *Br. Biotechnol. J.* 3, 1, 2013.
92. T. Collins, C. Gerday and G. Feller, *FEMS Microbiol. Rev.* 29, 3, 2005. http://www.ncbi.nlm.nih.gov/pubmed/15652973.
93. G. Paes, J.G. Perrin and J. Beaugrand, *Biotechnol. Adv.* 30, 564, 2012. http://www.ncbi.nlm.nih.gov/pubmed/22067746.
94. CAZypedia, Glycoside Hydrolase Family 10, 2012. http://www.cazypedia.org/index.php/Glycoside_Hydrolase_Family_10.
95. D. Tull, S.G. Withers, N.R. Gilkes, D.G. Kilburn, R.A. Warren and R. Aebersold, *J. Biol. Chem.* 266, 15621, 1991. http://www.ncbi.nlm.nih.gov/pubmed/1678739?dopt=Abstract and http://www.jbc.org/content/266/24/15621.long.
96. U. Derewenda, L. Swenson, R. Green, Y. Wei, R. Morosoli, F. Shareck, D. Kluepfel and Z.S. Derewenda, *J. Biol. Chem.* 269, 20811, 1994. http://www.ncbi.nlm.nih.gov/pubmed/8063693?dopt=Abstract and http://www.jbc.org/content/269/33/20811.long.
97. RCSB, PDB. http://www.rcsb.org/pdb/explore.do?structureId=1xas.
98. CAZypedia, Glycoside Hydrolase Family 11, 2012. http://www.cazypedia.org/index.php/Glycoside_Hydrolase_Family_11#bibkey_8.
99. W.W. Wakarchuk, R.L. Campbell, W.L. Sung, J. Davoodi and M. Yaguchi, *Protein Sci.* 3, 467, 1994. http://www.ncbi.nlm.nih.gov/pubmed/8019418?dopt=Abstract and http://www.ncbi.nlm.nih.gov/pmc/articles/PMC2142693/pdf/8019418.pdf.
100. RCSB, PDB. http://www.rcsb.org/pdb/explore.do?structureId=1BCX.
101. J.K. Yang, H.J. Yoon, H.J. Ahn, B.I. Lee, J.D. Pedelacq, E.C. Liong, J. Berendzen et al., *J. Mol. Biol.* 335, 155, 2004. http://www.ncbi.nlm.nih.gov/pubmed/14659747?dopt=Abstract.
102. D.J. Vocadlo, L.F. Mackenzie, S. He, G.J. Zeikus and S.G. Withers, *Biochem. J.* 335, 449, 1998. http://www.ncbi.nlm.nih.gov/pubmed/9761746.
103. CAZypedia, Glycoside Hydrolase Family 39, 2012. http://cazypedia.msl.ubc.ca/index.php/Glycoside_Hydrolase_Family_39.
104. D.J. Vocadlo, J. Wicki, K. Rupitz and S.G. Withers, *Biochemistry* 41, 9736, 2002. http://www.ncbi.nlm.nih.gov/pubmed/12146939?dopt=Abstract.

105. NCBI, Structure Summary MMDB, 2012. http://www.ncbi.nlm.nih.gov/Structure/mmdb/mmdbsrv.cgi?uid=25852.
106. RSCB Protein Data Bank. http://www.rcsb.org/pdb/explore.do?structureId=1UHV.
107. RCSB, PDB. http://www.rcsb.org/pdb/explore.do?structureId=2bs9.
108. EMBL-EBI, Protein Databank in Europe. http://www.ebi.ac.uk/pdbe-srv/view/entry/2bs9/summary.html.
109. C.R. Santos, C.C. Polo, J.M. Correa, R.C.J. Simao, F.A.V. Seixas and M.T. Murakami, *Acta Cryst.* D68, 1339, 2012. http://journals.iucr.org/d/issues/2012/10/00/dw5019/stdsup.html.
110. U.T. Bornscheuer, *FEMS Microbiol Rev.* 26, 73, 2002. http://envismadrasuniv.org/Microbes%20and%20Metals%20Interaction/pdf/Microbial%20carboxyl%20esterases.pdf.
111. B. Henrissat, P.M. Coutinho and G.J. Davies, *Crit. Rev. Biotechnol.* 47, 55, 2001.
112. http://www.cazy.org/Carbohydrate-Esterases.html
113. G. Degrassi, M. Kojic, G. Ljubijankic and V. Venturi, *Microbiology* 146, 1585, 2000. http://mic.sgmjournals.org/content/146/7/1585.full.
114. P. Biely, C.R. Mackenzie, J. Puls and H. Schneider, *Nat. Biotechnol.* 4, 731, 1986. http://www.nature.com/nbt/journal/v4/n8/full/nbt0886-731.html.
115. J. Zhang, M. Siika-Aho, M. Tenkanen and L. Viikari, *Biotechnol. Biofuels* 4, 60, 2011. http://www.ncbi.nlm.nih.gov/pmc/articles/PMC3259036.
116. M.C. Ravanal, E. Callegari and J. Eizaguirre, *Appl. Environ. Microbiol.* 76, 5247, 2010. http://aem.asm.org/content/76/15/5247.full.
117. H. Ichinose, M. Yoshida, Z. Fujimoto and S. Kaneko, *Appl. Microbiol. Biotechnol.* 80, 399, 2008. http://www.ncbi.nlm.nih.gov/pmc/articles/PMC2518083/.
118. S. Kaneko, M. Sano and I. Kusakabe, *Appl. Environ. Microbiol.* 60, 3425, 1994. http://www.ncbi.nlm.nih.gov/pmc/articles/PMC201823/.
119. A. Alhassid, A. Ben_David, O. Tabachnikov, D. Libster, E. Naveh, G. Zolotnitsky, Y. Shoham and G. Shoham, *Biochem. J.* 422, 73, 2009. http://www.biochemj.org/bj/422/0073/bj4220073.htm.
120. C.J. Yeoman, Y. Han, D. Dodd, C.M. Schroeder, R.I. Mackie and I.K.O. Cann, *Adv. Appl. Microbiol.* 70, 1, 2010. http://www.sciencedirect.com/science/article/pii/S0065216410700010.
121. http://www.chem.qmul.ac.uk/iubmb/enzyme/EC3/2/1/99.html
122. D. Nurizzo, J.P. Turkenburg, S.J. Charnock, S.M. Roberts, E.J. Dodson, V.A. Mckie, E.J. Taylor, H.J. Gilbert and G.J. Davies, *Nat. Struct. Biol.* 9, 665, 2002. http://www.ncbi.nlm.nih.gov/pubmed/12198486?dopt=Abstract.
123. http://www.ebi.ac.uk/pdbe-srv/view/entry/1gyd/summary.html
124. http://www.cazypedia.org/index.php/Glycoside_Hydrolase_Family_43#bibkey_10
125. Z. Fujimoto, H. Ichinose, T. Maehara, M. Honda, M. Kitaoka and S. Kaneko, *J. Biol. Chem.* 285, 34134, 2010. http://www.ncbi.nlm.nih.gov/pmc/articles/PMC2962512/.
126. http://www.brenda-enzymes.org/php/result_flat.php4?ecno=3.2.1.139&UniProtAcc=B3PC73&OrganismID=191848&ShowAll=True
127. D. Nurizzo, T. Nagy, H.J. Gilbert and G.J. Davies, *Structure* 10, 547, 2002. http://www.ncbi.nlm.nih.gov/pubmed/11937059?dopt=Abstract.
128. T. Nagy, D. Nurizzo, G.J. Davies, P. Biely, J.H. Lakey, D.N. Bolam and H.J. Gilbert, *J. Biol. Chem.* 278, 20286, 2003. http://www.jbc.org/content/278/22/20286.full.
129. http://www.rcsb.org/pdb/explore.do?structureId=1GQI
130. M.S.J. Centeno, C.I.P.D. Guerreiro, F.M.V. Dias, C. Morland, L.E. Tailford, A. Goyal, J.A.M. Prates et al., *FEMS, Microbiol. Lett.* 261, 123, 2006. http://onlinelibrary.wiley.com/doi/10.1111/j.1574-6968.2006.00342.x/full.

131. http://www.chem.qmul.ac.uk/iubmb/enzyme/EC3/2/1/78.html
132. E. Béki, I. Nagy, J. Vanderleyden, S. Jäger, L. Kiss, L. Fülöp, L. Hornok and J. Kukolya, *Appl. Environ. Microbiol.* 69, 1944, 2003. http://www.ncbi.nlm.nih.gov/pmc/articles/PMC154781/.
133. M. Couturier, J. Feliu, S. Bozonnet, A. Roussel and J.G. Berrin, *PLoS-One* 2013. doi: 10.1371/journal.pone.0079800.
134. J. Tao and R.J. Kazlauskas, Ed., *Biocatalysis for Green Chemistry and Chemical Process development*, John Wiley & Sons, Hoboken, NJ, 2011.
135. D. Hogg, E.J. Woo, D.N. Bolam, V.A. Mckie, H.J. Gilbert and R.W. Pickersgill, *J. Biol. Chem.* 276, 31186, 2001. http://www.jbc.org/content/276/33/31186.full.
136. L.E. Tailford, V.A. Money, N.L. Smith, C. Dumon, G.J. Davies and H.J. Gilbert, *J. Biol. Chem.* 282, 11291, 2007. http://www.jbc.org/content/282/15/11291.long.
137. http://www.rcsb.org/pdb/explore.do?structureId=2je8
138. CAZypedia, 2013. http://www.cazypedia.org/index.php/Glycoside_Hydrolase_Family_2.
139. http://en.wikipedia.org/wiki/Immunoglobulin_domain
140. http://en.wikipedia.org/wiki/Immunoglobulin_superfamily
141. http://www.brenda-enzymes.info/php/result_flat.php4?ecno=3.2.1.22&organism=
142. B. Henrissat and A. Bairoch, *Biochem. J.* 316, 695, 1996. http://www.ncbi.nlm.nih.gov/pmc/articles/PMC1217404/pdf/8687420.pdf.
143. D.O. Hart, S. He, C.J. 2nd Chany, S.G. Withers, P.F. Sims, M.L. Sinnott and H. 3rd Brumer, *Biochemistry* 39, 9826, 2000. http://www.ncbi.nlm.nih.gov/pubmed/10933800?dopt=Abstract.
144. http://www.cazypedia.org/index.php/Glycoside_Hydrolase_Family_GH27_%28GH27%29
145. http://www.cazy.org/Glycoside-Hydrolases.html
146. https://database.riken.jp/sw/en/Glycoside_hydrolase__clan_GH-D/crib124s1rib124u111i/
147. S.C. Garman, L. Hannick, A. Zhu and D.N. Garboczi, *Structure* 10, 425, 2002. http://www.ncbi.nlm.nih.gov/pubmed/12005440?dopt=Abstract.
148. A.M. Golubev, R.A.P. Nagem, J.R. Brandao Neto, K.N. Neustroev, E.V. Eneyskaya, A.A. Kulminskaya, K.A. Shabalin, A.N. Savel'ev and I. Polikarpov, *J. Mol. Biol.* 339, 413, 2004. http://www.sciencedirect.com/science/article/pii/S002228360400378X.
149. http://www.uniprot.org/manual/carbohyd
150. D. Grierson, Ed., *Biosynthesis and Manipulation of Plant Products*, Blackie Academic and Professional, Chapman & Hall, New York, 1993.
151. A.J. Domb, N. Kumar and A. Ezra, Eds., *Biodegradable Polymers in Clinical Use and Clinical Development*, John Wiley & Sons, Hoboken, NJ, 2011.
152. http://biocyc.org/META/new-image?object=Plant-Arabinogalactans-I
153. http://biocyc.org/META/new-image?object=Plant-Arabinogalactans-II
154. B. Seiboth and B. Metz, *Appl. Microbiol. Biotechnol.* 89, 1665, 2011. http://www.ncbi.nlm.nih.gov/pmc/articles/PMC3044236/.
155. D.L. Purich and R.D. Allison, *The Enzyme Reference: A Comprehensive Guidebook to Enzyme Nomenclature*, Academic Press, San Diego, CA, 2002.
156. T. Sakamoto, Y. Taniguchi, S. Suzuki, H. Ihara and H. Kawasaki, *Appl. Environ. Microbiol.* 76, 3109, 2007. http://aem.asm.org/content/73/9/3109.full.
157. N.X.Y. Ling, J. Lee, M. Ellis, M.-L. Liao, S.-L. Mau, D. Guest, P.H. Janssen, P. Kováč, A. Bacic and F.A. Pettolino, *Carbohydr. Res.* 352, 70, 2012. http://www.ncbi.nlm.nih.gov/pmc/articles/PMC3419940/.
158. http://www.brenda-enzymes.info/php/result_flat.php4?ecno=3.2.1.164&organism=
159. http://www.cazy.org/GH5.html
160. S. Shipkowski and J.E. Brenchley, *Appl. Environ. Microbiol.* 72, 7730, 2006. http://www.ncbi.nlm.nih.gov/pmc/articles/PMC1694227/.

161. P. Hagglund, Mannan-hydrolysis by hemicellulases, thesis, Lund University, 2002. http://www.lub.lu.se/luft/diss/sci_493/sci_493_without_paper_II_and_III.pdf.
162. L.R. Lynd, P.J. Weimer, W.H. van Zyl and I.S. Pretorius, *Appl. Environ. Microbiol.* 66, 506, 2002.
163. A. Ariza, J.M. Eklof, O. Spadiut, W.A. Offen, S.M. Roberts, W. Besenmatter, E.P. Friis et al., *J. Biol. Chem.* 286, 33890, 2011. http://www.ncbi.nlm.nih.gov/pubmed/21795708.
164. P. Kumar, D.M. Barrett, M.J. Delwiche and P. Stroeve, *Ind. Eng. Chem. Res.* 48, 3713, 2009. http://ucce.ucdavis.edu/files/datastore/234-1388.pdf.
165. S. Duttaa and K.C.-W. Wu, *Green Chem.* 16, 4615, 2014.

# 6 Structure and Biosynthesis of Lignin

## 6.1 INTRODUCTION

Lignin is a complex phenolic heteropolymer that imparts strength, rigidity, and hydrophobicity to plant secondary cell walls.[1] As found in all vascular plants, particularly within the woody tissues, lignin makes up a substantial fraction of the total organic carbon in the biosphere and is exceeded in abundance only by cellulose. Lignin is mainly deposited in terminally differentiated cells of supportive and water-conducting tissues, imparting on them the capacity to withstand the force of gravity, the mechanical stress, and the negative pressure generated by transpiration.

Lignin polymers make the cell wall rigid and impervious, allowing transport of water and nutrients through the vascular system and protecting plants against microbial invasion.[2] Lignification generally occurs after the polysaccharides have been laid down in the cell wall. Lignin is one of the main obstacles in the conversion of lignocellulosic biomass into fuels and chemicals.

The evolution of lignin and other hydrophobic polymers some 400 millions years ago was one of several critical developments in the colonization of terrestrial environments by vascular plants, allowing them to support increasingly large aerial structures, to supply these structures with water, and to protect them against desiccation.[1] Lignin is highly resistant to both mechanical disruption and enzymatic degradation, thus providing plant defense against herbivores and pathogens. The recalcitrance of lignin to degradation has driven scientific inquiry into its structure and biosynthesis, as the presence of lignin in animal feed and forage greatly reduces its nutritional quality, and removal of lignin from wood pulp in the process of paper production is costly and requires the use of harsh, polluting chemicals. More recently, lignin science has been the subject of increased interest as a result of the efforts toward developing lignocellulosic biorefineries. Just as it protects cellulose from digestion by herbivores, lignin limits the yield and increases the cost of generating fermentable sugars from plant biomass. The elucidation of lignin structure, biosynthesis, regulation, and function has major implications for potential bioengineering strategies to improve notably feed, fiber, and bioenergy crops.

## 6.2  LIGNIN PRECURSORS, THE MONOLIGNOLS

### 6.2.1  FUNDAMENTALS

Lignin is the generic term for a large group of aromatic polymers resulting from the *oxidative combinatorial coupling* of 4-hydroxyphenylpropanoids (*monolignols*).[3–5] It is the only naturally synthesized polymer with an aromatic backbone.[6] The name of *phenylpropanoids* is derived from the phenyl group and the propene tail of cinnamic acid, which is synthesized from the amino acid phenylalanine (Phe) in the first step of phenylpropanoid biosynthesis. The three most abundant monolignols are *p*-coumaryl (4-hydroxycinnamyl), coniferyl (3-methoxy 4-hydroxycinnamyl), and sinapyl (3,5-dimethoxy 4-hydroxycinnamyl) alcohols (Figure 6.1).[1]

Monolignol structures differ in the number of methoxy groups attached to the aromatic ring (in ortho to the hydroxyl group): *p*-coumaryl alcohol has no methoxy; coniferyl alcohol has one; and sinapyl alcohol has two. Upon incorporation into the lignin polymer, these monomers are referred to as *p*-hydroxyphenyl (**H**), guaiacyl (4-hydroxy-3-methoxyphenyl) (**G**), and syringyl (4-hydroxy-3,5-dimethoxyphenyl) (**S**) units, respectively. The units in the lignin polymer are linked by a variety of chemical bonds that have different chemical properties[2] (Figure 6.2).

During lignin deposition, monolignols are synthesized in the *cytoplasm*, translocated to the *apoplast*, and polymerized into lignin.[7] Over the last two decades, the biosynthesis of monolignols has been a major focus of research on lignification. The monolignol biosynthetic pathway has been now relatively well elucidated, at least for angiosperms.

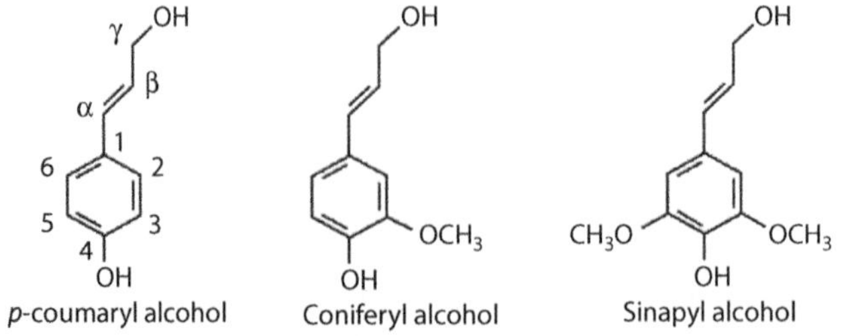

**FIGURE 6.1**  The three main monolignols.

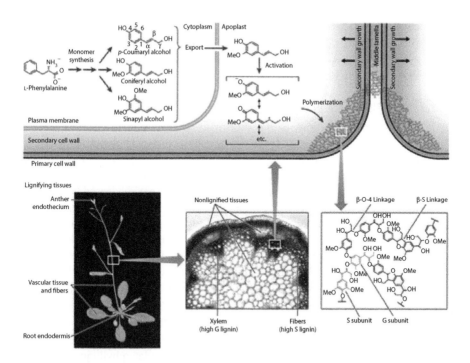

**FIGURE 6.2** Overview of lignin biosynthesis, structure, and distribution. Gray arrows follow the progression from the whole organism to the tissue, cellular, and molecular levels. (Reproduced from N.D. Bonawitz and C. Chapple, *Annu. Rev. Genet.*, 44, 337, http://www. annualreviews.org/doi/pdf/10.1146/annurev-genet-102209-163508, 2010. With permission.)

The monolignols are synthesized from the aromatic amino acid phenylalanine (Phe) through the general phenylpropanoid pathway supplying the precursor *p*-coumaroyl-CoA, and the monolignol-specific pathways beginning at *p*-coumaroyl-CoA.[3] Phe is derived from the shikimate pathway.[8] It is synthesized within the *plastids* in plants.[9]

## 6.2.2 THE SHIKIMATE AND AROMATIC AMINO ACID PATHWAYS

### 6.2.2.1 Introduction

Tryptophan (Trp), Phe, and tyrosine (Tyr) are aromatic amino acids (AAAs) that are required for protein biosynthesis in all living cells.[10,11] In plants, these AAAs also serve as precursors of a wide range of plant natural products that play crucial roles in plant growth, development, reproduction, defense, and environmental responses. Trp is a precursor of auxin as well as multiple secondary metabolites, whereas Tyr is a precursor of, for example, quinones. Phe is a precursor of numerous phenolic compounds, which include lignin, *lignans*, flavonoïdes, condensed tannins, and phenylpropanoid/benzenoid volatiles. Of the three AAAs, the highest carbon flux is often directed to Phe, as Phe-derived compounds can constitute up to 30% of organic matter in some plant species. All three AAAs are produced from the final product of the shikimate pathway, chorismate.

The AAA pathways consist of the shikimate pathway and individual postchorismate pathways leading to Trp, Phe, and Tyr.[10]

### 6.2.2.2   The Shikimate Pathway

The seven enzymatic reactions of the shikimate pathway connect central carbon metabolism and the AAA network by converting three-carbon phosphoenolpyruvate (PEP) and four-carbon D-erythrose 4-phosphate (4EP)—intermediates in glycolysis and the pentose phosphate pathways, respectively—to chorismate.

In the first step of the shikimate pathway, E4P and PEP are condensed to a seven-carbon six-membered heterocyclic compound, 3-deoxy-D-*arabino*-heptulosonate 7-phosphate (DAHP).[11] In the second step, the ring oxygen is exchanged for the exocyclic C7 of DAHP to form a highly substituted cyclohexane derivative, 3-dehydroquinate. The remaining five steps serve to introduce a side chain and two of the three double bonds that convert this cyclohexane into the benzene ring, the signature of aromatic compounds. Chorismate is the final product of the shikimate pathway.

### 6.2.2.3   The Tryptophan, Phenylalanine and Tyrosine Pathways

In plants, chorismate is a common precursor of at least four branches of metabolic pathways leading to the formation of Trp, Phe/Tyr, salicylate/phylloquinone, and folate.[10] Four enzymes catalyze the committed step of the respective pathways and compete for chorismate. The Trp pathway converts chorismate to Trp via six enzymatic reactions. In contrast to the Trp pathway, the knowledge of the plant Phe and Tyr pathways is still in its infancy. In the first step of the pathways, chorismate is converted by chorismate mutase (CM) to prephenate, of which subsequent conversion to Phe and Tyr may occur via two alternative pathways. In one route (the arogenate pathway), prephenate is first transaminated to L-arogenate followed by dehydration/decarboxylation to Phe or dehydrogenation/decarboxylation to Tyr. In the other route (the phenylpyruvate or 4-hydroxyphenylpyruvate pathway), these reactions occur in reverse order. Recent genetic evidence indicates that the arogenate pathway is the predominant route for Phe biosynthesis in plants.

### 6.2.3   Monolignol Biosynthetic Pathways

The synthesis of the monolignols from phenylalanine requires deamination, hydroxylation at one, two, or three positions of the aromatic ring, methylation of one or two of these hydroxyl groups, and two successive reductions of the monolignol side chain from a carboxylic acid first to an aldehyde and then to an alcohol (Figure 6.3).[1]

These reactions are mediated by

- A phenylalanine ammonia-lyase (PAL)
- Three different cytochrome P450-dependent monooxygenases: cinnamate 4-hydroxylase (C4H), *p*-coumarate 3-hydroxylase (C3H), and ferulate 5-hydroxylase (F5H)

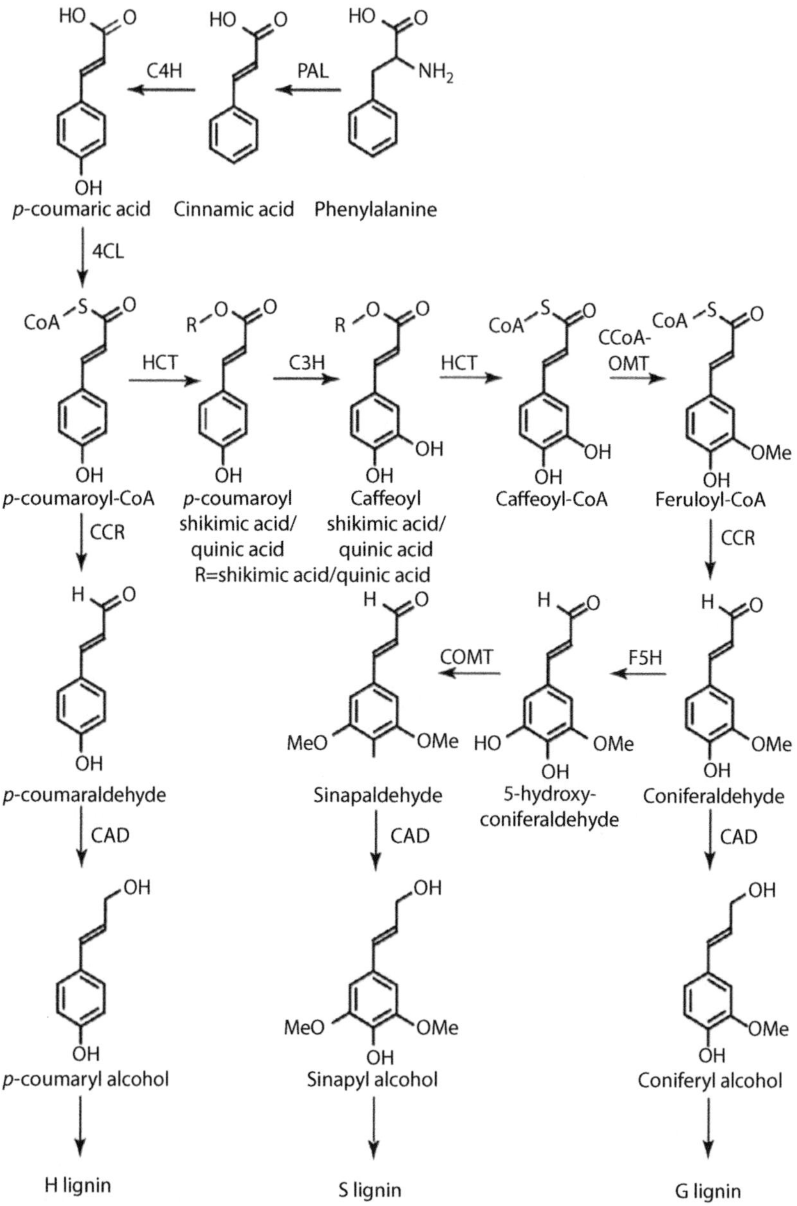

**FIGURE 6.3** The main biosynthetic routes toward the monolignols *p*-coumaryl, coniferyl, and sinapyl alcohols. PAL, phenylalanine ammonia-lyase; C4H, cinnamate 4-hydroxylase; 4CL, 4-coumarate:CoA ligase, CoA, coenzyme A; C3H, *p*-coumarate 3-hydroxylase; HCT, *p*-hydroxycinnamoyl-CoA:quinate/shikimate *p*-hydroxycinnamoyltransferase; CCoAOMT, caffeoyl-CoA *O*-methyltransferase; CCR, cinnamoyl-CoA reductase; F5H, ferulate 5-hydroxylase; COMT, caffeic acid *O*-methyltransferase; CAD, cinnamyl alcohol dehydrogenase. (Reproduced from H. Maeda and N. Dudareva, *Annu. Rev. Plant Biol.*, 63, 73, http://www.annualreviews.org/doi/full/10.1146/annurev-arplant-042811-105439, 2012. With permission.)

- Two methyltransferases: caffeoyl-CoA $O$-methyltransferase (CCoAOMT; CoA, coenzyme A) and caffeic acid $O$-methyltransferase (COMT)
- Two oxidoreductases: cinnamoyl-CoA reductase (CCR) and cinnamyl alcohol dehydrogenase (CAD)[1]

In addition to these eight enzymes, of which actions are apparent in the final pathways products, two more enzymes, 4-coumarate:CoA ligase (4CL) and $p$-hydroxycinnamoyl-CoA:quinate/shikimate $p$-hydroxycinnamoyltransferase (HCT), are required to synthesize pathway intermediates that serve as substrates for subsequent reactions.[1] 4CL is an ATP-dependent CoA ligase that catalyzes the synthesis of $p$-coumaroyl CoA. This $p$-coumaroyl CoA is then used as an acyl donor by acyltransferase HCT to synthesize $p$-coumaroyl shikimate, the substrate for 3-hydroxylation by C3H. Alternatively, $p$-coumaroyl CoA can serve as a substrate for chalcone synthase, diverting it to the flavonoid pathway, or CCR, committing it to H lignin biosynthesis.

The relatively minor pathway leading to $p$-coumaryl alcohol (H lignin) requires only a subset of these enzymes (PAL, C4H, 4CL, CCR, and CAD) whereas the synthesis of coniferyl alcohol (G lignin) requires these five plus HCT, C3H, and CCoAOMT, and the synthesis of sinapyl alcohol (S lignin) requires all ten enzymes, including F5H and COMT.[1] In many plants, some or all of these enzymes are found as multiple isoforms encoded by different genes.

## 6.3 LIGNIN DISTRIBUTION IN PLANTS AND TISSUES

The amount and composition of lignins vary among taxa, cell types, and individual cell wall layers and are influenced by development and environmental cues.[12] With some notable exceptions, lignin monomer composition varies substantially among plant species:

- Gymnosperms (softwood) lignins are composed mostly of G units with low levels of H units.
- Dicot angiosperm (hardwood) lignins are composed mainly of G and S units and traces of H units.
- Lignins from grasses (monocots) contain G and S units at comparable levels and more H units than dicots.[3,12]

The majority of secondary wall formation in most plants occurs in the water-conducting cells of the xylem, which tend to be enriched in G lignin, and in the structural *fibers*, which in angiosperms typically contain higher levels of S subunits.[1] Lignified cell walls are also found in *sclereid* cells, *endodermal* tissue in roots, and in specialized cells of *anthers* and some seed pods, in which they are important for the dehydration-driven release of pollen and seeds, respectively. The wood of both gymnosperm and angiosperm trees is composed largely of the secondary cell walls of vascular tissue and associated fibers, and as such, of lignin. It is important to note that lignin biosynthesis is also of crucial importance in herbaceous (nonwoody) species.

Other natural units than the three main monolignols have been identified from a variety of species and may be incorporated into the polymer at varying levels.

Particularly common other subunits include ferulates (which form crosslinks between lignin and hemicelluloses), coniferaldehyde, and acylated monolignols containing acetate, *p*-coumarate, or *p*-hydroxybenzoate moieties.[1] Lignin composition also varies among cell types and can even be different in individual call wall layers. Lignin composition is also influenced by environmental conditions; for example, lignin in compression wood is enriched in H units.[2] Hence, both developmental and environmental parameters influence the composition and thus the structure of the lignin polymer.

## 6.4 EVOLUTION OF LIGNIFICATION

It is commonly accepted that lignin evolved together with the adaptation of plants to a terrestrial life to provide them with the structural support needed for an erect growth habit.[3] The recent discovery of lignin in a bryophyte, the liverwort *Marchiantia polymorpha*, expands the distribution of lignification to nonvascular plants (Figure 6.4).[7] Even more strikingly, lignin was found in the cell wall of the red alga, *Calliathron cheilosporioides*, which shares a common ancestor with vascular plants over 1 billion years ago. These observations raise new questions on the evolutionary origin and history of lignification.

Similarly, recent research on the distribution of lignin monomers in the plant kingdom challenges our perception about the evolution of S lignin monomer biosynthesis.[7] S lignin (lignin that is rich in S units) has been generally considered to be

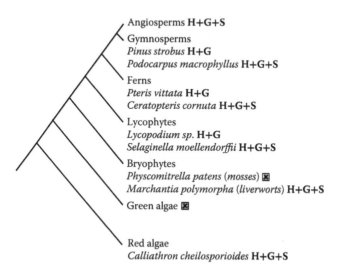

**FIGURE 6.4** Phylogenetic tree showing the distribution of lignification and lignin monomer composition through major plant lineages. Vascular plants include lycophytes, ferns, gymnosperms, and angiosperms. It is evident that S lignin is not restricted to angiosperms. The distribution of S lignin within lycophytes, ferns, and gymnosperms is not uniform. Crosses indicate no lignification. (Adapted from X. Li and C. Chapple, *Plant Physiol.*, 154, 449, http://www.plantphysiol.org/content/154/2/449.full, 2010.)

characteristic of the angiosperms; however, recent studies confirmed that S lignin is also present in gymnosperms and some basal vascular plants such as lycophytes and ferns. Furthermore, S lignin is also detected in liverworts and red algae that are more distantly related to angiosperms.

The occurrence of lignification in liverworts and red algae suggests that the genes required for lignin deposition evolved in their common ancestor before the divergence of specific lineages (e.g., red algae and liverworts) and were subsequently lost in certain lineages (e.g., green algae and mosses).[7] Alternatively, red algae and liverworts lineages could have independently evolved the biochemical pathways for lignification. Similar evolutionary scenarios can also be envisaged to explain the phylogenetic pattern of S lignin. *Convergent evolution* of S lignin biosynthesis via distinct pathways in the lycophyte *Selaginella* and angiosperms has been recently demonstrated.[13]

## 6.5  REGULATION OF LIGNIFICATION

Lignification is coordinated with the production of cellulose, hemicelluloses, and other polysaccharides during secondary wall formation.[1] The regulation of this process is overseen by a complex network of tissue-specific transcription factors that respond to as-yet-unidentified development signals triggering an expression cascade of other transcription factors and ultimately the enzymes and other proteins responsible for synthesizing the cell wall itself. Several different genetic approaches have contributed to the identification of this regulatory network.

The regulatory cascade explains why several of these transcription factors lead to enhanced or reduced lignification when misexpressed in plants while they do not directly regulate the lignin biosynthetic genes by binding to their promoters.[3]

Engineering the expression of transcription factors has the potential to alter lignification with fewer adverse effects on plant development for two reasons.[3] First, these transcription factors bind the promoters of multiple genes, thereby affecting the flux through the pathway in an orchestrated manner. Second, some might be specifically involved in the developmental lignification process and not in other processes, such as stress lignin formation, leaving the plant able to respond to environmental factors.

## 6.6  TRANSPORT AND POLYMERIZATION

### 6.6.1  Monolignol Transport

To be incorporated into the lignin polymer, monolignols must first be transported to the cell wall via a process that remains poorly understood.[1,3] In one model, monolignols are stored and transported over the plasma membrane as *glucosides* and then released for polymerization by the action of glucosidases. However, *Arabidopsis* mutants defective in the corresponding glucosyltransferases have normal lignin levels. In another model, monolignols are transported to the plasma membrane by Golgi-derived vesicles. However, this model has recently been challenged by Kaneda et al.[14] Furthermore, their work suggested that glucoside formation was not essential for monolignol transport or lignification. Finally, their data supported a model in which unknown membrane transporters, rather than Golgi vesicles, export monolignols.[14]

## 6.6.2 POLYMERIZATION

Lignin polymerization occurs via oxidative radicalization of phenols, followed by combinatorial radical coupling (without enzymatic activity).[2,3] In the first step, the monolignol phenol is oxidized, that is, dehydrogenated. The resulting phenolic radical is relatively stable due to delocalization of the unpaired electron in the conjugated system (Figure 6.5).

Subsequently, two monomer radicals may couple to form a dimer, establishing a covalent bond between both subunits (Figures 6.5 and 6.6).[3] Monolignol radicals favor coupling at their β positions, resulting primarily in the formation of a β-*O*-4 ether (β-aryl ether), a β-5 linked structure known as phenylcoumaran, and a β-β resinol structure.[1] The β-5 linked phenylcoumaran is a more rigid and more hydrophobic structure than that produced by β-*O*-4 coupling. This radical–radical coupling occurs in a chemical-combinatorial fashion; thus, the ratio of each of the possible coupling products depends largely on the chemical nature of each of the monomers and the conditions in the cell wall.

Once formed, the dimer needs to be dehydrogenated, again to a phenolic radical, before it can couple with another monomer radical.[3] This mode of action, in which

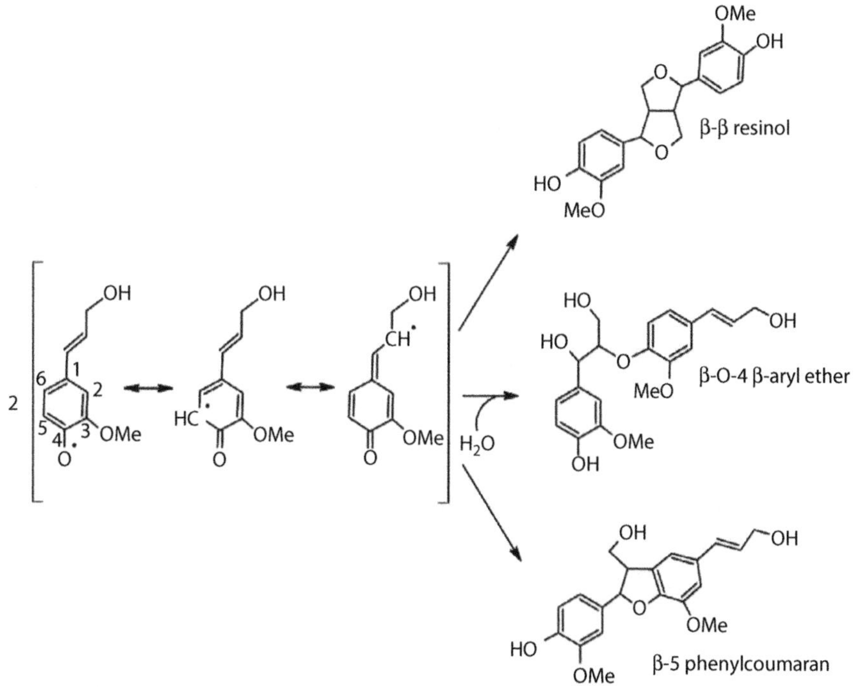

**FIGURE 6.5** Dimerization of two dehydrogenated coniferyl monomers. Resonance forms with the unpaired electron localized at C1 or C3 are not shown because radical coupling reactions do not occur at these positions. (Adapted from R. Vanholme et al., *Plant Physiol.*, 153, 895, http://www.plantphysiol.org/cgi/content/full/153/3/895, 2010.)

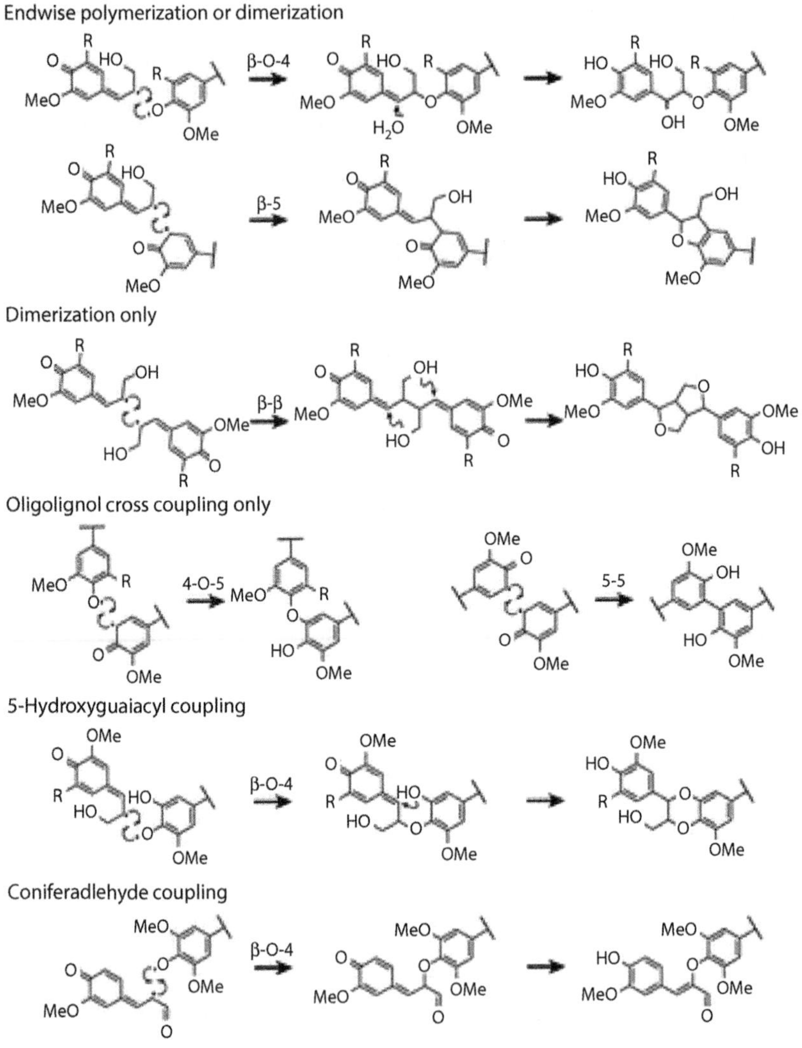

**FIGURE 6.6** Major structural units in the lignin polymer, including selected less abundant monomers, and the bimolecular coupling reactions that give rise to them. Half-headed arrows show single-electron mechanisms, and full arrows show two-electron processes, such as the position of the nucleophile attack in rearomatization reactions. (Reproduced from N.D. Bonawitz and C. Chapple, *Annu. Rev. Genet.*, 44, 337, http://www.annualreviews.org/doi/pdf/10.1146/annurev-genet-102209-163508, 2010. With permission.)

a monomer (radical) adds to the growing polymer, is called endwise coupling or endwise polymerization: the polymer grows one unit at a time. During each coupling reaction, two radicals are consumed as each single electron contributes to the newly formed bond, making this type of radical polymerization different from the radical chain reactions that occur in the polymerization of several fossil-based polymers.

The simplest case of endwise polymerization is the coupling of a monolignol radical to a syringyl radical at the end of a growing polymer (Figure 6.6).[1] By far, the most commonly formed bond is between the 4-O position of the syringyl terminal unit and the β-carbon of the monolignol radical. As shown in Figure 6.6, the resulting quinone methide intermediate is rearomatized by reaction with water.[15] Addition of a monolignol to a terminal guaiacyl unit can also proceed via formation of a β-O-4 ether. Alternatively, the β-carbon of the monolignol radical can react with the carbon at the 5-position of the guaiacyl subunit, which is not available in syringyl units, thereby forming a β-5 linked structure (Figure 6.6).

Because endwise polymerization always involves the β-carbon of the incoming monolignol, the common bonds remaining to be described only occur during the coupling of oligolignol end units, which do not have their β-carbon available for coupling (Figure 6.6).[1] Remarkably, if lignin polymerization was constrained to only β-O-4, β-5, and β-β bonds, then the lignin polymer would be entirely linear and devoid of any branches. In fact, branched may not be the best term to describe the cross-linked structures seen in the lignin polymer, given that these structures are not the result of one elongating polymer end giving rise to two new elongating ends but rather arise from the collision of two oligolignol ends to form one new end.

When such collisions do occur, there are two possible coupling modes available.[1] If at least one oligolignol end is a guaiacyl subunit, bond formation can proceed between the phenoxy radical at one end unit and the 5-carbon of the guaiacyl unit (4-O-5 coupling) (Figure 6.6). Alternatively, if both oligolignol ends are guaiacyl units, a bond can form between the carbons at the 5-positions of both subunits (5-5 coupling) (Figure 6.6). Syringyl subunits do not have a 5-carbon available for bonding, and thus two terminal syringyl units cannot couple with each other. Although 5-5 linked structures possess two phenolic hydroxyl groups, addition of a monolignol to one of them results in internal trapping of the other and formation of an eight-membered ring (dibenzodioxocin) (Figure 6.7).[4]

Thus, because at most one monolignol can be added to a 4-O-5 or 5-5 structure, the apparent three-way branching structures observed in the polymer at the eight-membered rings can be viewed as the collapse of two elongating ends into one. Nevertheless, lignin containing a high proportion of G subunits is more highly crosslinked than lignin rich in S subunits, presumably rendering it more rigid and more hydrophobic. It should be noted that coupling of two oligomers is rare in S/G lignins but relatively frequent in G lignins, where 5-5 coupling accounts for ~4% of the linkages.[3] The average length of a linear lignin chain in poplar (angiosperm; S/G lignin) is estimated to be between 13 and 20 units (Figure 6.8).[3]

Among the less abundant monomers, coniferaldehyde (in CAD-deficient plants) and 5-hydroxyconiferyl alcohol (in COMT-deficient plants) couple via similar mechanisms as G and S subunits, respectively.[1] However, the β-O-4 structures that each forms during polymerization are more rigid than β-O-4 structures formed by coniferyl and sinapyl alcohols. Because of its electronic structure, the quinone methide tail of a newly coupled coniferaldehyde subunit is able to rearomatize by β-proton abstraction, resulting in a new α-β double bond and only three rotatable bonds between the rings of the two lignin subunits instead of the four usually present. The benzodioaxane structure formed when a monolignol couples with a

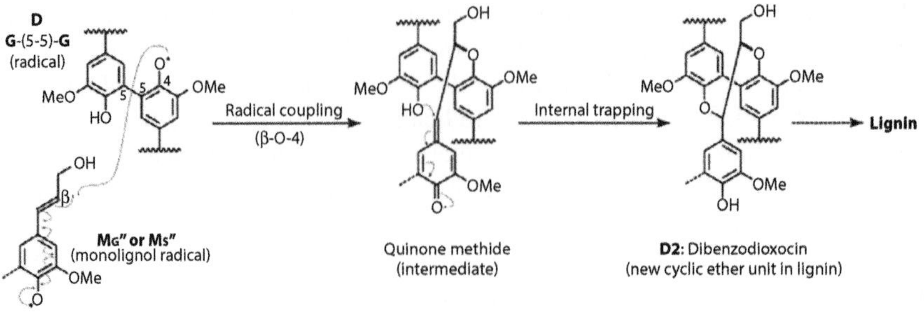

**FIGURE 6.7** The formation of dibenzodioxocin units in lignin. Following 5-5 coupling of two phenolic units, the next endwise coupling with a monolignol (reacting at its favored at its β-position, and coupling to the only available position, the 4-O position, of one of the phenolic moieties) produces the normal quinone methide intermediate. This quinone methide is internally trapped by the other phenol in the 5-5 moiety, producing an eight-membered ring, a dibenzodioxicin structure. Such structures are strikingly prevalent in high-guaiacyl lignins. The importance of both 5-5 and 4-O-5 structures is that they represent branch points in lignin. (With kind permission from Springer Science+Business Media: *Phytochem. Rev.*, Lignins: Natural polymers from oxidative coupling of 4-hydroxyphenyl-propanoids, 3, 2004, 29, Ralph. J. et al.)

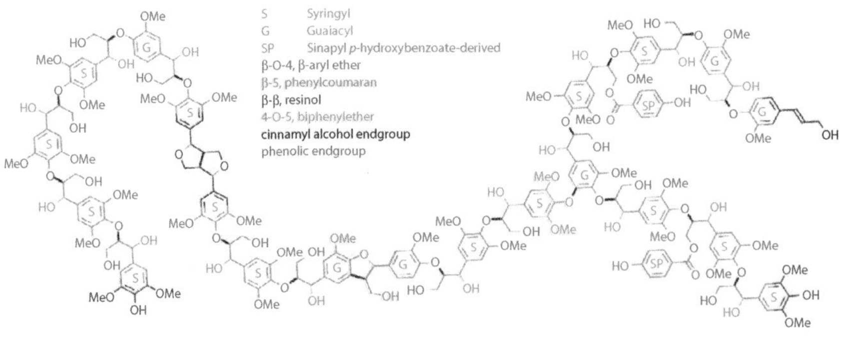

**FIGURE 6.8** Lignin polymer from poplar, as predicted from NMR (nuclear magnetic resonance)-based lignin analysis. (Reproduced from R. Vanholme et al., *Plant Physiol.*, 153, 895, http://www.plantphysiol.org/cgi/content/full/153/3/895, 2010. With permission.)

5-hydroxyguaiacyl end unit is much more rigid, containing only one rotatable bond between the two aromatic rings due to internal trapping of the quinone methide intermediate by the 5-hydroxyguaiacyl end unit.

Much less is known about the coupling and cross-coupling of *p*-coumaryl alcohol, which forms a minor component of the lignin polymer in most plants.[1]

In native lignin, there are some forms of reduced structures that cannot be explained only by oxidative coupling.[15] It has been shown via biomimetic model experiments that nicotinamide adenine dinucleotide (NADH), in an uncatalyzed process, reduces quinone methides to their benzyl derivates.

### 6.6.3 MONOLIGNOL DEHYDROGENATION

Monolignol dehydrogenation involves *peroxidases* and/or *laccases* (see Chapter 7).[3] Peroxidases use hydrogen peroxide as electron acceptor, whereas laccases use molecular oxygen to oxidize their metal centers to enable catalytic oxidation. Both types of enzymes belong to large gene families of which the members have overlapping activities, making the process difficult to study. Peroxidases may differ in their substrate specificities; some almost exclusively accept coniferyl alcohol, whereas others are highly specific toward sinapyl alcohol. Because the structure of lignin depends on the availability of monolignol radicals, peroxidases specificity may determine partly the structure of the final lignin polymers, opening possibilities for altering lignin structure by modified expression of specific peroxidase isoforms. Although monolignols can be dehydrogenated via direct interaction with an electron-removing enzyme, the radicals might also be generated by radical transfer.

As a conclusion on lignin polymerization, the present model of lignification involves peroxidases and/or laccases to provide the oxidative capacity in the cell wall.[3] All phenolic compounds that enter the cell wall will eventually have the potential to radicalize and incorporate into the lignin polymer, subject to chemical oxidation and coupling propensity. This lignification model also explains why many other phenolic molecules can be integrated into the growing lignin polymer and opens up the possibility of tailoring lignins for industrial applications.

## 6.7 STRUCTURE AND FUNCTION IN PLANT CELL WALLS

The secondary cell walls of woody tissues and grasses are composed predominantly of cellulose, lignin, and hemicelluloses.[16] Lignin is closely mixed with the other wall components. The covalent linkages between lignin and polysaccharides (predominantly hemicelluloses) form lignin–carbohydrate complexes (LCC).[17,18] Important LCCs involve quinone methide (QM) and phenolic acids.

Three major classes of LCCs via QM have been reported: benzyl ether linkages, benzyl ester linkages, and, to a lower extent, glycosidic linkages.[17,19] The two first are formed via the nucleophilic addition of polysaccharide hydroxyl or carboxylic groups to the α-carbon of the methylene quinone formed during monolignol polymerization (Figure 6.9). This mechanism competes with the addition of other nucleophilic agents such as water.

In grasses, another type of covalent linkage occurs via phenolic acids.[17,21,22] Phenolic acids, such as ferulic acid, are bifunctional molecules because they can form ester bonds via their carboxylic function, and ether bonds via their hydroxyl function. They are potential agents for wall polymer crosslinking.

Ferulates are implicated in crosslinking grass cell-wall polysaccharides with lignin.[22] Wall ferulates are involved in single-electron reactions as evidenced by their dimerization via radical processes. Radical cross-coupling products are formed from oxidative coupling of ferulate with coniferyl or sinapyl alcohols, or their oligomers. Certain ferulate coupling products are not produced in grass providing evidence for the role of feruloylated saccharides as initiation/nucleation for lignification.

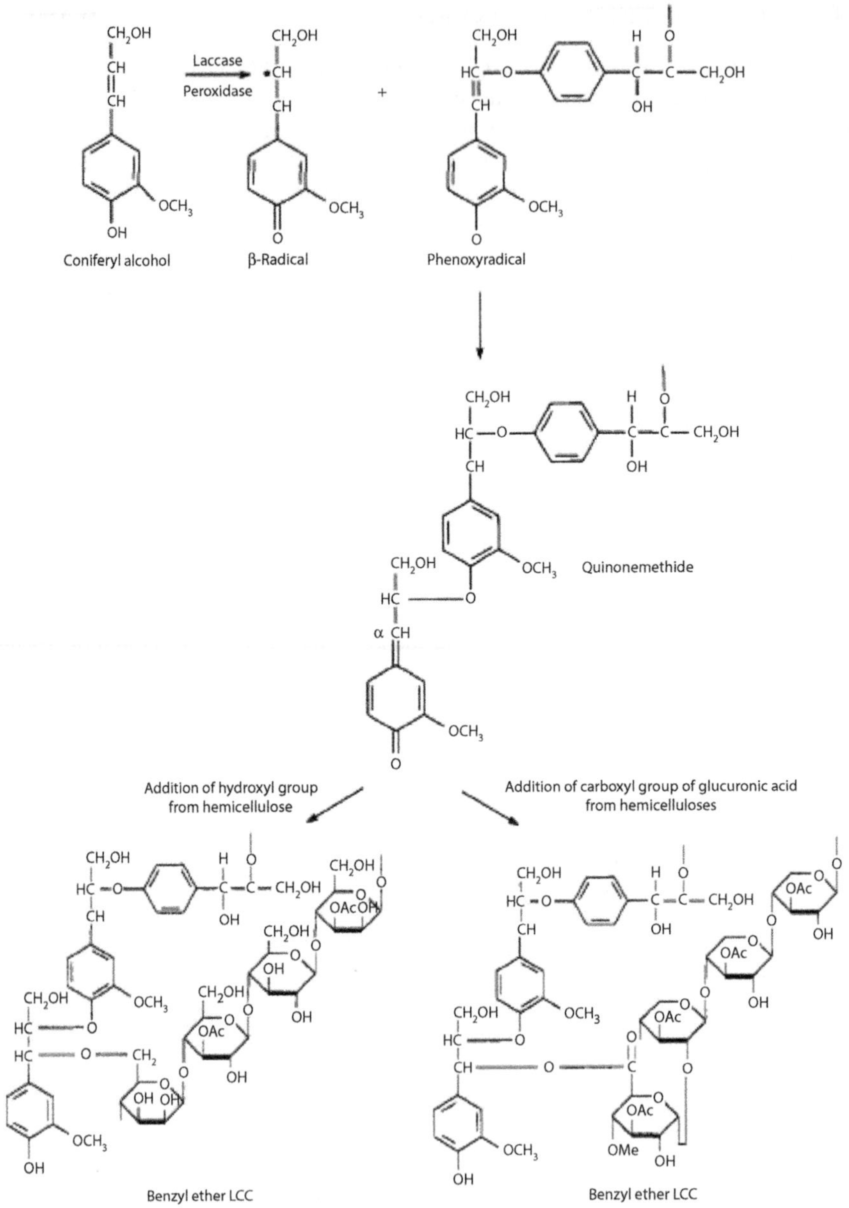

**FIGURE 6.9** Biosynthesis of benzyl ether and benzyl ester LCCs. (With kind permission from Springer Science+Business Media: Synthesis and oxidation of lignin-carbohydrate model compounds, Thesis, University of Maine, http://www.library.umaine.edu/theses/pdf/NguyenMTT2008.pdf, 2008, Nguyen, M.T.T.; Association Between Lignin and Carbohydrates in Wood and Other Plant Tissues, 2003. Koshijima, T. and Watanabe, T.)

Feruloylated polysaccharides in grasses are, therefore, critical entities in directing cell-wall crosslinking during plant growth and development.

Crosslinking of the lignin-hemicelluloses network in which cellulose microfibrils are embedded is believed to result in the elimination of water from the wall and the formation of a hydrophobic composite that limits accessibility of hydrolytic enzymes and is a major contributor to the structural characteristics of secondary walls.[16]

## REFERENCES

1. N.D. Bonawitz and C. Chapple, *Annu. Rev. Genet.* 44, 337, 2010. http://www.annual reviews.org/doi/pdf/10.1146/annurev-genet-102209-163508.
2. F.R.D. van Parijs, K. Moreel, J. Ralph, W. Boerjan and R.M.H. Merks, *Plant Physiol.* 153, 1332, 2010. http://www.plantphysiol.org/cgi/content/full/153/3/1332.
3. R. Vanholme, B. Demedts, K. Morreel, J. Ralph and W. Boerjan, *Plant Physiol.* 153, 895, 2010. http://www.plantphysiol.org/cgi/content/full/153/3/895.
4. J. Ralph, K. Lundquist, G. Brunow, F. Lu, H. Kim, P.F. Schatz, J.M. Marita et al., *Phytochem. Rev.* 3, 29, 2004. http://www.springerlink.com/content/lx20h1488802t565.
5. J. Ralph, Lignin structure: Recent developments, in *Proceedings of the 6th Brazilian Symposium Chemistry of Lignins and Other Wood Components*, 1999. http://www.dfrc. wisc.edu/DFRCWebPDFs/JR_Brazil99_Paper.pdf.
6. M. Dashtban, H. Schraft, T.A. Syed and W. Qin, *Int. J. Biochem. Mol. Biol.* 1, 36, 2010. http://www.ijbmb.org/files/IJBMB1004005.pdf.
7. X. Li and C. Chapple, *Plant Physiol.* 154, 449, 2010. http://www.plantphysiol.org/ content/154/2/449.full.
8. University of Maine, Department of Chemistry, 2011. http://chemistry.umeche.maine. edu/CHY431/Shikimate.html.
9. P. Rippert, J. Puyaubert, D. Grisollet, L. Derrier and M. Matringe, *Plant Physiol.* 149, 1251, 2009. http://www.plantphysiol.org/content/149/3/1251.full.
10. H. Maeda and N. Dudareva, *Annu. Rev. Plant Biol.* 63, 73, 2012. http://www.annualre-views.org/doi/full/10.1146/annurev-arplant-042811-105439.
11. K.M. Herrmann, *Plant Cell* 7, 907, 1995. http://www.plantcell.org/content/7/7/907.full.pdf.
12. W. Boerjan, J. Ralph and M. Baucher, *Annu. Rev. Plant Biol.* 54, 519, 2003. http://www. dfrc.wisc.edu/DFRCWebPDFs/2003-Boerjan-ARPB-54-519.pdf.
13. J.K. Weng, T. Akiyama, N.D. Bonawitz, X. Li, J. Ralph and C. Chapple, *Plant Cell* 22, 1033, 2010. http://www.plantcell.org/content/22/4/1033.full.
14. M. Kaneda, K.H. Rensing, J.C.T. Wong, B. Banno, S.D. Mansfield and A.L. Samuels, *Plant Physiol.* 147, 1750, 2008. http://www.plantphysiol.org/content/147/4/1750.full.
15. A. Holmgren, G. Brunow, G. Henriksson, L. Zhang and J. Ralph, *J. Org. Biomol. Chem.* 4, 3456, 2006. http://ddr.nal.usda.gov/bitstream/10113/14222/1/IND44051453.pdf.
16. Complex Carbohydrate Research Center, The University of Georgia. http://www.ccrc. uga.edu/~mao/intro/ouline.htm.
17. A. Barakat, Etude de la lignification de parois végétales de graminées par des assem-blages modèles : Réactivité, organisation et structure supramoléculaire, Thèse de doctorat, INRA, Université de Reims Champagne-Ardenne, 2007. http://ebureau.univ-reims.fr/slide/files/quotas/SCD/theses/exl-doc/GED00000531.pdf.
18. L. Christopher, in C.A. Okia, Ed., *Global Perspectives on Sustainable Forest Management*, InTech, Rijeka, Croatia, 2012.

19. M.T.T. Nguyen, Synthesis and oxidation of lignin-carbohydrate model compounds, Thesis, University of Maine, 2008. http://www.library.umaine.edu/theses/pdf/NguyenMTT2008.pdf.

20. T. Koshijima and T. Watanabe, *Association Between Lignin and Carbohydrates in Wood and Other Plant Tissues*, Springer, New York, 2003.

21. J.H. Grabber, R.D. Hatfield, J. Ralph, J. Zon and N. Amrhein, *Phytochemistry* 40, 1077, 1995. http://www.dfrc.ars.usda.gov/DFRCWebPDFs/1995-Grabber-Phyto-40-1077.pdf.

22. J. Ralph, J.H. Grabber and R.D. Hatfield, *Carbohydr. Res.* 275, 167, 1995. http://www.sciencedirect.com/science/article/pii/000862159500237N.

# 7 Biodegradation of Lignin

## 7.1 INTRODUCTION

Lignin is extremely recalcitrant to degradation.[1] By linking to both cellulose and hemicelluloses, it creates a barrier to any solutions or enzymes and prevents the penetration of lignocellulolytic enzymes into the lignocellulosic structure. However, some *basidiomycetes* white-rot fungi are able to degrade lignin efficiently using a combination of extracellular ligninolytic enzymes, organic acids, mediators, and accessory enzymes. White-rot fungi produce extracellular oxidative enzymes which catalyze the oxidation of lignin.[2] Due to their lignin-degrading capacity, whole cultures of various white-rot fungi cause extensive brightness gains and delignification of kraft pulp.

The role of fungi in delignification due to the production of extracellular oxidative enzymes has been studied more extensively than that of bacteria.[3] However, certain bacteria can degrade lignified cell walls of wood.[4]

## 7.2    FUNGI AND BACTERIA

Chemically and morphologically distinct types of decay result from attack of different microorganisms.[4] Wood decay fungi are usually separated into three main groups, causing white, soft, or brown rot. Bacterial degradation of wood has also been reported including erosion, tunneling, and cavity formation.

### 7.2.1    LIGNOCELLULOLYTIC ENZYME-PRODUCING FUNGI

Lignocellulolytic enzyme-producing fungi are widespread, and include species from the *ascomycetes* (e.g., *Trichoderma reesei*), and basidiomycetes phyla such as white-rot (e.g., *Phanerochaete chrysosporium*[5]) and brown-rot fungi (e.g., *Fomitopsis palustris*).[1]

In addition, a few anaerobic species are found to be able to degrade cellulose in the gastrointestinal tracts of ruminant animals. Biomass degradation by these fungi is performed by complex mixtures of cellulases, hemicellulases, and *ligninases.*

In nature, efficient lignin degradation during the process of wood decay becomes possible mainly by basidiomycetes white-rot fungi.[1] Many white-rot fungi simultaneously attack lignin, hemicelluloses, and cellulose, whereas some other white-rot fungi preferentially work on lignin in a selective manner. Selective lignin degraders may have potential biotechnological applications when the removal of lignin is required to obtain pure cellulose such as in biopulping and in processes where the objective is to provide an unprotected carbohydrate for subsequent use.

In contrast to white-rot fungi, brown-rot fungi are able to circumvent the lignin barrier, removing the hemicelluloses and cellulose with only minor modification to lignin.[1,5] Ascomycetes are mostly able to degrade cellulose and hemicelluloses, while their ability to degrade lignin is limited.[6] Various plant pathogenic fungi are able to degrade lignin by production of laccase and lignin peroxidase.[1,7]

### 7.2.2    LIGNOCELLULOLYTIC ENZYME-PRODUCING BACTERIA

Filamentous bacteria belonging to *Streptomyces*, a genus of Actinomycetes, are well-known degraders of lignin and can mineralize up to 15% of [14]C-labeled lignins but usually much less.[4] Typically, *Streptomyces* spp. solubilize part of lignin to yield a water soluble but acid-precipitable polymeric lignin. Actinomycetes produce extracellular peroxidases, for example, lignin peroxidase-type enzymes. Lignin model compound studies demonstrated that monomeric compounds, that is, vanillic acid and protocatechuic acid were formed, indicating the cleavage of $C_\alpha$–$C_\beta$ bonds.

Nonfilamentous bacteria usually mineralize less than 10% of lignin preparations and can degrade only the low-molecular weight part of lignin as well as degradation products of lignin.[4]

## 7.3    FUNGAL EXTRACELLULAR LIGNINASES

Fungi degrade lignin by secreting enzymes collectively termed ligninases or lignin-modifying enzymes.[1,8] Ligninases can be classified into two categories:

- Heme peroxidases
- Laccases (phenol oxidases)

In general, laccases use molecular oxygen as the electron acceptor while peroxidases use hydrogen peroxide. Heme peroxidases include mainly:

- Lignin peroxidases (LiP)[9–11]
- Manganese peroxidases (MnP)[12]
- Versatile peroxidases (VP)[3,13]

However, a new family of heme peroxidases, known as *dye-decolorizing peroxidases (DyP)*,[14] has been described recently.[15] These latter enzymes constitute an independent group of heme peroxidases and seem to offer attractive catalytic properties.

In addition to the peroxidases and laccases, fungi produce other accessory oxidases such as aryl-alcohol oxidase and the glyoxal oxidase, which generate the hydrogen peroxide required by the peroxidases.

It has been shown that *P. chrysosporium* produces several LiP and MnP *isoenzymes* but no laccase.[1] Although LiP is able to oxidize the nonphenolic part of lignin (which forms 80%–90% of lignin composition), it is absent from many lignin-degrading fungi. As oxidative ligninolytic enzymes are too large to penetrate into the wood cell wall micropores, it has been suggested that prior to enzymatic attack, low-molecular-weight oxidative compounds must initiate changes to the lignin structure. Figure 7.1 shows the major steps and enzymes involved in lignin degradation mainly by basidiomycetes white-rot fungi.

Lignin-degrading enzymes have attracted the attention for their valuable biotechnological applications especially in the pretreatment of recalcitrant lignocellulosic biomass for biofuel production.[3] The use of lignin-degrading enzymes has been studied in various applications such as paper industry, textile industry, wastewater treatment, and the degradation of herbicides.

## 7.3.1 Phenol Oxidases (Laccases)

Laccases are found in plants, fungi, and bacteria and belong to the multi-copper oxidase superfamily.[3]

Laccases (benzenediol:oxygen oxidoreductases, EC 1.10.3.2), phenol oxidases with a molecular mass of ~70 kDa, are glycosylated blue multi-copper oxidoreductases that use molecular oxygen to oxidize various aromatic and nonaromatic compounds through a radical-catalyzed reaction mechanism (Equation 7.1).[1,9,16,17]

$$4 \text{ benzenediol} + O_2 \leftrightarrow 4 \text{ benzosemiquinone} + 2 H_2O \qquad (7.1)$$

Laccases represent one of the oldest enzymes ever described.[9] The enzyme seems ubiquitous in all white-rot fungi, and its presence is less frequent in plants. The better-characterized enzymes are from the fungus *Trametes versicolor* and the Japanese lacquer tree *Rhus vernicifera* (hence the name laccase). The plant enzyme participates together with peroxidases in the biosynthesis of lignin.

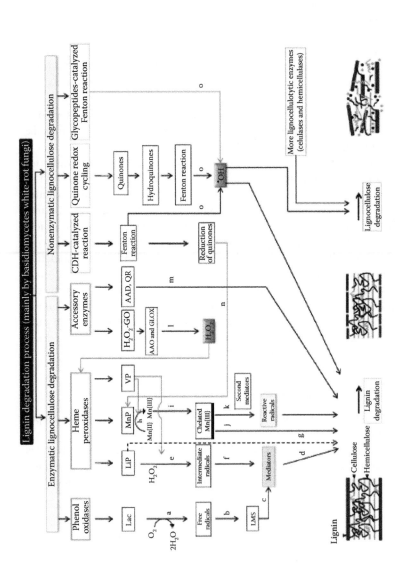

**FIGURE 7.1** Schematic diagram of lignin degradation by basidiomycetes white-rot fungi: the major steps and enzymes involved (refer to text). Lac, laccase; LMS, laccase-mediator system; LiP, lignin peroxidase; MnP, manganese peroxidase; VP, versatile peroxidase; AAD, aryl-alcohol dehydrogenases; AAO, aryl-alcohol oxidase; GLOX, glyoxal oxidase; $H_2O_2$-GO, $H_2O_2$-generating oxidases; QR, quinone reductases; CDH, cellulose dehydrogenase, LMS, laccase-mediator system. (From Dashtban, M. et al., *Int. J. Biochem. Mol. Biol.*, 1, 36, http://www.ijbmb.org/files/IJBMB1004005.pdf, 2010.)

### 7.3.1.1 Molecular Structure

Fungal laccases occur often as isoenzymes with monomeric or dimeric protein structures.[9] Both intracellular and extracellular isoenzymes may be produced from a single microorganism. The monomeric proteins have a molecular mass ranging from 50 to 110 kDa. Laccases are ~10%–45% glycosylated with the fungal enzyme less substituted than the plant enzyme. The three-dimensional structures of several laccases have been reported. All fungal laccases show a similar architecture consisting of three β-barrel domains assembled to form three spectroscopically distinct copper binding sites.[18]

Their active site is well conserved.[9] It contains four copper ions that mediate the redox process: one type-1 (T1) copper ion, one type-2 (T2) copper ion, and two type-3 (T3) copper ions, based on the copper's coordination and spectroscopic properties. The T2 and T3 copper centers are close together forming a trinuclear copper cluster and the T1 site contains a single type-1 copper ion.[19] The electrons are transferred from the mononuclear center where oxidation takes place to the trinuclear center where reduction of oxygen to water occurs.[20,21]

From an electrochemical point of view, laccases from different sources exhibit a wide range of redox potentials.[9,22,23] The T1 site has a high-redox potential reaching 780–800 mV for the enzymes from the white-rot fungus *T. versicolor* and the ascomycetes fungus *Neurospora crassa*, whereas the plant *R. vernicifera* enzyme has a value of 420 mV. The redox potential of T2 and T3 sites for the *R. vernicifera* laccase are, respectively, 390 and 460 mV at pH 7.5. The *T. versicolor* laccase T3 site has been reported to be 782 mV. In general, the T1 sites in fungal laccases are much higher than those of plant laccases.

### 7.3.1.2 Mechanism of Catalysis

Laccases couple the four-electron reduction of dioxygen into two molecules of water to the four one-electron oxidation processes of a variety of substrates, such as phenols, arylamines, anilines, thiols, and lignins (Figure 7.1a).[1,21] The oxidation reactions catalyzed by laccases lead to the formation of free radicals which act as intermediate substrates for the enzymes (Figure 7.1b).[1,21,24] These mediators can leave the enzyme site and react with high-redox potential substrates and thus create nonenzymatic routes of oxidative polymerizing or depolymerizing reactions (Figure 7.1c). Ultimately, laccase-mediator system (LMS) becomes involved in a range of physiological functions such as lignolysis (Figure 7.1d), lignin synthesis, morphogenesis, pathogenesis, and detoxification.

Laccases catalyze four $1e^-$ oxidations of a reducing substrate with concomitant two $2e^-$ reductions of dioxygen to water. The stoichiometry is four molecules of reducing substrates for each molecular oxygen, involving a total transfer of four electrons.[9]

The first step of laccase catalysis is the oxidation of the electron donor substrate by the copper ($Cu^{2+}$ to $Cu^+$) at the T1 site, which is the primary electron acceptor.[9] The electrons extracted from the substrate are transferred to the T2/T3 trinuclear site, resulting in the conversion of the fully oxidized form of the enzyme to a fully reduced state. A successive four-electron oxidation (from four substrate molecules) is required to fully reduce the enzyme (Figure 7.2).

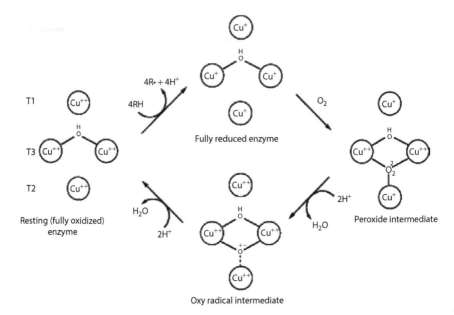

**FIGURE 7.2** Catalytic cycle of laccases. (Reprinted from *Appl. Biochem. Biotechnol.*, 2009, 157, Wong, D.W.S., Structure and action mechanism of ligninolytic enzymes, 174–209, Copyright 2016, with permission from Elsevier.)

The reduction of dioxygen takes place in two steps via the formation of bound oxygen intermediates. The dioxygen molecule first binds to the T2/T3 site, and two electrons are transferred from the T3 cuprous ions, resulting in the formation of a peroxide intermediate. The peroxide intermediate decays to an oxy radical and undergoes a two-electron reductive cleavage of the O-O bond with the release of a water molecule. The decay of peroxide intermediate is facilitated by the final electron transfer from the T2 cuprous ion. In the last step, all four copper centers are oxidized, and $O^{2-}$ is released as a second water molecule.

### 7.3.1.3 Oxidation of Phenolic Substrates

Laccases promote the subtraction of one electron from phenolic hydroxyl groups of lignin to form phenoxy radicals, which generally undergo polymerization via radical coupling.[9] The reaction is also accompanied by demethylation, formation of quinone, resulting in ring cleavage. The degradation of phenolic β-1 lignin substructure models occurs via the formation of phenoxy radicals, leading to Cα-Cβ cleavage, Cα oxidation, alkyl-aryl cleavage, and aromatic ring cleavage (Figure 7.3).[25] Laccase-catalyzed oxidation of phenols, anilines, and benzenethiols correlates with the redox potential difference between laccases T1 copper site and the substrate.

### 7.3.1.4 Oxidation of Nonphenolic Substrates

Laccases have been found to oxidize nonphenolic model compounds and β-1 lignin dimers in the presence of a mediator, indicating that the enzymes play a significant role in the depolymerization of lignin.[6] One of the most studied mediators is

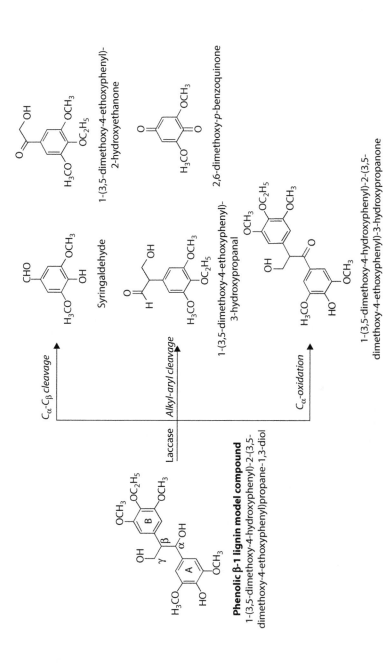

**FIGURE 7.3** Laccase-catalyzed oxidation of phenolic β-1 lignin model compound. (Reprinted from *Appl. Biochem. Biotechnol.*, 157, Wong, D.W.S., Structure and action mechanism of ligninolytic enzymes, 174–209, Copyright 2016, with permission from Elsevier.)

1-hydroxybenzotriazole (HBT), which is oxidized to its nitroxide radical by laccases.[26] The degradation of nonphenolic $\beta$-$O$-4 model compounds, which represent the major substructure in lignin, has been studied using laccase-mediator systems. Four types of reactions, $\beta$-ether cleavage, C$\alpha$-C$\beta$ cleavage, C$\alpha$ oxidation, and aromatic ring cleavage, are catalyzed by laccase-HBT coupled system (Figure 7.4).

In summary, laccases catalyze a one-electron oxidation with the concomitant four-electron reduction of molecular oxygen to water. Fungal laccases can oxidize phenolic lignin model compounds and have higher redox potential than bacterial laccases. In the presence of redox mediators, fungal laccases can oxidize nonphenolic lignin model compounds.

As laccases work efficiently on many substrates without cofactors, they may be valuable in many biotechnological industrial applications, such as pulp biobleaching, biosensors, food industries, textile industries, soil remediation, and in the production of complex polymers.[1]

## 7.3.2  HEME PEROXIDASES

As a common trait, most peroxidases, though phylogenetically unrelated, contain a heme B (iron protoporphyrin IX) molecule as the redox cofactor.[27]

The heme peroxidases have been classified into two distinct groups, termed the animal (found only in animals) and plant (found in plants, fungi, and prokaryotes) superfamilies.[28] The plant peroxidases, which share similar overall protein folds and specific features, such as catalytically essential histidine and arginine residues in their active sites, have been subdivided into three classes on the basis of sequence comparison. Class I consists of intracellular enzymes of prokaryotic origin. Class II consists of the secretory fungal peroxidases such as lignin peroxidases. Class III contains the secretory plant peroxidases. Another family of heme peroxidases, which constitutes an independent group, is the DyP-type peroxidase family.[29,30]

LiP, MnP, and VP are class II extracellular fungal peroxidases.[3,31] LiPs are strong oxidants with high-redox potential that oxidize the major nonphenolic structures of lignin (80%–90% of lignin composition). MnP is a Mn-dependent enzyme that catalyzes the oxidation of various phenolic substrates but is not capable of oxidizing the more recalcitrant nonphenolic lignin. VP enzymes combine the catalytic activities of both MnP and LiP and are able to oxidize $Mn^{2+}$ like MnP, and nonphenolic compounds like LiP. DyPs (Dye-decolorizing peroxidases), which occur in both fungi and bacteria, do not belong to plant peroxidase superfamily but are members of a new superfamily of heme peroxidases called DyPs.[32] DyPs oxidize high-redox potential anthraquinone dyes and were recently reported to oxidize nonphenolic lignin model compounds such as veratryl alcohol (Figure 7.5) and adlerol, a nonphenolic $\beta$-$O$-4 lignin model dimer.[33]

With regard to the oxidation potential of heme peroxidases, LiP possesses the highest redox potential ($E_0 \sim 1.2$ V, pH 3), followed by MnP ($E_0 \sim 0.8$ V, pH 4.5).[34]

The crystal structures of a number of these heme peroxidases show that they share two all-$\alpha$ domains between which the heme group is embedded.[35] The crystal structures of dye-decolorizing peroxidases reveal a $\beta$ barrel fold.[31]

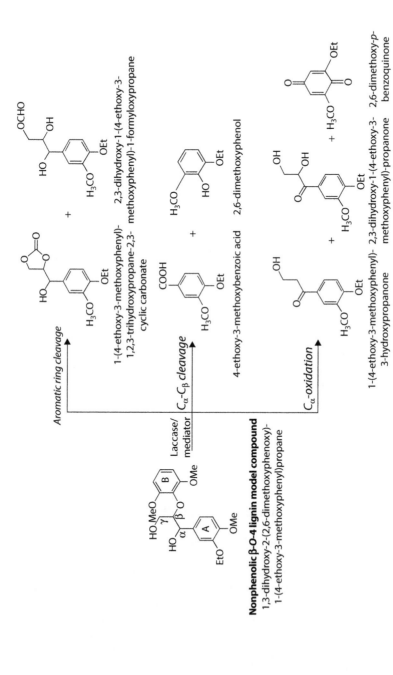

**FIGURE 7.4** Laccase-catalyzed oxidation of nonphenolic β-O-4 lignin model compound. (Reprinted from *Appl. Biochem. Biotechnol.*, 157, Wong, D.W.S., Structure and action mechanism of ligninolytic enzymes, 174–209, Copyright 2016, with permission from Elsevier.)

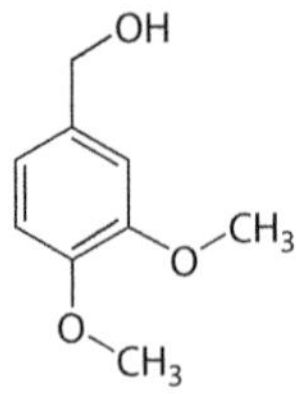

**FIGURE 7.5**   Veratryl alcohol (3,4-dimethoxybenzyl alcohol).

The general catalytic cycle of peroxidases is summarized by the following reactions (Equations 7.2 through 7.4)[28]:

$$\text{Enzyme} + H_2O_2 \rightarrow \text{Compound I} + H_2O \tag{7.2}$$

$$\text{Compound I} + XH_2 \rightarrow \text{Compound II} + XH\cdot \tag{7.3}$$

$$\text{Compound II} + XH_2 \rightarrow \text{Enzyme} + XH\cdot + H_2O \tag{7.4}$$

In a first step, $H_2O_2$ enters the heme cavity of the enzyme in resting state where it displaces a water molecule that occupies the sixth ferric iron coordination site of the protoporphyrin IX system. A distal basic amino acid residue mediates the rearrangement of a proton in $H_2O_2$. In plant peroxidases, this base typically is a histidine, while in DyPs this residue is substituted by an aspartate. The heme molecule is then oxidized to the radical-cationic oxoferryl species compound I by twofold single-electron transfer, releasing a water molecule. Two electrons are successively drawn from substrate molecules $(XH_2)$ leading to their oxidized counterparts. Concomitantly, the heme is stepwise reduced back to its initial oxidation state leading to the enzyme resting state.

### 7.3.2.1   Lignin Peroxidases

Lignin peroxidases (LiP) (1,2-bis(3,4-dimethoxyphenyl)propane-1,3-diol:hydrogen peroxide oxidoreductases, also called diarylpropane peroxidases, EC 1.11.1.14) are heme-containing glycoproteins and play a central role in the biodegradation of lignin (Equation 7.5).[1,9]

$$1,2\text{-bis}\left(3,4\text{-dimethoxyphenyl}\right)\text{propane-1,3-diol} + H_2O_2 \leftrightarrow 3,4\text{-dimethoxy-}$$
$$\tag{7.5}$$
$$\text{benzaldehyde} + 1\text{-}\left(3,4\text{-dimethoxyphenyl}\right)\text{ethane-1,2-diol} + H_2O$$

Lignin peroxidases catalyze the $H_2O_2$-dependent oxidative depolymerization of non-phenolic lignin compounds (e.g., diarylpropane), β-$O$-4 nonphenolic lignin model compounds, and many phenolic compounds (e.g., guaiacol, vanillyl alcohol, catechol, syringic acid, acetosyringone) with redox potentials up to 1.4 V. Lignin peroxidases oxidize the substrates in multistep electron transfers and form intermediate radicals, such as phenoxy radicals and nonphenolic veratryl alcohol radical cations (Figure 7.1e).

These intermediate radicals undergo nonenzymatic reactions, such as radical coupling and polymerization, side-chain cleavage, demethylation and intramolecular addition, and rearrangement (Figure 7.1f). Unlike the other peroxidases, lignin peroxidase is able to oxidize nonphenolic aromatic substrates and does not require the participation of mediators due to its unusually high-redox potential (Figure 7.1g).

### 7.3.2.2 Molecular Structure

Lignin peroxidase was first discovered in *P. Chrysosporium*, and various isoforms are known to exist with this microorganism and other white-rot fungi.[9] The lignin peroxidase isozymes are glycoproteins of 38–46 kDa. Lignin peroxidase has a distinctive property of an unusually low pH optimum near pH 3. The enzyme contains one mole of iron protoporphyrin IX per mole of protein.

The crystal structure of *P. Chrysosporium* lignin peroxidase has been described in details.[36] The enzyme is globular with a dimension of 50 Å × 40 Å × 40 Å comprising a proximal (C-terminal) and distal (N-terminal) domain (Figure 7.6).

LiP has a rather globular shape and is divided by the heme into a proximal (C-terminal) and distal (N-terminal) domain.[37] The heme is completely embedded in the protein but is accessible from the solvent via two small channels.[37] The monomeric LiP molecule contains eight major and eight minor α-helices. In addition, three short

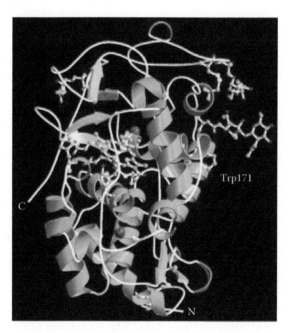

**FIGURE 7.6** Three-dimensional structure of *P. chrysosporium* lignin peroxidase.[37] Secondary structure elements are indicated by blue helices and orange arrows. The heme with the proximal and distal histidine and the two distal water molecules, the two calcium ions (purple spheres), His82 at the entrance of the active-site channel (left side), four carbohydrate molecules, Trp171 and the disulphide bridges (S atoms as yellow spheres) are represented as ball and stick models. (Reprinted from *J. Mol. Biol.*, 286, Choinowski, T. et al., The crystal structure of lignin peroxidase at 1.70 Å resolution reveals a hydroxyl group on the Cβ of tryptophan 171: A novel radical site formed during the redox cycle, 809–827, Copyright 2016, with permission from Elsevier.)

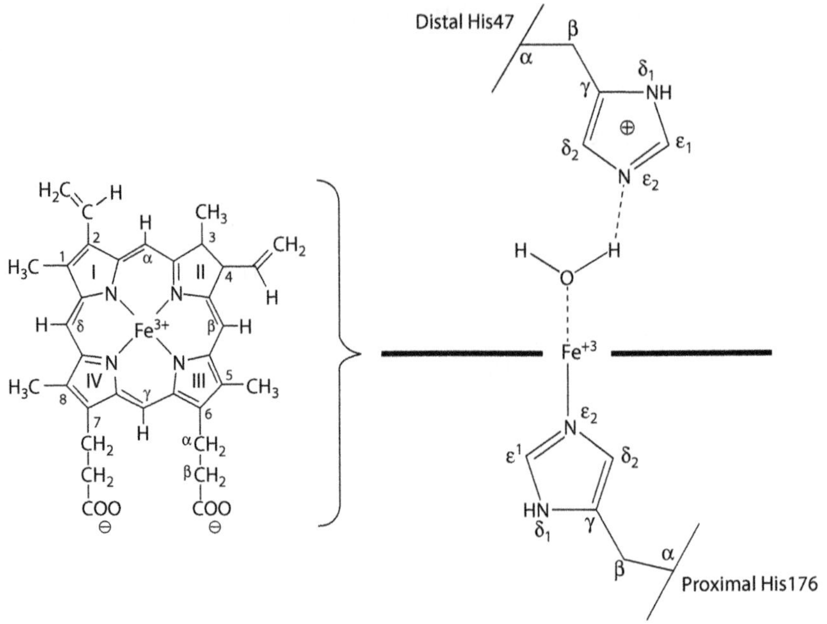

**FIGURE 7.7** The heme iron environment in lignin peroxidase. (Reprinted from *Appl. Biochem. Biotechnol.*, 157, Wong, D.W.S., Structure and action mechanism of ligninolytic enzymes, 174–209, Copyright 2016, with permission from Elsevier.)

antiparallel β-sheets complete the fold. An extended C-terminal segment of about 50 amino acids traverse over the surface with little contact to the core of the protein.[9] There are eight cysteine residues in LiP, all of them forming disulphide bridges. LiP has two calcium-binding sites, one in each domain, with possible function of maintaining the topology of the active site. Several *N*- and *O*-glycosylation sites can be identified in the crystal structure. The carbohydrate chains may play a role in the protection of the C-terminal peptide from proteolysis (breakdown of proteins). The heme iron is predominantly pentacoordinated with histidine at the proximal side as the fifth ligand, and water hydrogen-bonded to the distal histidine (Figure 7.7).

The peroxide-binding pocket is located on the distal side of the heme, with a channel extending to the exterior of the protein.[6]

### 7.3.2.3 Mechanism of Catalysis

LiP has a typical peroxidase catalytic cycle.[9] The mechanism of lignin peroxidase-catalyzed reaction includes two steps (Figure 7.8):

- A two-electron oxidation of the native ferric enzyme [Fe(III), LiP] to yield compound I intermediate that exists as a ferryl iron (Fe(IV)=O) porphyrin radical cation [Fe(IV)=O$^{\bullet+}$, LiP-I], with the peroxide substrate cleaved at the O-O bond (Equation 7.6):

$$LiP[Fe(III)] + H_2O_2 \rightarrow LiP\text{-}I[Fe(IV) = O^{\bullet+}] + H_2O \qquad (7.6)$$

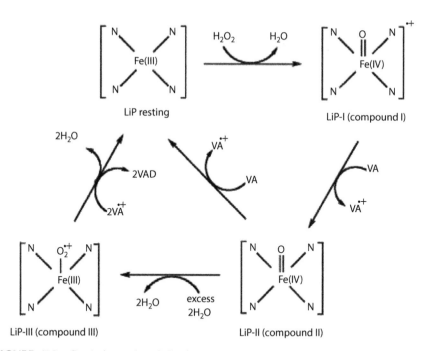

**FIGURE 7.8** Catalytic cycle of lignin peroxidase. (Reprinted from *Appl. Biochem. Biotechnol.*, 157, Wong, D.W.S., Structure and action mechanism of ligninolytic enzymes, 174–209, Copyright 2016, with permission from Elsevier.)

- A two consecutive single-electron reduction of LiP-I by electron donor substrates to the native enzyme. The first reduction of LiP-I by a reducing substrate (AH), such as veratryl alcohol, yields compound II [Fe(IV)=O, LiP-II], and a radical cation. A second reduction returns the enzyme to the ferric oxidation state, completing the catalytic cycle (Equations 7.7 and 7.8).

$$\text{LiP-I} + \text{AH} \; \rightarrow \; \text{LiP-II}\left[\text{Fe(IV)}{=}\text{O}\right] + \text{A}^{\bullet+} \tag{7.7}$$

$$\text{LiP-II} + \text{AH} \; \rightarrow \; \text{LiP} + \text{A}^{\bullet+} \tag{7.8}$$

At ph 3.0, in the presence of excess $H_2O_2$ and the absence of a reducing substrate, LiP-II reacts with $H_2O_2$ to form a catalytic inactive form of the enzyme, known as compound III (LiP-III).[9] The heme of LiP-III exists as a ferric-superoxo complex. LiP-III can be converted to the resting enzyme by spontaneous autoxidation or by oxidation with a veratryl alcohol radical cation.

### 7.3.2.4 Oxidation of Phenolic and Nonphenolic Substrates

Like other peroxidases, LiP is capable of oxidizing a wide variety of phenolic substrates.[9] LiP catalyzes the oxidation of phenolic compounds preferentially at a much higher rate compared to nonphenolic substrates. In the reduction of LiP-I and LiP-II, phenolic substrates are converted to phenoxy radicals. In the presence of oxygen, the phenoxy radical may react to form ring-cleavage products, or they may lead to coupling and polymerization.

Oxidation of nonphenolic diarylpropane and β-*O*-4 lignin model compounds involves initial formation of radical cation, followed by side-chain cleavage, demethylation, intramolecular addition, and rearrangements. In the mechanism, only the formation of the radical cation is enzyme-catalyzed.

Veratryl alcohol is a metabolite produced at the same time as LiP by *P. chrysosporium*.[9] The addition of veratryl alcohol is known to increase LiP activity and the rate of lignin mineralization. It is well established that veratryl alcohol (standard redox potential = 1.36) is oxidized by LiP-I and LiP-II, resulting in the formation of a veratryl alcohol radical cation intermediate. The radical cation decays by deprotonation of Cα to form veratraldehyde. It has been proposed that veratryl alcohol radical cation is a redox mediator in the oxidation of lignin.

### 7.3.2.5 Manganese Peroxidases

Manganese peroxidases (MnP) (Mn(II):hydrogen peroxide oxidoreductases, EC 1.11.1.13) are extracellular glycoproteins and are secreted in multiple isoforms which contain one molecule of heme as iron protoporphyrin IX.[1] The enzymes catalyze the peroxide-dependent oxidation of Mn(II) to Mn(III) (Figure 7.1h and Equation 7.9), which is then released from the enzyme surface in complex with oxalate or other chelators:

$$2\ \mathrm{Mn(II)} + 2\ \mathrm{H}^+ + \mathrm{H_2O_2} \leftrightarrow 2\ \mathrm{Mn(III)} + 2\ \mathrm{H_2O} \tag{7.9}$$

Chelated Mn (III) complex acts as a reactive, low-molecular weight, diffusible redox mediator (Figure 7.1j) of phenolic substrates including simple phenols, amines, dyes, phenolic lignin substructures, and dimers. The oxidation potential of Mn(III) chelate is only limited to phenolic lignin structures. However, for the oxidation of nonphenolic substrates by chelated Mn(III), reactive radicals must be formed in the presence of a second mediator (Figure 7.1k). Organic acids, such as oxalic and malonic acids, are the primary compounds that act as second mediators in the production of reactive radicals.

### 7.3.2.6 Molecular Structure

MnP contains one molecule of heme as iron protoporphyrin IX.[9] The heme iron in the native protein is in the pentacoordinate, ferric state with a histidine residue coordinated as the fifth ligand. The overall structure of *P. chrysosporium* MnP is similar to LiP, consisting of two domains with the heme sandwiched in-between (Figure 7.9).

The protein molecule contains ten major helices and one minor helix as found in LiP.[9] MnP has five rather than four disulfide bridges, with the additional Cys (cysteine)—Cys bond located near the C terminus of the polypeptide chain. This additional disulfide bond helps to form the Mn(II)-binding site and is responsible for pushing the C terminus segment away from the main body of the protein. The Mn(II) is located in a cation-binding site at the surface of the protein and coordinates to the carboxylate oxygens of three amino residues, the heme propionate oxygen, and two water oxygens. Two heptacoordinate structural calcium ions are important for thermal stabilization of the active site of the enzyme.

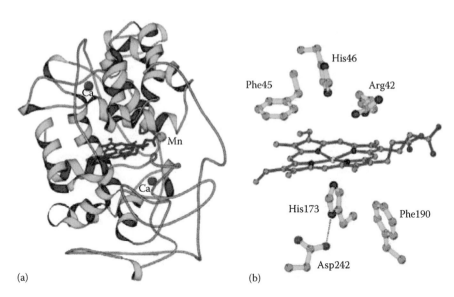

(a)                                                           (b)

**FIGURE 7.9**  (a) The overall structure of manganese peroxidases (MnP). The red spheres are structural $Ca^{++}$ ions conserved in extracellular heme peroxidases. The location of the substrate, Mn(II), near the heme, is indicated. (b) The active-site structure of MnP. This architecture is highly conserved in heme peroxidases. The main variations are the Phe (phenylalanine) residues which are Trp (tryptophan) in the intercellular peroxidases, cytochrome c and ascorbate peroxidase. The Asp (aspartic acid) 242-His (histidine) 173 pair is conserved. (Reprinted from *J. Inorg. Biochem.*, 104, Sundaramoorthy, M. et al., Ultrahigh (0.93Å) resolution structure of manganese peroxidase from *Phanerochaete chrysosporium*: Implications for the catalytic mechanism, 683–690, Copyright 2016, with permission from Elsevier.)

### 7.3.2.7  Mechanism of Catalysis

The characteristics of the cycle are very similar to that of LiP.[6,9] Addition of $H_2O_2$ to the native ferric enzyme yields compound I (MnP-1), which is a Fe(IV)-oxo-porphyrin radical cation [Fe(IV)=O$^+$]. The peroxide bond of $H_2O_2$ is cleaved subsequent to a two-electron transfer from the enzyme (Figure 7.10, Equation 7.10).

$$MnP + H_2O \rightarrow MnP\text{-}I + H_2O \tag{7.10}$$

Addition of MnP-1 to Mn(II) reduces compound I to compound II (MnP-II), which is a Fe(IV)-oxo-porphyrin complex [Fe(IV)=O] (Equation 7.11).

$$MnP\text{-}I + Mn^{2+} \rightarrow MnP\text{-}II + Mn^{3+} \tag{7.11}$$

The native enzyme is then regenerated from the further reduction of MnP-II by $Mn^{2+}$ and the release of another water molecule (Equation 7.12).

$$MnP\text{-}II + Mn^{2+} \rightarrow MnP + Mn^{3+} + H_2O \tag{7.12}$$

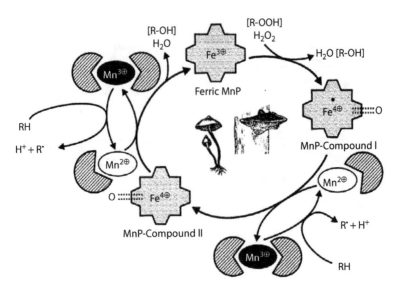

**FIGURE 7.10** Catalytic cycle initiated by manganese peroxidase. (Reprinted from *Enzyme Microb. Technol.*, 30, Hofrichter, M., Review: Lignin conversion by manganese peroxidase (MnP), 454–466, Copyright 2016, with permission from Elsevier.)

The formed $Mn^{3+}$, a strong oxidant, then oxidizes phenolic structures by single-electron oxidation (Equation 7.13).

$$Mn^{3+} + RH \rightarrow Mn^{2+} + R^{\bullet} + H^+ \qquad (7.13)$$

The Mn(III) formed is dissociated from the enzyme and stabilized by forming complexes with α-hydroxy acids at a high positive redox potential of 0.8–0.9 V.[9] Oxalate and malonate are optimal ligands that are secreted by the fungus.

### 7.3.2.8 Oxidation of Phenolic and Nonphenolic Substrates

The Mn(III) chelate acts as a diffusible oxidant of phenolic substrates involving single-electron oxidation of the substrate to form a phenoxy radical intermediate, which undergoes rearrangements, bond cleavages, and nonenzymatic degradation to yield various breakdown products.[9]

For nonphenolic substrates, the oxidation by Mn(III) involves the formation of reactive radicals in the presence of a second mediator, in contrast to the LiP-catalyzed reaction, which involves electron abstraction from the aromatic ring forming a radical cation.[9]

Manganese peroxidase may be capable of rivaling the potential applications of laccases in biotechnology.[1] Studies have shown that the presence of manganese peroxidase can increase the degree of dye decolorization. In addition, manganese peroxidase from white-rot fungi is considered as the primary enzyme responsible for biobleaching of kraft pulps.

## 7.3.2.9 Versatile Peroxidases

Versatile peroxidases (VPs) (Mn(II):hydrogen-peroxidase oxidoreductases/ diarylpropane:oxygen, hydrogen peroxide oxidoreductases; EC 1.11.1.16[40]) are glycoproteins with hybrid properties capable of oxidizing typical substrates of other basidiomycetes peroxidases including Mn(II) and also veratryl alcohol, MnP and LiP substrate, respectively (Figure 7.1).[1] VPs form an attractive ligninolytic enzyme group due to their dual oxidative ability to oxidize Mn(II) and also phenolic and non-phenolic aromatic compounds. This makes versatile peroxidases more efficient than both lignin peroxidases and manganese peroxidases, which are not able to efficiently oxidize phenolic compounds in the absence of veratryl alcohol or oxidize phenols in the absence of Mn(II), respectively. Similar to the manganese peroxidase mechanism, Mn(III) is released from versatile peroxidases and acts as a diffusible oxidizer of phenolic lignin and free phenol substrates (Figure 7.1h–j). Like other members of heme peroxidases, heme is buried in the interior of the protein and has access to the outer medium through two channels. The function of the first channel is similar to that described for lignin peroxidase and is conserved among all heme peroxidases. Conversely, the second channel is found to be specific to versatile peroxidase and manganese peroxidase, and is where the oxidation of Mn(II) to Mn(III) takes place.

## 7.3.2.10 Molecular Structure

Molecular characterization of VPs reveals structures that are closer to the LiPs than to the MnP isoenzymes of *P. chrysosporium*.[9,41] A Mn(II) binding site formed by three acidic residues typical of MnP enzymes is found near the heme internal propionate, which accounts for the ability to oxidize Mn(II) (Figure 7.7).[42] Furthermore, residues involved in lignin peroxidase interaction with varatryl alcohol and other aromatic substrates are also found in the protein structure. The molecular model of the *Pleurotus eryngii* versatile peroxidase reveals 12 helices, 4 disulfide bonds, a heme pocket containing the characteristic proximal histidine and the distal histidine, 2 structural $Ca^{2+}$ sites and a Mn(II) binding site (Figure 7.11).[42,43]

## 7.3.2.11 Mechanism of Catalysis

The catalytic cycle for VP is similar to that of LiP and MnP, in that the enzyme catalyzes the electron transfer from an oxidizable substrate, involving the formation and reduction of compound I and compound II intermediates (Figure 7.12).[9,42] Compound I is a two-oxidizing equivalent intermediate, with one-oxidizing equivalent localized in the ferryl state of the iron, and the second localized as a porphyrin radical cation. Furthermore, a protein-centered radical has been detected, implicating that the second oxidizing equivalent may also localize on a tryptophan (Trp) residue near the heme prosthetic group during the catalytic cycle of *Bjerkandera adusta* VP (Figure 7.13). In *P. eryngii* VP, it has been shown that the surface Trp exists in the form of a neutral radical.[43] The oxidation of aromatic substrates by direct electron transfer to the heme and long-range electron transfer from a surface accessible Trp residue is analogous to that observed in *P. chrysosporium* LiP, but the reactive form in the latter exists as a Trp radical cation.[9]

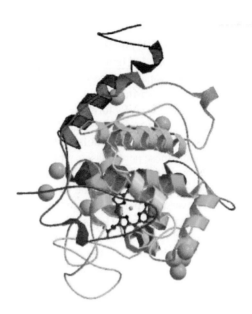

**FIGURE 7.11**  Crystal structure of *Pleurotus eryngii* versatile peroxidase. (From RCSB PDB Structure Summary—2BOQ; From Ruiz-Duenas, F.J. et al., *Biochemistry*, 46, 66, http://pubs.acs.org/doi/full/10.1021/bi061542h, 2007; http://www.pdb.org/pdb/explore.do?structureId=2BOQ.)

### 7.3.2.12  Summary of Plant Peroxidases and Laccases

The white-rot fungi produce an array of extracellular oxidative enzymes that degrade lignin.[45] The two major groups of ligninolytic enzymes include peroxidases and laccases (phenol oxidases). The peroxidases are heme-containing enzymes with catalytic cycles that involve the activation by $H_2O_2$ and substrate reduction of compound I and compound II intermediates. Laccases are multi-copper oxidoreductases. Key characteristics of the three subgroups of peroxidases and laccases include:

- Lignin peroxidases have the unique ability to catalyze oxidative cleavage of C–C bonds and ether (C-O-C) bonds in nonphenolic aromatic substrates of high-redox potential.
- Manganese peroxidases oxidize Mn(II) to Mn(III), which facilitates the degradation of phenolic compounds or, in turn, oxidizes a second mediator for the breakdown of nonphenolic compounds.
- Versatile peroxidases combine catalytic properties of lignin peroxidases and manganese peroxidases.
- Laccases are able to oxidize a wide variety of phenolic and nonphenolic compounds with the reduction of molecular oxygen to water.

### 7.3.2.13  Dye-Decolorizing Peroxidase

Dye-decolorizing peroxidase (DyP) (EC 1.11.1.19), a unique dye-decolorizing enzyme from the fungus *Thanatephorus cucumeris* Dec 1 (renamed *B. adusta*),

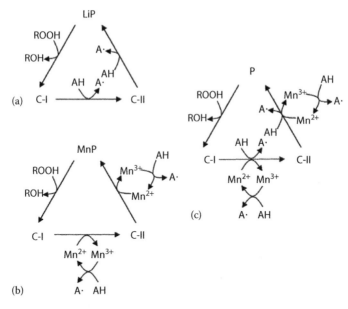

**FIGURE 7.12** Scheme of catalytic cycles of *P. chrysosporium* Lip (a) and MnP (b) compared with the *P. eryngii* VP (c). All cycles include two-electron oxidation of enzyme by hydroperoxidases (ROOH) to compound I, a complex of oxo-iron, and porphyrin cation radical ($[Fe^{4+}=OP^{•+}]$) followed by two one-electron reductions of compound I to compound II ($[Fe^{4+}=OP]$) and native enzyme ($[Fe^{3+}P]$), producing two oxidations of the electron donor. They differ in peroxidase electron donors, which could be (i) aromatic compounds (AH) for LiP; (ii) $Mn^{2+}$ for MnP; and $Mn^{2+}$ and phenolic or nonphenolic compounds for the *P. eryngii* VP. (From Ruiz-Dueñas, F.J. et al.: Molecular characterization of a novel peroxidase isolated from the ligninolytic fungus Pleurotus eryngii. *Mol. Microbiol.* 1999. 31. 223–239. Copyright Wiley-VCH Verlag GmbH & Co. KGaA. Reproduced with permission.)

has been classified as a peroxidase but lacks homology to almost all other known plant peroxidases.[33] The primary structure of DyP shows moderate sequence homology to only two known proteins: the peroxide-dependent phenol oxidase, TAP, and the hypothetical peroxidase, cpop21. In 2007, Sugano et al.[32] showed the first crystal structure of DyP and revealed that this protein has a unique tertiary structure with a distal heme region that differs from that of most other peroxidases.

DyP contains one heme with an iron at the center of the molecule.[33] The full structure has dimensions of ~62 Å × 66 Å × 48 Å. The size and heme existence are consistent with other peroxidase structures. Of 442 total residues in DyP, 192 residues form 18 α-helices and 15 β-strands.

One unique motif found in the secondary structure of DyP is two sets of antiparallel β-sheets (β1 and β4, β2 and β3) located between two α-helices (α3 and α8). This motif has not been identified in any other registered protein. Consistent with the structure of other peroxidases, the fifth ligand of the heme iron of DyP is histidine. However, DyP lacks an important distal histidine residue known to assist in the formation of compound I ($[Fe^{4+}=OP^+]$) during the action of ubiquitous peroxidases.

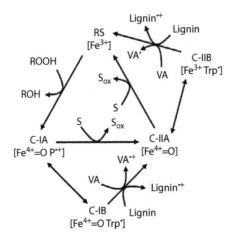

**FIGURE 7.13**  Extended catalytic cycle proposed for ligninolytic peroxidases (LiP and VP). In addition to normal compound-I and compound-II of Figure 7.3 (now C-IA and C-IIA), C-IB (containing $Fe^{4+}=O$ and tryptophan radical) and C-IIB (containing $Fe^{3+}$ and tryptophan radical) are included, being involved in oxidation of high-redox potential compounds such as veratryl alcohol (VA) and lignin units to their corresponding cation radicals. C-IB and C-IIB are formed by LRET to the activated haem cofactor. (From Ruiz-Dueñas, F.J. and Martínez, A.T.: Microbial degradation of lignin: How a bulky recalcitrant polymer is efficiently recycled in nature and how we can take advantage of this. *Microb. Biotechnol.* 2009. 2. 164–177. Copyright Wiley-VCH Verlag GmbH & Co. KGaA. Reproduced with permission.)

Instead, an aspartic acid and an arginine were suggested to be involved in the formation of compound I. It was proposed that DyP from *Thanatephorus cucumeris* represents a novel heme peroxidase family.[46]

In 2007, Zubieta et al.[30] reported the crystal structures of a DyP from *Bacteroides thetaiotaomicron* (Figure 7.14) and TyrA from *Shewanella oneidensis*, two novel bacterial dye-decolorizing peroxidases. The crystal structures reveal a β-barrel fold with a conserved heme-binding motif.

BtDyP assembles into a hexamer (Figure 7.15), while TyrA assembles into a dimer; the dimerization interface is conserved between the two proteins. Each monomer exhibits a two-domain, α+β ferredoxin-like fold [(β-α-β) × 2].[48] All of these proteins contain duplicated ferredoxin-like folds arranged as a β-barrel.[49] A site for heme binding was identified. Comparisons with other DyPs demonstrated a conservation of putative heme-binding residues, including an absolutely conserved histidine (Figure 7.16).

In 2013, E. Strittmatter et al.[27] determined the first crystal structure of a fungal high-redox potential dye-decolorizing peroxidase: AauDyPI from *Auricularia auricula-judae* (Figure 7.17).

The mostly helical structure shows a β-sheet motif typical for DyPs structures and includes the complete polypeptide chain. At the distal side of the heme molecule, a flexible aspartate residue (Asp168) plays a key role in catalysis. It guides incoming hydrogen peroxide toward the heme iron and mediates proton rearrangement in the

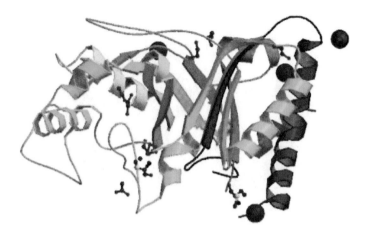

**FIGURE 7.14** Crystal structure of a DyP from *Bacteroides thetaiotaomicron* (asymmetric unit). (From RCSB PDB Structure Summary—2GVK; From http://www.rcsb.org/pdb/explore. do?structureId=2GVK.)

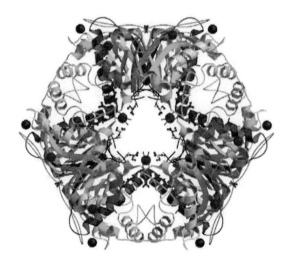

**FIGURE 7.15** Crystal structure of *Bacteroides thetaiotaomicron* DyP (hexameric assembly). In BtDyP, three dimers pack around the central threefold axis to form the hexamer. (From RCSB PDB Structure Summary—2GVK; From http://en.wikipedia.org/wiki/ Ferredoxin_fold.)

process of compound I formation (first step in the reaction mechanism of plant heme peroxidases; Figure 7.12). Afterward, its side chain changes its conformation, now pointing toward the protein backbone. It appears that Asp168 acts like a gatekeeper by altering the width of the heme cavity access channel. Tyr337 is identified as a surface-exposed substrate interaction site and therefore a residue directly involved in AauDyPI-catalysis.

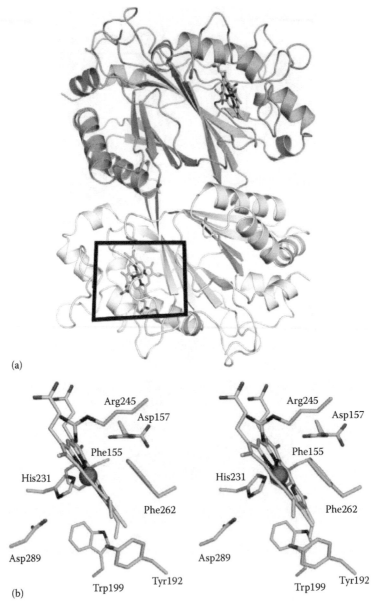

(a)

(b)

**FIGURE 7.16**  Heme-binding model for BtDyP: (a) BtDyP dimer showing two independent heme-binding sites. (b) Stereo view of putative heme-binding residues. The heme is
colored by atom, with carbons (cyan), nitrogens (blue), oxygens (red), and iron (orange).
Residues from BtDyP are labeled and colored by atom as for the heme, with carbons in green.
Side-chain conformations were not optimized for heme binding. (From Zubieta, C. et al.:
Crystal structures of two novel dye-decolorizing peroxidases reveal a β-barrel fold with a
conserved heme-binding motif. *Proteins.* 2007. 69. 223–234. Copyright Wiley-VCH Verlag
GmbH & Co. KGaA. Reproduced with permission.)

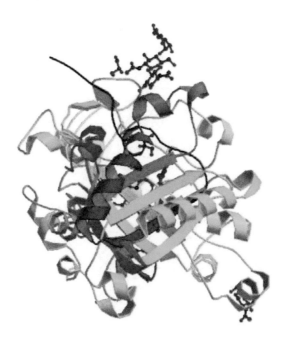

**FIGURE 7.17** Crystal structure of AauDyPI, a fungal DyP-type peroxidase from *Auricularia auricula-judae* (monomer). (From RCSB PDB Structure Summary—4AU9; From http://www.rcsb.org/pdb/explore.do?structureId=4au9.)

From a structural viewpoint, DyPs are best considered members of a highly diverse superfamily of proteins sharing a ferredoxin-like core of β-sheets as a common feature.[28] Actually, the term *DyP* circumscribes a polyphyletic group that can be roughly divided into four different entities. The groups DyPA-C comprise primarily bacterial enzymes, whereas DyPD is a fungal group.

In 2014, Colpa et al.[46] wrote a review on DyP-type peroxidases described as promising and versatile class of enzymes.[47] DyP-type peroxidases are unrelated at the primary sequence level to peroxidases of the plant and animal superfamily. They lack the typical heme-binding motif of plant peroxidases, comprising one proximal histidine, one distal histidine, and one crucial arginine (Figure 7.18).

DyPs are bifunctional enzymes displaying not only oxidative activity but also hydrolytic activity. Moreover, these enzymes are able to oxidize a variety of organic compounds of which some are poorly converted by established peroxidases, including dyes, β-carotenes, and aromatic sulfides. In addition, growing evidence shows that microbial DyPs play a crucial role in the degradation of lignin.

### 7.3.3 Accessory Enzymes

Besides ligninases, other fungal extracellular enzymes, which act as accessory enzymes, are involved in lignin degradation.[1] These include oxidases generating hydrogen peroxide required by peroxidases, and mycelium-associated

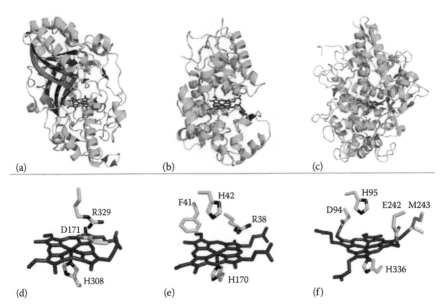

**FIGURE 7.18**  Structural comparison of DyP from *Bjerkandera adusta* (a), plant peroxidase HRP from *Armoracia rusticana* (b), and human myeloperoxidase from *Homo sapiens* (c). α-Helices are shown in green; β-sheets are in blue; and the heme cofactor is in red. Close-up of key amino acids in the heme-surrounding region of DyP (d), HRP (e), and human myeloperoxidase (f). (With kind permission from Springer Science+Business Media: *J. Ind. Microbiol. Biotechnol.*, DyP-type peroxidases: A promising and versatile class of enzymes, 41, 2013, 1–7, Colpa, D.I. et al., Copyright [2016].)

dehydrogenases, which reduce lignin-derived compounds (Figure 7.1l). Oxidases generating hydrogen peroxide include aryl-alcohol oxidase (AAO) found in various fungi, such as *P. eryngii*, and glyoxal oxidase (GLOX, a copper-radical protein) found in *P. chrysosporium*. Furthermore, aryl-alcohol dehydrogenases (AAD) and quinone reductases (QR) are also involved in lignin degradation by fungi (Figure 7.1m). Moreover, it has been shown that cellulose dehydrogenase (CDH), which is produced by many different fungi under cellulolytic conditions, is also involved in lignin degradation in the presence of hydrogen peroxide and chelated Fe ions. It has been proposed that the effect of CDH on lignin degradation is through the reduction of quinones, to be used by ligninolytic enzymes or the support of a Mn-peroxidase reaction (Figure 7.1n).[1]

## 7.4  OXIDATIVE (NONLIGNOCELLULOLYTIC) LIGNOCELLULOSE DEGRADATION

Studies over the last few decades have provided evidence in support of the involvement of nonenzymatic mechanisms in plant cell wall polysaccharide degradation.[1] These mechanisms are mostly assisted by oxidation through the production of free hydroxyl radicals (Figure 7.1o). These hydroxyl radicals are powerful oxidants that can both depolymerize polysaccharides via hydrogen abstraction and attack lignin by demethylation/demethoxylation.[51] Many white- and brown-rot fungi produce hydrogen peroxide which enters the Fenton reaction and results in the release of hydroxyl radicals (Equation 7.14):

$$Fe^{2+} + H_2O_2 + H^+ \rightarrow Fe^{3+} + H_2O + OH^{\cdot} \tag{7.14}$$

By attacking polysaccharides and lignin in a nonspecific manner, these radicals create a number of cleavages which facilitate penetration of the cell wall by lignocellulolytic enzymes. The pathways by which fungi generate free hydroxyl radicals are as follows:

- CDH-catalyzed reactions
- Low-molecular-weight peptides/quinone redox cycling
- glycopeptide-catalyzed Fenton reactions (Figure 7.1o)

### 7.4.1 CELLOBIOSE DEHYDROGENASE

Cellobiose dehydrogenase (CDH) (cellobiose:acceptor 1-oxidoreductase; EC 1.1.99.18), an extracellular protein, is produced in a number of wood- and cellulose-degrading fungi, including basidiomycetes (mostly white-rot fungi) and ascomycetes, growing on cellulosic medium.[1,52] CDH is a monomeric enzyme containing an N-terminal heme domain and a C-terminal flavin domain.[52] This enzyme catalyzes the two-electron oxidation of cellobiose and more generally cellodextrins, mannodextrins, and lactose to the corresponding lactones. Oxidation takes place in the flavin domain where electrons are transferred to *flavin adenine dinucleotide* (*FAD*) and redistributed using electron acceptors such as dioxygen, quinones, phenoxy radicals, or a subsequent one-electron transfer to the heme domain. The heme part is also involved in electron transfer to a wide variety of substrates acting as electron acceptors including quinones, metal ions, and organic dyes. Reduced heme iron may reduce dioxygen to hydrogen peroxide or participate in *Fenton reaction.*

CDH participates to a wide variety of reactions including cellulose and lignin degradation.[31] It is now well accepted that CDH is able to degrade and modify the three major components of biomass by producing free hydroxyl radicals in a Fenton-type reaction.[1]

### 7.4.2 LOW MOLECULAR WEIGHT COMPOUNDS

A variety of low-molecular-weight compounds have been proposed as potential diffusible agents in mechanisms supporting the direct attack of both polysaccharides and lignin or as mediators of ligninolytic enzymes for the indirect attack of lignin.[52]

It has been shown that white- and brown-rot fungi produce low-molecular-weight compounds which are able to penetrate into the cell wall.[1] For example, *G. trabeum* produces a low-molecular-weight peptide, known as short fiber generating factor, which can degrade cellulose into short fibers by an oxidative reaction. Some of these low-molecular-weight compounds are quinones, which must first be converted to hydroquinones by particular fungal enzymes before free hydroxyl radicals can be produced through the Fenton reaction (Figure 7.1o).

### 7.4.3 GLYCOPEPTIDES

*Glycopeptides* have been found in many brown- and white-rot fungi.[1] Similar to other mechanisms, glycopeptides are able to catalyze redox reactions and thus to produce free hydroxyl radicals (Figure 7.1o). It has been postulated that a small glycopeptide secreted by fungi reduces $O_2$ and $Fe^{3+}$ to $H_2O_2$ and $Fe^{2+}$.[53]

## 7.5 SUMMARY

White-rot fungi degrade lignin by secreting extracellular enzymes collectively called lignin-modifying enzymes (LME). Lignin-modifying enzymes are oxidoreductases. They can be classified as:

- Phenol oxidases (laccases; EC 1.10.3.2)
- Ligninolytic peroxidases, which include lignin peroxidases (LiP; EC 1.11.1.14), manganese peroxidases (MnP; 1.11.1.13), and versatile peroxidases (VP; 1.11.1.16)

Phenol oxidases and ligninolytic peroxidases show the following features (Table 7.1).

In general, laccases use diphenols and related substances as donors and molecular oxygen as acceptor while peroxidases use a peroxide as the acceptor.

In addition to lignin-modifying enzymes, other fungal extracellular enzymes which act as accessory enzymes have bound found to be involved in lignin degradation. These include $H_2O_2$-generating oxidases and dehydrogenases.

### TABLE 7.1
#### Features of the Two Main Groups of Fungal Ligninolytic Enzymes

| Enzyme | Reaction |
|---|---|
| Phenol oxidase | 4 benzenediol + $O_2$ ⇆ 4 benzosemiquinone + 2 $H_2O$ |
| Lignin peroxidase | 1,2-bis(3,4-dimethoxyphenyl)propane-1,3-diol + $H_2O_2$ ⇆ 3,4-dimethoxybenzaldehyde + 1-(3,4-dimethoxyphenyl)ethane-1,2-diol + $H_2O$ |
| Manganese peroxidase | 2 Mn(II) + 2 $H^+$ + $H_2O_2$ ⇆ 2 Mn(III) + 2 $H_2O$ |
| Versatile peroxidase | Donor + $H_2O_2$ ⇆ oxidized donor + 2 $H_2O$ |

*Source:* Hatakka, A., *Biodegradation of Lignin*, Biopolymers online, Wiley online, http://www.wiley-vch.de/books/biopoly/pdf/v01_kap05.pdf, 2005.

## 7.6   PERSPECTIVES

Lignin is the most abundant source of aromatic polymer in nature, and its decomposition is indispensable for carbon recycling.[46] The microbial decomposition and recycling of natural polymers such as lignin and polysaccharides have contributed to the development of innovative biotechnological processes in different industries (pulp and paper, textile, chemical synthesis…).[54]

The interest in utilizing ligninolytic enzymes for biotechnological applications has increased rapidly since the discovery of these enzymes in white-rot fungi.[9] Emerging technologies include

- Selective delignification for production of cellulosics in pulp bleaching.
- Conversion of lignocellulosics into multiple energetic and nonenergetic products in a second-generation biorefinery.[55]
- Conversion of lignin extracted from lignocellulosics into low- and high-molecular weight chemicals.[55]
- Treatment of environmental pollutants and toxicants generated in various industrial processes.

Research on enzyme applications has shifted from focusing on a single enzyme preparation to a selective combination of different enzymes or whole organisms.[9]

The understanding of the degradation process has been hampered by substrate complexity and by the multiplicity of enzymes involved. Local environmental factors, such as temperature, moisture, pH, nitrogen, and oxygen levels exert critical influence on the ability of the fungus to degrade lignin. It is also clear that lignin degradation involves not only the four major ligninolytic oxidases, but is a multi-enzymatic process with an array of accessory enzymes. Elucidation of the role and interaction of enzyme combination in lignin degradation are essential to providing a scientific base for the development of application technologies. The advent of DNA technology may provide a powerful tool for solving some of the challenges and allow access to the tremendous biotechnological potential of lignocellulolytic fungi, and particularly white-rot fungi.

A review of the applications of lignin and products derived from its degradation is given in Chapter 9.

## REFERENCES

1. M. Dashtban, H. Schraft, T.A. Syed and W. Qin, *Int. J. Biochem. Mol. Biol.* 1, 36, 2010. http://www.ijbmb.org/files/IJBMB1004005.pdf.
2. M.T. Moreira, G. Feijoo, T. Mester, P. Mayorga, R. Sierra-Alvarez and J.A. Field, *Appl. Environ. Microbiol.* 64, 2409, 1998. http://aem.asm.org/content/64/7/2409.full.
3. A.M. Abdel-Hamid, J.O. Solbiati and I.K. Cann, *Adv. Appl. Microbiol.* 82, 1, 2013. http://www.ncbi.nlm.nih.gov/pubmed/23415151.
4. A. Hatakka, *Biodegradation of Lignin*, Biopolymers online, Wiley online, 2005. http://www.wiley-vch.de/books/biopoly/pdf/v01_kap05.pdf.

5. S.M. Geib, T.R. Filley, P.G. Hatcher, K. Hoover, J.E. Carlson, M. del Mar Jimenez-Gasco, A. Nakagawa-Izumi, R.L. Sleighter and M. Tien, *Proc. Natl. Acad. Sci. USA* 105, 12932, 2008. http://www.pnas.org/content/105/35/12932.full.

6. S. Shary, S.A. Ralph and K.E. Hammel, *Appl. Environ. Microbiol.* 73, 6691, 2007. http://aem.asm.org/cgi/content/full/73/20/6691.

7. J.A. Hoff, N.B. Klopfenstein, J.R. Tonn, G.I. McDonald, P.J. Zambino, J.D. Rogers, T.L. Peever and L.M. Carris, *Roles of Woody Root-Associated Fungi in Forest Ecosystem Processes: Recent Advances in Fungal Identification*, USDA, Research Paper, RMRS-RP-47, 2004. http://www.fs.fed.us/rm/pubs/rmrs_rp047.pdf.

8. D.W.S. Wong, *Appl. Biochem. Biotechnol.* 157, 174, 2009. http://www.udea.edu.co/portal/page/portal/bibliotecaSedesDependencias/unidadesAcademicas/Corporacion AcademicaAmbiental/BilbiotecaDiseno/Archivos/StructureActionMechanismLigninolytic Enzymes.pdf.

9. A.M. Pirttila and A.C. Frank (Eds.), *Endophytes of Forest Trees: Biology and Applications*, Springer, New York, 2011.

10. http://www.brenda-enzymes.org/php/result_flat.php4?ecno=1.11.1.14&UniProtAcc=P 11543&OrganismID=4722&ShowAll=True

11. K.T. Swe, Screening of potential lignin-degrading microorganisms and evaluating their optimal enzyme producing culture conditions, Thesis, Chalmers University of Technology, Göteborg, Sweden, 2011. http://publications.lib.chalmers.se/records/fulltext/154250.pdf.

12. http://www.usbio.net/item/M2195

13. http://www.worldofchemicals.com/wochem/pub/chemversatile-peroxidase.html

14. http://en.wikipedia.org/wiki/Dye_decolorizing_peroxidase

15. D. Salvachua, A. Prieto, A.T. Martinez and M.J. Martinez, *Appl. Environ. Microbiol.* 79, 4316, 2013. http://www.ncbi.nlm.nih.gov/pubmed/23666335 and http://aem.asm.org/content/early/2013/05/06/AEM.00699-13.full.pdf.

16. P.M. Hanna, D.R. Mcmillin, M. Pasenkiewicz-Gierula, W.E. Antholine and B. Reinhammar, *Biochem. J.* 253, 561, 1988. http://www.ncbi.nlm.nih.gov/pmc/articles/PMC1149334.

17. K. Piontek, M. Antorini and T. Choinowski, *J. Biol. Chem.* 277, 37663, 2002. http://www.jbc.org/content/277/40/37663.short?cited-by=yes&legid=jbc;277/40/37663.

18. A.B. Taylor, C.S. Stoj, L. Ziegler, D.J. Kosman and P.J. Hart, *Proc. Natl. Acad. Sci. USA* 102, 15459, 2005. http://www.pnas.org/content/102/43/15459.long.

19. U.C. Davis, ChemWiki, University of California, Laccase, 2008. http://chemwiki.ucdavis.edu/Wikitexts/UCD_Chem_124A%3A_Berben/Laccase/Laccase_1.

20. M. Ferraroni, N.M. Myasoedova, V. Schmatchenko, A.A. Leontievsky, L.A. Golovleva, A. Scozzafava and F. Briganti, *BMC Struct. Biol.* 7, 60, 2007. http://www.biomedcentral.com/content/pdf/1472-6807-7-60.pdf.

21. I. Bento, C.S. Silva, Z. Chen, L.O. Martins, P.F. Lindley and C.M. Soares, *BMC Struct. Biol.* 10, 28, 2010. http://www.biomedcentral.com/1472-6807/10/28.

22. S. Sadhasivam, S. Savitha, K. Swaminathan and F.H. Lin, *Process Biochem.* 43, 736, 2008. http://ntur.lib.ntu.edu.tw/bitstream/246246/128038/1/83.pdf.

23. B.R.M. Reinhammar, *Biochim. Biophys. Acta* 275, 245, 1972. http://www.sciencedirect.com/science/article/pii/000527287290045X.

24. B. Brogioni, D. Biglino, A. Sinicropi, E.J. Reijerse, P. Giardina, G. Sannia, W. Lubitz, R. Basosi and R. Pogni, *Phys. Chem. Chem. Phys.* 10, 7284, 2008. http://pubs.rsc.org/en/Content/ArticleLanding/2008/CP/b812096j.

25. T. Satyanarayana, B.N. Johri and A. Prakash, Eds., *Microorganisms in Sustainable Agriculture and Biotechnology*, Springer, Dordrecht, the Netherlands, 2012.

26. E. Srebotnik and K.E. Hammel, *J. Biotechnol.* 81, 179, 2000. http://www.fpl.fs.fed.us/documnts/pdf2000/srebo00a.pdf.

27. E. Strittmatter, C. Liers, R. Ullrich, S. Wachter, M. Hofrichter, D.A. Plattner and K. Piontek, *J. Biol. Chem.* 288, 4095, 2013. http://www.jbc.org/content/288/6/4095 and http://www.jbc.org/content/early/2012/12/12/jbc.M112.400176.full.pdf.

28. A.N.P. Hiner, J.H. Ruiz, J.N.R. Lopez, F.G. Canovas, N.C. Brisset, A.T. Smith, M.B. Arnao and M. Acosta, *J. Biol. Chem.* 277, 26879, 2002. http://www.jbc.org/content/277/30/26879.long.

29. C. Zubieta, R. Joseph, S.S. Krishna, D. McMullan, M. Kapoor, H.L. Axelrod, M.D. Miller et al., *Proteins* 69, 234, 2007. http://www.ncbi.nlm.nih.gov/pubmed/17654547.

30. C. Zubieta, S.S. Krishna, M. Kapoor, P. Kozbial, D. McMullan, H.L. Axelrod, M.D. Miller et al., *Proteins*, 69, 223, 2007. http://onlinelibrary.wiley.com/doi/10.1002/prot.21550/abstract and http://www.ncbi.nlm.nih.gov/Structure/mmdb/mmdbsrv.cgi?uid=2HAG.

31. N. Busse, D. Wagner, M. Kraume and P. Czermak, *Am. J. Biochem. Biotechnol.* 9, 365, 2013. http://www.thescipub.com/pdf/10.3844/ajbbsp.2013.365.394.

32. Y. Sugano, R. Muramatsu, A. Ichiyanagi, T. Sato and M. Shoda, *J. Biol. Chem.* 282, 36652, 2007. http://www.jbc.org/content/282/50/36652.full#xref-ref-16-1 and http://www.jbc.org/content/282/50/36652.full.pdf.

33. C. Liers, C. Bobeth, M. Pecyna, R. Ullrich and M. Hofrichter, *Appl. Microbiol. Biotechnol.* 85, 1869, 2010. http://link.springer.com/article/10.1007%2Fs00253-009-2173-7.

34. E. Marco-Urrea and C.A. Reddy, *Microbial Degradation of Xenobiotics*, S.N. Singh (Ed.), Springer-Verlag, Berlin, Germany, 2012.

35. http://en.wikipedia.org/wiki/Haem_peroxidase

36. T. Choinowski, W. Blodig, K.H. Winterhalter and K. Piontek, *J. Mol. Biol.* 286, 809, 1999. http://www.sciencedirect.com/science/article/pii/S0022283698925074.

37. K. Piontek, A.T. Smith and W. Blodig, *Biochem. Soc. Trans.* 29, 111, 2001. http://www.biochemsoctrans.org/bst/029/0111/bst0290111.htm.

38. M. Sundaramoorthy, M.H. Gold and T.L. Poulos, *J. Inorg. Biochem.* 104, 683, 2010. http://www.ncbi.nlm.nih.gov/pubmed/20356630.

39. M. Hofrichter, *Enzyme Microb. Technol.* 30, 454, 2002.

40. http://www.jenabioscience.com/images/207bdf791d/EN-203.pdf

41. F.J. Ruiz-Duenas, M.J. Martinez and A.T. Martinez, *Mol. Microbiol.* 31, 223, 1999. http://onlinelibrary.wiley.com/doi/10.1046/j.1365-2958.1999.01164.x/pdf.

42. M. Perez-Boada, F.J. Ruiz-Duenas, R. Pogni, R. Basosi, T. Choinowski, M.J. Martinez, K. Piontek and A.T. Martinez, *J. Mol. Biol.* 354, 385, 2005. http://www.sciencedirect.com/science/article/pii/S0022283605011137.

43. F.J. Ruiz-Duenas, M. Morales, M. Perez-Boada, T. Choinowski, M.J. Martinez, K. Piontek and A.T. Martinez, *Biochemistry* 46, 66, 2007. http://pubs.acs.org/doi/full/10.1021/bi061542h.

44. http://www.pdb.org/pdb/explore.do?structureId=2BOQ

45. D.W. Wong, *Appl. Biochem. Biotechnol.* 157, 174, 2009. http://www.ncbi.nlm.nih.gov/pubmed/18581264.

46. D.I. Colpa, M.W. Fraaije and E. van Bloois, *Ind. Microbiol. Biotechnol.* 41, 1, 2014. http://www.deepdyve.com/lp/springer-journals/dyp-type-peroxidases-a-promising-and-versatile-class-of-enzymes-e4hiL0QVJa and http://link.springer.com/article/10.1007/s10295-013-1371-6#page-1.

47. http://www.rcsb.org/pdb/explore.do?structureId=2GVK

48. http://en.wikipedia.org/wiki/Ferredoxin_fold

49. http://www.ebi.ac.uk/thornton-srv/databases/cgi-bin/pdbsum/GetPage.pl?pdbcode=2gvk

50. http://www.rcsb.org/pdb/explore.do?structureId=4au9

51. L. Jouanin and C. Lapierre, Eds., *Advances in Botanical Research, Lignins: Biosynthesis, Biodegradation and Bioengineering*, J.P. Jaquot and P. Gadal, Series Eds., Academic Press, London, 2012.

52. S.D. Mansfield, E. de Jong and J.N. Saddler, *Appl. Environ. Microbiol.* 63, 3804, 1997. http://www.ncbi.nlm.nih.gov/pmc/articles/PMC1389261/pdf/hw3804.pdf.
53. L. Jouanin and C. Lapierre, Eds., *Advances in Botanical Research, Lignins: Biosynthesis, Biodegradation and Bioengineering*, Academic press, Elsevier, London, 2012.
54. C. Liers, T. Arnstadt, R. Ullrich and M. Hofrichter, *FEMW, Microbiol. Ecol.* 78, 91, 2011. http://onlinelibrary.wiley.com/doi/10.1111/j.1574-6941.2011.01144.x/full.
55. J.L. Wertz and O. Bedue, *Lignocellulosic Biorefineries*, EPFL Press, Lausanne, Switzerland, 2013.

# 8 Pretreatments of Lignocellulosic Biomass

## 8.1 INTRODUCTION

Biofuels and biobased products generated from various lignocellulosic materials, such as wood, agricultural, or forest residues, have the potential to be valuable substitutes for fossil-based products.[1] Many structural and compositional factors hinder the hydrolysis of polysaccharides present in biomass to sugars that can later be converted to useful energy and nonenergy products.[2] Overcoming the recalcitrance of lignocellulosic biomass is a key step in the production of biofuels and biochemicals.[3] The recalcitrance is mainly due to the crystalline structure of cellulose that is embedded in a matrix of lignin and hemicelluloses. The main goal of pretreatment is to overcome this recalcitrance, to separate cellulose, hemicelluloses, and lignin, and to make polysaccharides more accessible not only for enzymatic hydrolysis but also for obtaining high added value bioproducts.[3,4] This goal is achieved by (1) degrading and removing hemicelluloses and lignin, (2) reducing the crystallinity of cellulose, and (3) increasing the porosity of the lignocellulosic material (Figure 8.1).[5]

Pretreatment technologies aiming at producing, for example, bioethanol must meet the following requirements: (1) improve the formation of sugars or the ability to subsequently form sugars by hydrolysis, (2) avoid the degradation or loss of carbohydrates, (3) avoid the formation of byproducts that are inhibitory to the subsequent

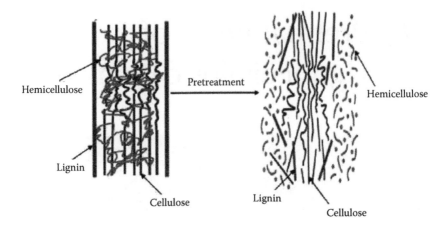

**FIGURE 8.1** Schematic of the role of pretreatment in the conversion of lignocellulosic biomass to biofuels and biobased chemicals. (From Brodeur, G. et al., *Enzyme Res.*, 2011, 787532, http://www.ncbi.nlm.nih.gov/pmc/articles/PMC3112529/#!po=3.12500, 2011.)

hydrolysis and fermentation processes, (4) allow lignin recovery to generate valuable coproducts, and (5) be cost effective.[1,4]

During the last years, a variety of pretreatment methods that change the physical and chemical structures of the components of lignocellulosic biomass and improve hydrolysis rates have been developed. Many methods have been shown to result in high sugar yields, above 90% of the theoretical yield for lignocellulosic biomasses such as woods, grasses, corn, and so on. In this chapter, the objective is to give an overview of the various available pretreatment technologies, and we review the recent literature on the use of these technologies for pretreatment of lignocellulosic biomass.

Among the various pretreatment methods, liquid hot water (hydrothermal), organosolv (organic solvents), steam explosion (used in DuPont's and Poet/DSM's cellulosic biorefineries in the United States), acid hydrolysis, and alkaline hydrolysis are involved at a demonstration or industrial scale.

## 8.2 THE DIFFERENT PRETREATMENT METHODS

### 8.2.1 MECHANICAL COMMINUTION

The purpose of comminution and radiation pretreatments (see Section 8.2.2 for radiations) is mainly the increase of the accessible surface area and the size of pores in the lignocellulosic structures, and the decrease of crystallinity degree and polymerization degree of cellulose.[4] The power requirement of these pretreatments is relatively high and depends on the type of biomass and on the final particle size.

Reduction of particle size, that is, comminution, is often needed to make material handling easier and to increase surface/volume ratio.[5] Biomass materials can be comminuted by various chipping, grinding, and milling.[6] Mechanical pretreatment is usually carried out before a following processing step, and the desired particle size is dependent on these subsequent steps. The comminution is time consuming, energy intensive, and expensive; furthermore, it does not remove lignin, which is known to restrict access of enzymes to cellulose.

## 8.2.2 RADIATIONS

Digestibility of cellulosic biomass has been enhanced by the use of radiation methods, including $\gamma$-rays, electron beam, pulsed electrical field, UV, and microwave.[6] The action mode behind radiation could be one or more changes in biomass structure, such as increase in specific surface area, decrease in polymerization degree and crystallinity of cellulose, hydrolysis of hemicelluloses and partial depolymerization of lignin. However, these radiation methods are usually slow, energy intensive, and expensive. They also appear to be substrate specific.

Microwaves are electromagnetic waves with frequencies ranging from 300 MHz to 300 GHz.[7] When these waves interact with organic matter, they get absorbed by water, fats, and sugars, and their energy gets transferred to organic molecules, generating heat. In this way, microwaves exhibit heating effect. Microwaves are also employed in the pretreatment of biomass. Microwave treatment of biomass can lead to high lignin removal and increased sugar yields after hydrolysis. However, the process generates high temperature and nonuniform heating of biomass, which leads to the formation of inhibitors.

Comminution and radiation pretreatments are often inadequate in providing complete fractionation to a readily digestible and fermentable product.[8] The cell structure of lignocellulosic biomass is by nature complex and difficult to penetrate, so fractionation requires chemical reactions in addition to physical destructuring. Sections 8.2.3 through 8.2.8 describe chemical pretreatments responsible for rupture of lignocellulosic biomass.

## 8.2.3 LIQUID HOT WATER

Liquid hot water processes (or hydrothermal processes) are biomass pretreatments with water at high temperature (160°C–240°C) and pressure.[5] Water treatments use pressure to maintain the water in the liquid state at elevated temperature.[2] This type of pretreatment by liquid water has been termed hydrothermolysis, aqueous fractionation, *aquasolv*, *autohydrolysis*, *solvolysis*, hydrothermal pretreatment, subcritical water, and so on.[4,9,10]

Mok et al. studied solvolysis of biomass samples with hot compressed liquid water for 0–15 min at 200°C–230°C.[11] Between 40% and 60% of the total biomass were solubilized. In all cases, 100% of the hemicelluloses were solubilized, of which 90% were recoverable as monomeric sugars. Between 4% and 20% of the cellulose and between 35% and 60% of the lignin were also solubilized. Variability in results was related to the biomass type. High lignin solubilization apparently reduced recovery of hemicellulose sugars.

Liquid hot water pretreatments are both helped and hindered by the cleavage of *O*-acetyl and *uronic acid* substitutions from hemicelluloses to generate acetic acid and other organic acids.[2] The release of these acids helps catalyze the hydrolysis into oligosaccharides, monosaccharides, and their degradation products. However, the

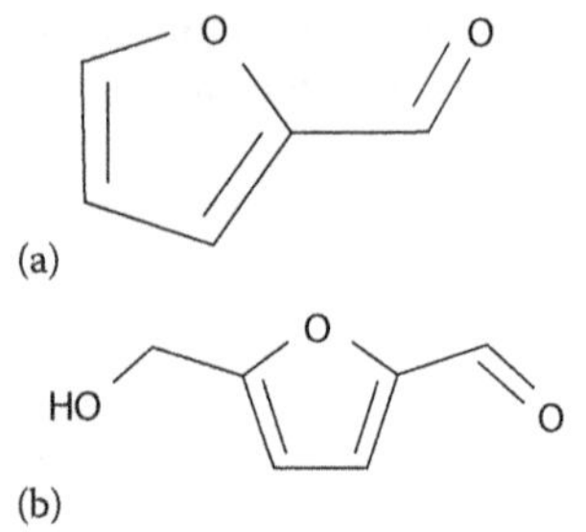

(a)

HO

(b)

**FIGURE 8.2**  2-Furfural structure (a) and 5-hydroxymethylfurfural (b), inhibitors of fermentation.

monomeric sugars are partially degraded to aldehydes if acid is used. These aldehydes, mainly *furfural* from dehydration of pentoses and 5-hydroxymethylfurfural from dehydration of hexoses, are inhibitory to microbial fermentation (Figure 8.2).

There are three types of liquid hot water reactor configurations: cocurrent, countercurrent, and flow through (Figure 8.3).

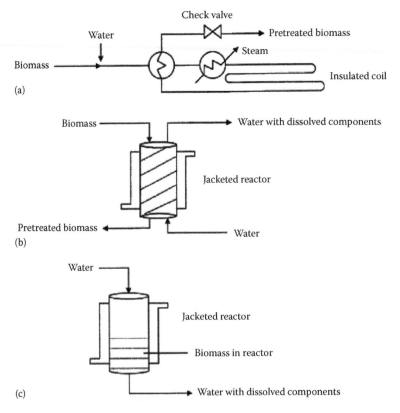

**FIGURE 8.3**  Schematic illustration of (a) cocurrent liquid hot water pretreatment, (b) countercurrent reactor, and (c) flow-through reactor. (From Mosier, N. et al., *Biores. Technol.* 96, 673, http://www.sciencedirect.com/science/article/pii/S0960852404002536, 2005.)

In cocurrent pretreatment, slurry of biomass and water is heated to the desired temperature and held at the pretreatment conditions for a controlled residence time before being cooled. Countercurrent pretreatment is designed to move water and lignocellulose in opposite directions through the pretreatment reactor. In a flow-through system, hot water passes over a stationary bed of lignocellulose, hydrolyzes, dissolves lignocellulose components, and carries them out of the reactor.

In summary, the characteristics of liquid hot water can be described as follows:

- *Effect on cellulose*: Minor effect.
- *Effect on hemicelluloses*: Removal (major).
- *Effect on lignin*: Alteration (minor).

The Danish company Inbicon, subsidiary of Dong Energy, has developed a patented hydrothermal pretreatment, which is integrated in a three-stage process: mechanical conditioning of the biomass, hydrothermal pretreatment, and enzymatic hydrolysis.[12] Core Inbicon technology solubilizes lignin and unlocks cellulose and hemicelluloses to be converted into sugars.

Another important feature of the Inbicon's pretreatment is to separate the alkali content (particularly Na and K) and a portion of the inhibitors formed through partial degradation of the hemicelluloses. The process operates at high (30%–40%) dry matter content, minimizing water, and energy consumption. The biomass is fed into the pressure reactor with proprietary particle pumps.

The Inbicon demonstration biorefinery at Kalundborg (DK) can be described by the following facts[13]:

Raw materials:

- 4 t/h equivalent to ~30,000 tons of straw a year.
- Enzymes supplied by DuPont, Novozymes, and DSM.

Annual production:

- 5.4 million liters (1.4 million gallons) of cellulosic ethanol.
- 11,400 tons of lignin pellets.
- 13,900 tons of C5 molasses.

Employees: 30
Total construction costs: ~54 million EUR.

## 8.2.4 Acid Hydrolysis

The dilute acid pretreatment of lignocellulosic biomass has received considerable research attention.[2,14] Hemicelluloses are the first of the constituents of biomass to break down during acid hydrolysis. For example, when sulfuric acid is used, 0.01 M is generally enough to break down the hemicelluloses into their monomers.

Dilute sulfuric acid has been added to lignocellulosic materials for some years to commercially produce furfural.[15] Dilute sulfuric acid is mixed with biomass to hydrolyze hemicelluloses to xylose and other sugars and then to dehydrate xylose to form furfural. The furfural is recovered by distillation.

The biomass/acid mixture is held at temperatures of 160°C–220°C for periods ranging from seconds to minutes.

The addition of sulfuric acid has been initially applied to hydrolyze hemicelluloses either in combination with breakdown of cellulose to glucose or prior to acid hydrolysis of cellulose.[2] Hemicelluloses are removed when sulfuric acid is added, and this enhances digestibility of cellulose in the residual solids. Most approaches are based on dilute sulfuric acid, but nitric acid, hydrochloric acid, and phosphoric acid have also been tested. Furthermore, it is worth mentioning the autocatalyzed acid hydrolysis, a process in which no acid is added. The process solely depends on the hydrolysis of acetic esters present in biomass which form acetic acid. The latter becomes the acid during the hydrolysis of polysaccharides.

Acid hydrolysis releases oligomers and monosaccharides and has historically been modeled as a homogeneous reaction in which acid catalyzes breakdown of cellulose to glucose followed by breakdown of the glucose to form hydroxymethylfurfural and other degradation products.[16] This reflects the approximately equal reactivity of glycosidic bonds with respect to hydrolysis. The model was adapted to describe the hydrolysis of hemicelluloses and formation of furfural and other decomposition products.

As an alternative to inorganic acids, organic acids, such as maleic acid and fumaric acid, can be used for dilute acid pretreatment.

Dilute acid treatment is one of the most effective pretreatment methods for lignocellulosic biomass.[1,5] Two types of dilute-acid treatment processes are typically used:

1. A high-temperature (T > 160°C), continuous-flow process for low solids loadings (5–10 wt% substrate concentration).
2. A low-temperature (T < 160°C) batch process for high solids loadings (10–40 wt% substrate concentration).

Dilute acid (mostly sulfuric acid) is mixed with lignocellulosic material, and the mixture is held at the proper temperature for short periods up to a few minutes.[5] Hydrolysis of hemicelluloses then occurs, releasing monomeric sugars and soluble oligomers from the cell walls into the hydrolysate. Hemicelluloses removal increases porosity and enhances enzymatic digestibility, with maximum digestibility usually coinciding with complete hemicellulose removal.[17]

The treatment offers good performance in terms of recovering hemicellulose sugars but has some important limitations including corrosion that mandates expensive material of construction.[2,5] The acid must be neutralized before the sugars proceed to fermentation. Furthermore, the formation of degradation products and release of biomass fermentation inhibitors are other characteristics of acid treatment. The pretreatment is especially suitable for biomass with low lignin content, as almost no lignin is removed by the pretreatment.

In summary, the characteristics of dilute acid hydrolysis can be described as follows:

- *Effect on cellulose*: Minor effect.
- *Effect on hemicelluloses*: Removal (major).
- *Effect on lignin*: Alteration (minor).
- Formation of fermentation inhibitors.

## 8.2.5   ALKALINE HYDROLYSIS

The general principle behind alkaline pretreatment methods is the removal of lignin, whereas cellulose and a portion of hemicelluloses remain in the solid fraction. Alkaline treatment of woody biomass is well known in the pulp and paper industry as Kraft pulping (also known as Kraft process or sulfate process).

### 8.2.5.1   Kraft Process

After grinding, and sometimes steam activation, wood chips are cooked with sodium hydroxide and sodium sulfide (known as *white liquor*) at temperatures above 150°C.[18,19] Kraft treatment conditions degrade and dissolve readily lignin molecules by alkaline cleavage of ether bonds.[20] The reaction starts with an alkaline rearrangement of a phenolate to a quinone methide with cleavage of α-aryl ether bonds (Figure 8.4).[21]

A nucleophile ($HS^-$ in Kraft pulping) adds to the quinone methide with the subsequent cleavage of the adjacent β-*O*-4 ether bond (Figures 8.5 and 8.6).

However, the cooking conditions for global solubilization of lignin are also favorable to cellulose and glucomannan (hemicellulose) degradation, mainly by peeling oxidation reaction.[21,23] Therefore, delignification goes together with carbohydrate decomposition, and the final yield of the reaction is significantly lower than the theoretical target.[21,24] Additives like anthraquinone (AQ) are known to favor lignin solubilization.

To avoid environmental impact and reduce cost, the Kraft process is carried out with a sodium sulfide recycling loop. In the process, about half of the wood is dissolved, and together with the spent pulping chemical, forms a liquid stream called weak *black liquor*.[26] This black liquor, which is separated from the pulp by washing, contains lignin fragments, carbohydrates, sodium carbonate, sodium sulfate, and other inorganic salts. The weak black liquor is concentrated by evaporation to 65 or even 80% solids (heavy black liquor) and burned in the *recovery boiler* to recover the inorganic chemicals for reuse in the pulping process (Equation 8.1).

$$Na_2SO_4 + 2\,C \text{ (organic carbon in the mixture)} \rightarrow Na_2S + 2\,CO_2 \qquad (8.1)$$

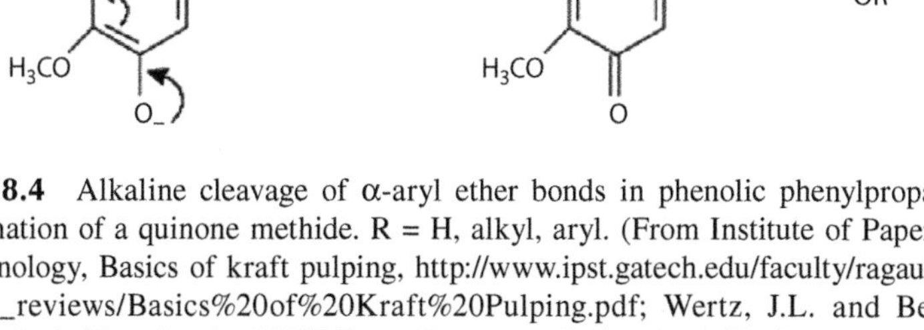

**FIGURE 8.4**   Alkaline cleavage of α-aryl ether bonds in phenolic phenylpropane units with formation of a quinone methide. R = H, alkyl, aryl. (From Institute of Paper Science and Technology, Basics of kraft pulping, http://www.ipst.gatech.edu/faculty/ragauskas_art/technical_reviews/Basics%20of%20Kraft%20Pulping.pdf; Wertz, J.L. and Bedue, O., *Lignocellulosic Biorefineries*, EPFL Press, Lausanne, Switzerland, 2013.)

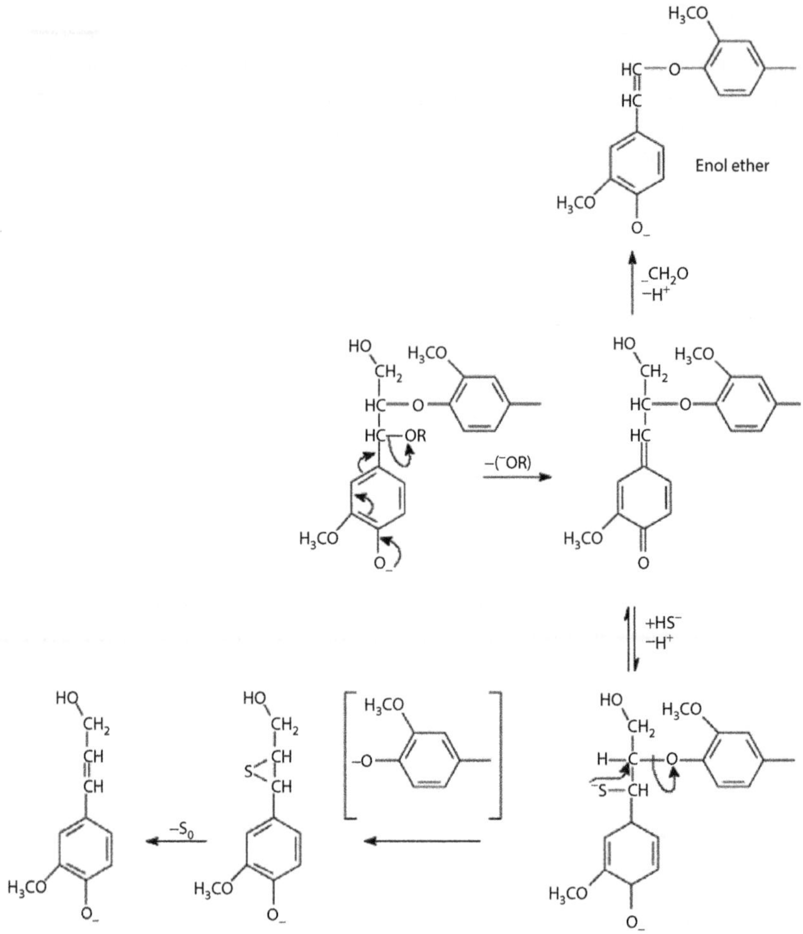

**FIGURE 8.5** Mechanism of lignin degradation in kraft pulping: sulfidolytic cleavage of β-aryl ether bonds in phenolic phenylpropane units and conversion into enol-ether units. (From Institute of Paper Science and Technology, Basics of kraft pulping, http://www.ipst.gatech.edu/faculty/ragauskas_art/technical_reviews/Basics%20of%20Kraft%20Pulping.pdf.)

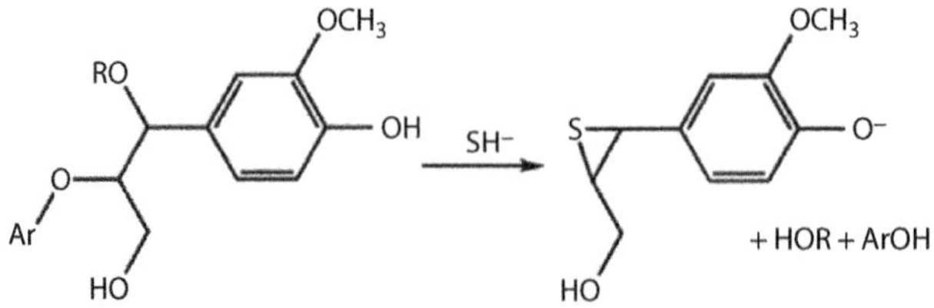

**FIGURE 8.6** Net reaction in depolymerization of lignin by SH⁻ (Ar = aryl; R = alkyl groups). (Courtesy of R.J.A. Gosselink, https://www.researchgate.net/profile/Richard_Gosselink/publication/254833784_Lignin_as_a_Renewable_Aromatic_Resource_for_the_Chemical_Industry/links/55c3205f08aebc967defeeb2/Lignin-as-a-Renewable-Aromatic-Resource-for-the-Chemical-Industry.pdf.)

NaOH that remains in the black liquor will create sodium carbonate when it reacts with carbon dioxide. The molten salts ($Na_2S$ and $Na_2CO_3$) from the recovery boiler are dissolved in a process water to form *green liquor* which consists mostly of $Na_2S$ and $Na_2CO_3$. This liquid is mixed with CaO (lime), which becomes $Ca(OH)_2$ (slake lime) in solution, to regenerate the white liquor used in the pulping process (Equation 8.2).

$$Na_2S + Na_2CO_3 + Ca(OH)_2 \leftrightarrow Na_2S + 2\,NaOH + CaCO_3 \qquad (8.2)$$

The precipitated $CaCO_3$ (lime mud) from the causticizing reaction is washed and sent to a lime kiln where it is heated to regenerate CaO for reuse (Equation 8.3).

$$CaCO_3 \rightarrow CaO + CO_2 \qquad (8.3)$$

Finally, CaO is reacted with $H_2O$ to regenerate $Ca(OH)_2$ (Equation 8.4).

$$CaO + H_2O \rightarrow Ca(OH)_2 \qquad (8.4)$$

These recovery steps can be strongly perturbed by the precipitation of sodium silicate (amphoteric salt) obtained from silica present in lignocellulosic raw materials. As silica content goes from 1 kg/t in wood, to about 20 kg/t in bamboo, and 50–150 kg/t in wheat or rice straw, kraft process is not applicable for most nonwoody lignocellulosic residues.

Today, the idea of transforming a standard kraft pulping mill into an integrated biorefinery becomes a reality.[27,28] An advantage is that the conversion of a kraft mill from 100% pulp making to 100% biorefinery can be done in a stepwise fashion. One biorefinery concept involves extracting hemicelluloses, which normally end up in the black liquors of kraft pulp mills, prior to pulping, and using the extract for the production of acetic acid and ethanol.[29]

### 8.2.5.2   LignoBoost

Developed by Chalmers University of Technology (Sweden) and Innventia (Sweden), the patented LignoBoost process is designed to extract lignin from pulp mill black liquor (Figure 8.7).[30] The process is now owned by Valmet, previously Metso's Pulp, Paper, and Power businesses.

LignoBoost responds to the growing interest for utilizing biobased raw materials. By treating the black liquor with carbon-dioxide and a strong acid, the lignin is precipitated, then washed, and dried. Lignin has a heating value similar to carbon. It can be used as fuel in the lime kilns. It can also be sold for new material applications to external customers.

Metsos's Pulp, Paper, and Power businesses have supplied a LignoBoost lignin separation plant to Swedish-Finish Stora Enso's Sunila mill in Finland. The order is one part of Stora Enso's €32 million biorefinery project. The start-up was scheduled for the first quarter of 2015, and Stora Enso has now started lignin production.[31] The LignoBoost plant is to be integrated with the pulp mill to separate and collect lignin from the black liquor.

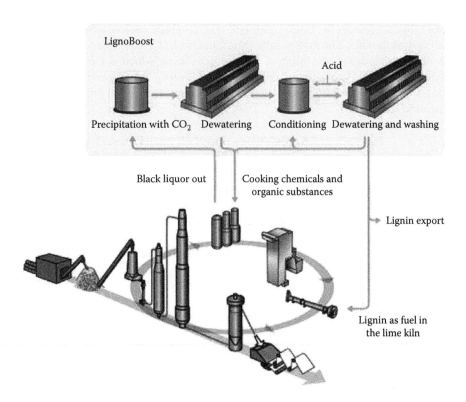

**FIGURE 8.7** Extraction of lignin from kraft pulp mill black liquor by the LignoBoost process. (From http://www.valmet.com/pulp/chemical-recovery/lignin-separation/lignoboost-process/; Courtesy of Jean-Luc Wertz.)

### 8.2.5.3 Alkaline Pretreatment

Several authors evaluated the ability of sodium hydroxide solutions in different concentrations to enhance enzymatic digestibility of corn stover, sorghum straw, wheat straw, cotton stalks, or switch grass.[32-34] Treatment is applied alone or with addition of irradiation, such as microwave (300 MHz–300 GHz) and ultrahigh frequency (UHF, 300 MHz–3 GHz).[35,36]

Pedersen et al.[37] assessed the influence of pretreatment pH, temperature, and time, and their interactions on the subsequent enzymatic glucose and xylose yields from mildly pretreated wheat straw. They concluded that pretreatment pH exerted significant effects and factor interactions on the enzymatic glucose and xylose releases. Quite extreme pH values (pH 1 or pH 13) were necessary with mild thermal pretreatment conditions (T < 140°C, t < 10 min). Alkaline pretreatments generally induced higher sugar release during enzymatic hydrolysis and did so at lower temperatures than required with acidic pretreatments.

As glucomannans are much less stable in alkaline media than xylans, the proportion of remaining hemicelluloses in the solid part may vary significantly with the composition of lignocellulosic raw material. Alkaline treatments are well known to be more efficient on hardwoods and straw than on softwood. In any case, further enzymatic treatment requires hemicellulases both acting on dissolved and solid hemicelluloses fractions.[38]

By using sodium hydroxide or calcium hydroxide (slaked lime), salts are formed, which may be incorporated in the biomass and need to be removed or recycled.[5] Slaked lime has the additional benefits of low reagent cost and safety and being recoverable from water as insoluble calcium carbonate by reaction with carbon dioxide. The carbonate can then be converted to lime (CaO or burnt lime) using established lime kiln technology ($CaCO_3$ + heat → CaO + $CO_2$).

Slaked lime treatments have been applied to different cellulosic materials (poplar, straw, bagasse, corn stover, and switchgrass), showing subsequent enzymatic digestion similar to or even better than after acidic treatments.[39–43] Like for soda, treatment efficiently solubilizes lignins and removes acetyl and uronic acid substitutions on hemicellulosic compounds.

Essentially developed at the laboratory scale by Holtzapple et al.[44–46] the lime activation process (MixAlco process), including a pretreatment phase, was deployed by the U.S. company Terrabon which declared bankruptcy in 2012.[47,48] The process converts biomass such as municipal solid waste, sewage sludge, forest product residues, and energy crops into valuable chemicals such as acetic acid, ketones, and alcohols that can be processed into renewable hydrocarbon fuels.

The biomass is treated with lime and air to enhance digestibility.[50] It is then fed to a mixture culture of microorganisms that produce carboxylic acids, which are transformed into salts by addition of calcium carbonate. After drying, the salts are acidified and thermally converted into ketones. Finally, the ketones are hydrogenated to alcohols. Two different versions of the MixAlco process are available. The original process produces mixed alcohol fuels. The second version produces carboxylic acids and primary alcohols (ethanol).

As a conclusion, the characteristics of alkaline pretreatment can be summarized as follows:

- *Effect on cellulose*: No.
- *Effect on hemicelluloses*: Removal (minor).
- *Effect on lignin*: Removal (major).

## 8.2.6 ORGANOSOLV/SOLVENT FRACTIONATION

Organosolv pretreatment processes use an organic solvent or mixtures of organic solvents with water to remove lignin before enzymatic hydrolysis of the cellulosic fraction.[5] Common solvents for the processes include ethanol, methanol, acetone, and ethylene glycol. Temperatures used for the processes can be as high as 200°C, but lower temperatures can be sufficient depending on, for example, the type of biomass and the use of a catalyst. Possible catalysts include inorganic and organic acids.

Originally developed in the years 1980–1990 as an *environmentally friendly* alternative to kraft and sulfite pulping, this process received regain of interest recently with the development of biorefineries. Main advantages are the absence of odoriferous sulfur compounds, increase of yield compared with kraft pulp process, and possibility to apply to all feedstocks; main drawbacks are energy consumption and complicated recovery systems for solvent.

Main reactions during organosolv cooking are hydrolysis of ether linkages in lignin and hemicellulose–lignin bonds.[21,49] $\alpha$-ether bonds are more readily broken, but it is likely that $\beta$-aryl ether bonds are also broken under the conditions of many processes. Most of the organosolv processes use either a neutral solvent, with or without an acid added catalyst, or an acidic solvent. In cases where no acid is added, the liquor becomes acidic as a result of release of acetic acid from hemicelluloses. Under these conditions, neutral or acidic solvolysis reactions of lignin may be expected to occur.

Model compounds have shown that $\alpha$-aryl ether linkages are more easily split that $\beta$-aryl ether linkages, especially when they occur in a lignin structural unit containing a free hydroxyl group in the para position.[52] In this case, the formation of a quinone methide intermediate is possible (Figure 8.8). Otherwise, a nucleophilic substitution reaction occurs at the benzylic position (Figure 8.9). In the presence of added acid, the hydrolysis of benzyl ether linkages is facilitated by acid catalysis as shown in Figure 8.9.

The likelihood of $\beta$-ether cleavage is greater in more acidic systems.[52] Formation of either a resonance-stabilized benzyl carbocation or a similarly stabilized transition state is followed by elimination of water to form the readily hydrolyzed enol ether (Figure 8.10). This releases guaiacol and forms $\beta$-hydroxyconiferyl alcohol, which exists in equilibrium with its keto form. In general, hardwoods are delignified more easily than softwoods, mainly due to differences in $\beta$-ether reactivity, $\alpha$-ether concentration, lignin concentration, and propensity to undergo condensation reactions.

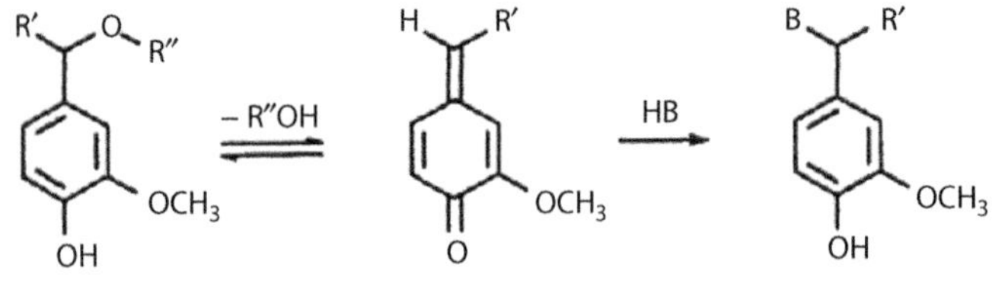

**FIGURE 8.8**  Solvolytic cleavage of a phenolic $\alpha$-aryl ether linkage via a quinone methide intermediate. B = OH, $OCH_3$. (From Alvira, P. et al., *Bioresour. Technol.*, 101, 4851, http://www.scribd.com/doc/53790135/Pre-Treatment-Technologies-for-an-Efficient-Bio-Ethanol-Production-Process, 2010.)

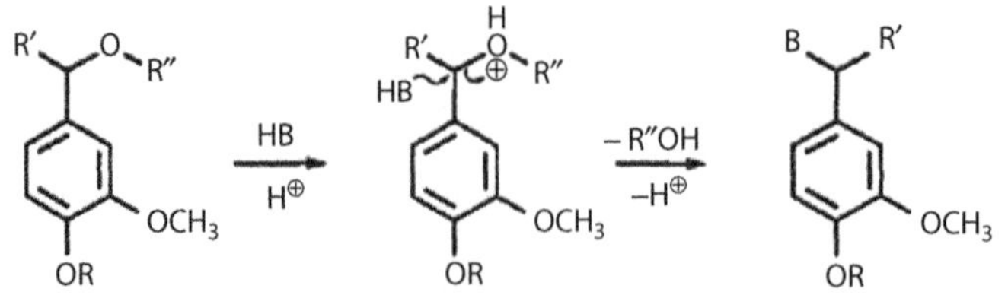

**FIGURE 8.9**  Solvolytic cleavage of an $\alpha$-aryl ether linkage by nucleophilic substitution. R = H or $CH_3$; B = OH, OCH3, and so on. (From Alvira, P. et al., *Bioresour. Technol.*, 101, 4851, http://www.scribd.com/doc/53790135/Pre-Treatment-Technologies-for-an-Efficient-Bio-Ethanol-Production-Process, 2010.)

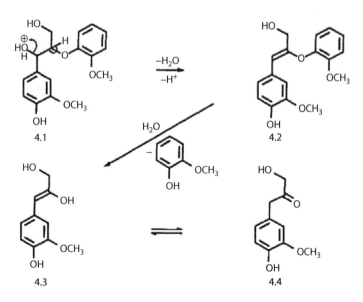

**FIGURE 8.10** Solvolytic cleavage of a β-aryl ether linkage to form β-hydroxyconiferyl alcohol, which is in equilibrium with its keto form. (From Alvira, P. et al., *Bioresour. Technol.*, 101, 4851, http://www.scribd.com/doc/53790135/Pre-Treatment-Technologies-for-an-Efficient-Bio-Ethanol-Production-Process, 2010.)

Organosolv pulping in alkaline systems is less distinct from conventional processes than are the autocatalyzed and acidic organosolv processes.[52] Unlike the latter, the alkaline systems would remain viable pulping processes if the nonaqueous component were omitted. In alkaline systems, α-ether is readily cleaved if they occur in units containing free phenolic groups but not otherwise. Alkaline cleavage of β-ethers also occurs, especially in the presence of nucleophilic additives such as hydrosulfide ion or an anthraquinone derivative.

In a review related to organosolv pulping, Muurinen[50] concluded in 2000 that organosolv pulping processes are still in a developing stage and are not yet ready to seriously threat the leading position of the kraft process. This is still true today. Distillation seems to be the main option as a process for recovering the solvent in organosolv pulping. The main organosolv processes are summarized in Table 8.1. Some of them were exploited industrially in the 1990s.

Recent reviews show the growing interest of organosolv processes for pretreatment of biomass.[51–53] The organosolv processes with acetic acid (Acetosolv), formic acid (Formosolv), or peroxyformic acid (Milox) have been tested on miscanthus, showing efficient routes for fractionation.[54]

Some studies utilize high-boiling-point organic solvents, such as glycerol, but most of the literature is based on ethanol–water mixes, eventually with acidic catalysis and sometimes coupled with steam explosion or mechanical treatments.[55] Ethanol–water treatment at high temperature (~200°C) under pressure produces spontaneously carboxylic acids, but addition of sulfuric or hydrochloric acid increases delignification rate and xylan degradation.[56] Studies using the ethanol organosolv process show efficient conversion to biofuel and valuable coproducts.[57,58]

## TABLE 8.1
## Main Organosolv Processes Developed for Pulp Production

| Process Name | Solvent System |
|---|---|
| Asam | Water + alkaline sulfide + anthraquinone + methanol |
| Organocell | Water + sodium hydroxide + methanol |
| Alcell (APR[a]) | Water + low aliphatic alcohol such as ethanol (Lignol process) |
| Milox | Water + formic acid + hydrogen peroxide (forming peroxyformic acid) |
| Acetosolv | Water + acetic acid + hydrochloric acid |
| Acetocell | Water + acetic acid |
| Formacell | Water + acetic acid + formic acid (CIMV process) |
| Formosolv | Water + formic acid + hydrochloric acid |
| Battelle-Geneva | Water + phenol + acid catalyst |
| Acos | Water + acetone + mineral acid |
| MEA | Water + monoethanolamine |

*Source:* Chen, H. and Qiu, W., *Biotechnol. Adv.* 28, 556, http://www.ncbi.nlm.nih.gov/pubmed/20546879, 2010.

[a] Alcohol pulping and recovery.

### 8.2.6.1  Lignol

Canadian Lignol is an emerging producer of biofuels, biochemicals, and renewable materials from waste biomass.[59] It is undertaking the development of biorefining technologies based on former Alcell (ethanol/water) process (Table 8.1). The objective is to build a completely flexible biorefinery, able to treat different feedstocks and produce fuel ethanol and valuable chemicals.

### 8.2.6.2  CIMV

The CIMV company (Compagnie Industrielle de la Matière Végétale) has patented acetic acid/formic acid/water process for the fractionation of lignocellulosic biomass.[61,62] The CIMV organosolv process cogenerates, besides cellulose, partially depolymerized hemicelluloses and sulfur-free, low-molecular-weight lignins. In preliminary studies, cellulose extracted from the CIMV process appeared to be highly amenable to cellulase liquefaction/saccharification, with more than 90% of glucose being recovered in monomeric form. The CIMV process, such as biomass fractionation, sugar and lignin extraction, and separation of sugars and lignin, is illustrated in Figure 8.11.[63,64]

CIMV biomass fractionation process has been at the heart of the biorefinery concept developed by the EU Biocore project.[66,67]

As a summary, the effects of organosolv treatments on biomass can be described as follows:

- *Effect on cellulose*: No.
- *Effect on hemicelluloses*: Removal (minor).
- *Effect on lignin*: Removal (major).
- High-quality lignin produced.

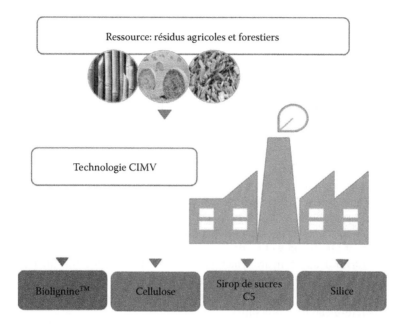

**FIGURE 8.11**  CIMV organosolv process. (From CIMV, http://www.cimv.fr/technologie/technologie-cimv/technologie-cimv.html.)

### 8.2.7  OXIDATIVE DELIGNIFICATION

Delignification of lignocellulose can be achieved by a treatment with an oxidizing agent.[5] The effectiveness of delignification can be attributed to the high reactivity of oxidizing agents with the aromatic ring. The oxidative opening of the aromatic rings leads to compounds such as carboxylic acids. On account of these acids act as inhibitors in the fermentation step, they have to be neutralized or removed. In addition to an effect on lignin, oxidative treatment also affects the hemicellulose component of the lignocellulosic biomass. A significant part of the hemicelluloses might be degraded and cannot be used anymore for sugar production.

#### 8.2.7.1  Hydrogen Peroxide

An oxidative agent commonly used is hydrogen peroxide.[5] Dissolution of ~50% of lignin and most of the hemicelluloses has been achieved in a solution of 2% hydrogen peroxide at 30°C. The yield of subsequent enzymatic hydrolysis can be as high as 95%.

There is not much literature on peroxide treatment, but it is generally applied in alkaline conditions (pH 10–14). Alkaline hydrogen peroxide treatment has been mostly studied on nonwoody materials (corn stover,[68] different types of grass and straw,[69,70] miscanthus[71]). Conditions may change slightly but are generally mild: temperature of 20°C–25°C, biomass loading of 1%–15%, hydrogen peroxide loading of 0.1–0.5 g/g biomass, and durations of 24–48 h. Authors report good results concerning subsequent enzymatic digestibility improvement, even compared with dilute acid, lime, or AFEX pretreatments.[72,73]

### 8.2.7.2   Ozonolysis

Ozone treatment degrades preferentially lignin by attacking and cleaving aromatic ring structures.[5] It can be used to disrupt the structure of many lignocellulosic materials, such as wheat straw, bagasse, pine, peanut, cotton straw, and poplar sawdust.

### 8.2.7.3   Wet Oxidation

Wet oxidation operates with oxygen or air in combination with water at elevated temperature and pressure.[5] It was presented as an alternative to steam explosion, dilute acid hydrolysis, and hydrothermal pretreatment, operating in the same temperature range (170°C–220°C) and reaction time (1–30 min).[74] Bjerre et al.[75] discovered that combination of alkali and wet oxidation (alkaline wet oxidation) prevented the formation of furfural and 5-hydroxymethylfurfural.[76] Industrially, wet air oxidation processes have been used for the treatment of wastes with a high organic content by oxidation of soluble or suspended materials by using oxygen in aqueous phase at high temperature and high pressure.[5]

Palonen et al.[77] investigated the wet oxidation pretreatment of softwood for enhancing enzymatic hydrolysis. Six different combinations of reaction time, temperature, and pH were applied, and the compositions of solid and liquid fractions were analyzed. The solid fraction after wet oxidation pretreatment contained 58%–64% cellulose, 2%–16% hemicelluloses, and 24%–30% lignin. Important factors affecting enzymatic hydrolysis are the temperature of the pretreatment and the residual hemicellulose content of the substrate. The highest sugar yield, 79% of theoretical, was obtained using a pretreatment at 200°C for 10 min at neutral pH.

Lissens et al.[78] investigated the wet oxidation pretreatment of woody yard waste with high lignin content (22% of dry matter). The effects of temperature (185°C–200°C), oxygen pressure (3–12 bar), and addition of sodium carbonate (0–3.3 g per 100 g dry matter biomass) on enzymatic cellulose and hemicelluloses (xylan) convertibility were studied. The enzymatic cellulose conversion was the highest after wet oxidation for 15 min at 185°C with addition of 12 bars of oxygen and 3.3 g sodium carbonate per 100 g waste. Total carbohydrate recoveries were high (91%–100% for cellulose and 72%–100% for hemicelluloses) and up to 49% of the original lignin and 79% of the hemicelluloses could be solubilized during wet oxidation pretreatment and converted into carboxylic acids mainly.

Klinke et al.[77] investigated the alkaline wet oxidation pretreatment (water, sodium carbonate, oxygen, high temperature, and pressure) of wheat straw. They found that the pretreatment resulted in solid fractions with high cellulose recovery (96%) and high enzymatic convertibility to glucose (67%). Carbonate and temperature were the most important factors for fractionation of wheat straw by wet oxidation. Optimal conditions were 10 min at 195°C with addition of 12 bar oxygen and 6.5 g/l sodium carbonate.

The effects of wet oxidation on biomass can be summarized as follows:

- *Effect on cellulose*: Decrystallization.
- *Effect on hemicelluloses*: Dissolution.
- *Effect on lignin*: Removal (major).
- Formation of fermentation inhibitors.

## 8.2.8 Room-Temperature Ionic Liquids

Ionic liquids (defined as salts in the liquid state, ILs) have been recognized as a promising way to fractionate lignocellulosic biomass.[79,80]

ILs have been presented for more than 10 years as very promising solvents for catalysis, analysis, and organic synthesis.[81–85] Due to the very high number of possible combinations between an organic cation and an organic/inorganic anion. The main cations mentioned in the literature include imidazolium, pyridinium, quaternary ammonium, quaternary phosphonium, and pyrazolium; the main anions of generally smaller size include chloride, bromide, iodide, nitrate, and phosphate (Figure 8.12).[85]

During recent years, a number of publications have introduced a variety of technical developments and solvent systems based on several types of ILs to fractionate lignocellulose into individual components, after full or partial dissolution. Ideally, there are two ways to fractionate lignocellulose in ILs: (1) complete dissolution of biomass followed by selective precipitation of the sought components as purified fractions, by addition of a nonsolvent, or (2) selective extraction of components from the biomass.[82]

Ionic liquids are considered as *green solvents*. They are liquid at room temperature (typically below 100°C). Their vapor pressure is extremely low, so that they are nonvolatile, do not evaporate, and are nonflammable. They could be an environmentally

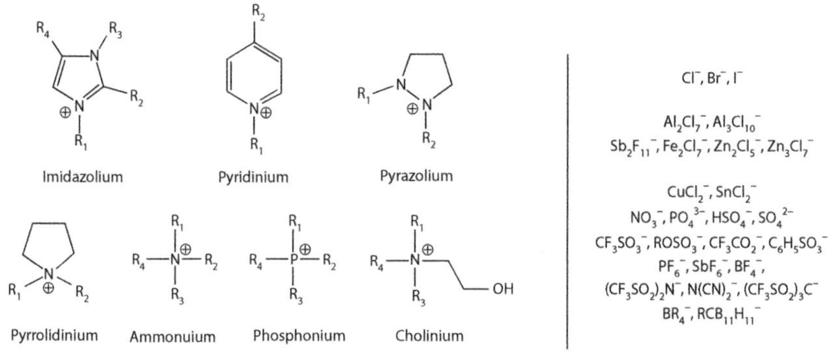

**FIGURE 8.12** Main cations and anions in ionic liquids (see Lignocellulosic Biorefineries book for resolution). (From Earle, M.J. and Seddon, K.R., *Pure Appl. Chem.*, 72, 1391, http://www.iupac.org/publications/pac/72/7/1391/, 2000.)

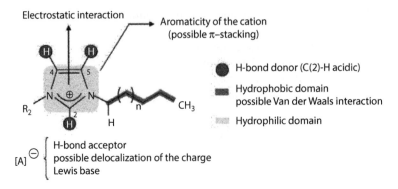

**FIGURE 8.13** Different types of interactions present in imidazolium-based ionic liquids ([A]⁻, anion). (From Earle, M.J. and Seddon, K.R., *Pure Appl. Chem.*, 72, 1391, http://www.iupac.org/publications/pac/72/7/1391/, 2000.)

friendly alternative to classical organic solvents.[86] The solvent properties are mainly linked not only to polarity but also to abundant different interactions with solutes (Figure 8.13).[85]

Concerning the ability of ionic liquids to dissolve cellulose, it is generally accepted that the results obtained in University of Alabama by Rogers et al.[87] were the starting point: high DP cellulose solutions (up to 25%) were obtained in 1-butyl-3-methylimidazolium chloride (BmimCl) at temperatures of 70°C–110°C.

As shown by several studies, most ionic liquids tend to inactivate cellulases, imposing cumbersome precipitation/purification steps before enzymatic treatment.[88] Therefore, some research teams have been exploring the possibility of direct enzymatic hydrolysis of cellulose in specific mixtures of cellulases and ionic liquids.[89,90]

As a summary, the dissolution of biomass into ILs has been shown to be a promising alternative biomass pretreatment technology, facilitating faster breakdown of cellulose through the disruption of lignin and the decrystallization of cellulose. The effect of ILs on lignocellulosic biomass can be described as follows:

- *Effect on cellulose*: Partial or total dissolution (through decrystallization).
- *Effect on hemicelluloses*: Partial or total dissolution.
- *Effect on lignin*: Partial or total dissolution.

### 8.2.9 STEAM EXPLOSION

#### 8.2.9.1 General

Steam explosion is the most commonly used pretreatment of biomass and uses both physical and chemical methods to break the structure of the lignocellulosic material.[4] In general terms, a steam explosion is a violent boiling or flashing of water into steam.[91] The biomass/steam mixture is held for a period of time under pressure and temperature (typically at 160°C–260°C for a few seconds to a few minutes) to promote hemicellulose hydrolysis and other chemical and physical changes. This period is followed by an explosive decompression. Pressure vessels that operate at above atmospheric pressure can also provide the conditions for a rapid boiling event which can be characterized as steam explosion.

### 8.2.9.2 Mode of Action

In the first stage, steam under high-pressure and high-temperature penetrates within the structure of the material.[98] Consequently, steam condenses and wets the material. The condensed water, together with high temperature, initiates the hydrolysis of acetyl and methylglucuronic acid groups attached to hemicelluloses. The released organic acids catalyze the depolymerization of hemicelluloses. The application of more severe conditions not only results in the formation of monosaccharides but also increases the concentration in furfural and 5-hydroxymethylfurfural.

In the second stage, the explosive decompression results in the instantaneous evaporation of the condensed moisture within the structure.[98] This expansion of the water vapor exerts a shear force on the surrounding structure. If the shear force is high enough, the vapor will cause the mechanical breakdown of the lignocellulosic structures. Combined effects of both stages include modification of the physical properties of the material (specific surface area, porosity, water retention, coloration, and cellulose crystallinity), hydrolysis of hemicellulosic components, and modification of the chemical structure of lignin, facilitating their extraction. These phenomena permit the opening of the lignocellulosic material and facilitate the extraction of its components.

According to Chornet and Overend,[92] steam explosion for biomass pretreatment is a physicochemical process in which breakdown of structural components of lignocellulosic materials is aided by heat in the form of steam, shear forces due to expansion of moisture, and hydrolysis of glycosidic bonds.[98]

The two governing parameters controlling the effect of steam explosion are reaction temperature and retention time.[93,98] The amount of time the biomass spends in the reactor helps to determine the extent of hemicellulose hydrolysis by the organic acids.[97] However, long retention times also increase the production of degradation products, which must be minimized for the subsequent fermentation process. Temperature governs the steam pressure within the reactor. Higher temperatures translate to higher pressures, therefore, increasing the difference between reactor pressure and atmospheric pressure. The pressure difference is in turn proportional to the shear force during decompression.

### 8.2.9.3 Process Layout

The apparatus for steam explosion consists of a boiler (steam generator) and a reactor. A schematic of the steam explosion process is shown in Figure 8.14.[94] The reactor is filled with biomass through valve 1. Valve 1 is then closed, and steam is let into the reactor through valve 2. The reactor is allowed to reach target temperature before time begins. Typically, ~20 seconds are required to attain the desired temperature. At the end of the allotted steaming time, valve 2 is closed, and valve 3 is opened for the explosive decompression to occur. The steam-exploded material shoots through the connecting piping and collects in the collection bin.

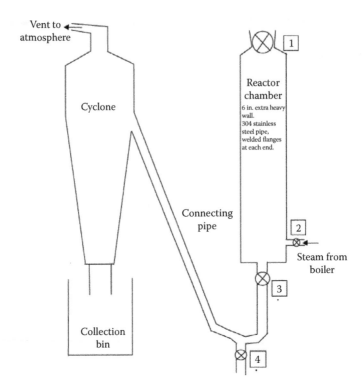

**FIGURE 8.14** (see Lignocellulosic Biorefineries book for redrawing) A schematic of the steam explosion process. Valve 1: sample charging valve. Valve 2: Saturated steam supply valve. Valve 3: discharge valve. Valve 4: condensate drain valve. (From Jeoh, T., Thesis, Virginia Polytechnic Institute and State University, 1998; Tanahashi, M., Characterization and degradation mechanisms of wood components by steam explosion and utilization of exploded wood, abstract of the PhD Thesis by the author, Kyoto University, http://repository. kulib.kyoto-u.ac.jp/dspace/bitstream/2433/53271/1/KJ00000017898.pdf, 1989.)

Another example of a steam explosion pilot plant, as developed by Gembloux Agro-Bio Tech, University of Liège (Belgium), is shown in Figure 8.15.[95]

### 8.2.9.4 Severity Factor

With the success of steam explosion came a need to standardize the process parameters to facilitate comparisons.[97] For example, it is important to be able to relate the net product yields to the pretreatment severity. From the observation that treatment temperatures and times are interchangeable, Overend and Chomet defined the severity of a steam explosion pretreatment in terms of the combined effect of both temperature and residence time. The model is based on the assumptions that the process kinetics is first order and obeys Arrhenius' law. In so doing, they were able to develop the reaction ordinate (R0) (Equation 8.5):

$$R0 = \int_0^t \exp\left[\frac{(Tr - Tb)}{14.75}\right] dt \qquad (8.5)$$

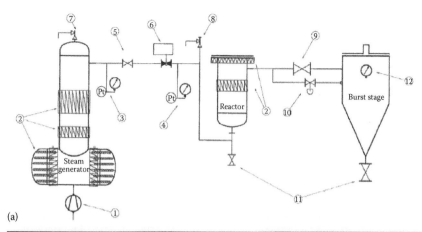

**FIGURE 8.15** (a) (see Biorefineries book for redrawing) Steam explosion/vapocracking pretreatment process flow diagram including a steam generator, a reactor, and a burster as developed by Gembloux Agro-Bio Tech, University of Liège. (1) High-pressure pump. (2) Heating collars. (3) Generator pressure and temperature probes. (4) Reactor pressure and temperature probes. (5) Isolation valve. (6) Loading valve. (7) Generator safety valve. (8) Reactor safety valve. (9) Bursting disk. (10) Purge valve. (11) Product recovery valves. (12) Burst manometer. (From Tanahashi, M. et al., *Wood Res.*, 75, 1, http://ci.nii. ac.jp/els/110000012875.pdf?id=ART0000336908&type=pdf&lang=en&host=cinii&order_no=&ppv_type=0&lang_sw=&no=1310116757&cp=, 1998.) (b) Photograph of the steam explosion/vapocracking equipment at Gembloux Agro-Bio Tech, University of Liège. (Courtesy of Nicolas Jacquet, GxABT, Gembloux, Belgium.)

where:

Tr is the reaction temperature (°C)

Tb is the base temperature (boiling point of water at atmospheric pressure; 100°C)

$t$ is the retention time (min)

14.75 is the conventional energy of activation assuming that the overall process is hydrolytic and the overall conversion is first order

The $\log_{10}$ value of the reaction ordinate gives the severity factor (or severity) that is used to map the effects of steam explosion pretreatment on biomass (Equation 8.6):

$$\text{Severity} = \log_{10}(R0) \qquad (8.6)$$

Chornet and Overend demonstrated the application of the reaction ordinate model using previously documented steam explosion data.[97,98] The data used by the authors were based on wood feedstocks. A study using steam-exploded sugarcane bagasse, however, concluded that the reaction ordinate model does not apply universally. In particular, the authors found that glucose yields from enzyme hydrolysis of steam-exploded sugarcane bagasse were not constant at a given severity over a range of temperatures.

### 8.2.9.5 Specific Effects on Lignocellulosic Substrates

#### 8.2.9.5.1 Uncatalyzed Steam Explosion

Tanahashi et al.[96] studied the effects of steam explosion on the morphology and physical properties of wood.[97] They found that at pressures greater than 28 kg/cm$^2$ (230°C) and 16 minute residence time, the microfibrils become completely separated from each other. The microfibrils were found to be thicker and shorter with increased steaming time.[97] Furthermore, the cellulose crystallinity was found to increase. This led the authors to conclude that amorphous cellulose becomes crystalline during the steaming process. Steam explosion at moderate severities was also shown to promote delignification. In another study, Tanahashi et al.[98] observed the chemical effects of the steam explosion process on wood. The hemicelluloses were found to be readily hydrolyzed to oligosaccharides by steaming, at lower severities. Higher severities further not only hydrolyzed the hemicelluloses to monosaccharides, but also increased the concentration of furfural and 5-hydroxymethyl furfural.

Similarly, Excoffier et al.[99] found that the degree of crystallinity of cellulose increases due to the steam treatment.[97] This observation was attributed to the crystallization of amorphous regions during the heat treatment. The authors also found that as the hemicelluloses are removed by hydrolysis, lignin softens, and depolymerizes.

Focher et al.[100] studied steam exploded wheat straw by scanning electron microscopy (SEM) and observed the formation of droplets on the cellulose fibers at high severities believed to be a physically modified form of lignin. Marchessault et al.[101] observed similar globular deposits on steam exploded pulp.

The effects of steam explosion on lignocellulose can be summarized as follows:

- Lignocellulosic materials are physically defibrillated/disintegrated.
- Crystallinity of cellulose is increased due to crystallization of amorphous portions.
- Hemicelluloses are easily hydrolyzed.
- Delignification is promoted due to modifications in lignin structure.

### 8.2.9.5.2   Catalyzed Steam Explosion

Numerous studies have also been carried out to investigate the possibilities of combining steam explosion pretreatment with catalysts.

Morjanoff et al.[102] were the first to show that the addition of sulfuric acid during the pretreatment improves the subsequent enzymatic hydrolysis and decreases the production of inhibitors such as furfurals and hydroxymethylfurfurals.[98] Optimum conditions for pretreatment of sugarcane bagasse were treatment for 30 seconds with saturated steam at 220°C and the addition of 1 g sulfuric acid/100 g dry bagasse.

Later, Vignon et al.[103] investigated hemp core samples impregnated in acid solution (0.1% w/w sulfuric acid) and steamed at temperatures between 200°C and 240°C for 180 seconds. Treatment at 200°C and 210°C led to samples that were difficult to delignify because the destructuring and disintegration of lignocellulosic materials were insufficient. A temperature of ~220°C–230°C is required to obtain well-separated fibers. However, at a temperature of 240°C, degradation and fiber damage were noted.

Viola et al.[104] pretreated the marine eelgrass *Zostera marina* by steam explosion, and selected temperature, time of pretreatment, and oxalic acid load as experimental factors. The best results were obtained at 180°C, 300 seconds, and 2 wt% of oxalic acid.

## 8.2.10   AMMONIA PRETREATMENTS

### 8.2.10.1   Ammonia Fiber Explosion

In the ammonia fiber/lignocellulose explosion/expansion (AFEX) pretreatment, lignocellulosic biomass is exposed to liquid ammonia at high temperature and pressure for a period of time, and then the pressure is suddenly reduced (Figure 8.16).[1,5,7]

The AFEX process is very similar to steam explosion. The operating conditions of a typical AFEX process are a dosage of liquid ammonia of 1–2 kg of ammonia/kg of dry biomass, a temperature of 90°C, a pressure of 21 atm, and a residence time of 30 min. AFEX pretreatment can significantly improve the fermentation rate of various herbaceous crops and grasses (Figure 8.17).[1]

The technology has been used for the pretreatment of many lignocellulosic materials including aspen wood, wheat straw, wheat chaff, and alfalfa stems.[105] During pretreatment, only a small amount of the solid material is solubilized. The hemicelluloses are degraded to oligomeric sugars and deacetylated.[106] The structure of the material is changed, resulting in higher digestibility. Over 90% hydrolysis of cellulose and hemicelluloses was obtained after AFEX treatment of biomass with low lignin content (<15% lignin).[107] However, the AFEX process

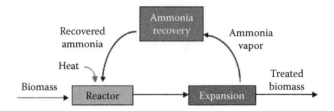

**FIGURE 8.16**    AFEX process. AFEX™ is a trademark of MBI. (Courtesy of MBI, Lansing, MI; Bruce Dale, Michigan State University, East Lansing, MI.)

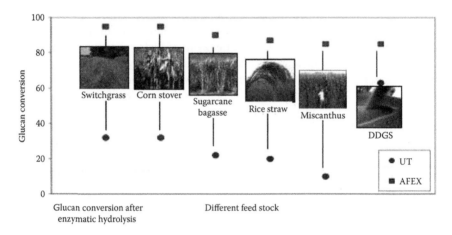

**FIGURE 8.17**    Efficiency of enzymatic hydrolysis of different feedstocks without pretreatment (UT) and after AFEX. (UT–no pretreatment, AFEX–ammonia pretreatment.) (Courtesy of MBI, Lansing, MI; Bruce Dale, Michigan State UniversityEast Lansing, MI.)

was not very effective for biomass with higher lignin content such as newspaper (18%–30% lignin) and aspen chips (25% lignin).[108] Hydrolysis yields of AFEX-pretreated newspaper and aspen chips were reported to be 40% and below 50%, respectively.[109]

To reduce the cost and protect environment, ammonia must be recycled after the pretreatment. In an ammonia recovery process, a superheated ammonia vapor with a temperature up to 200°C was used to vaporize and strip the residual ammonia in the pretreated biomass and the evaporated ammonia was then withdrawn from the system by a pressure controller for recovery.[111]

### 8.2.10.2  Ammonia Recycle Percolation

Another type of process using ammonia is the ammonia recycle percolation (ARP).[1,110] In this process, aqueous ammonia (10–15 wt%) passes through biomass at elevated temperatures (150°C–170°C) and high pressures (9–17 atm) with a fluid velocity of 1 cm/min and a residence time of 14 min, after which the ammonia is recovered. In the ARP method, the ammonia is separated and recycled (Figure 8.18).

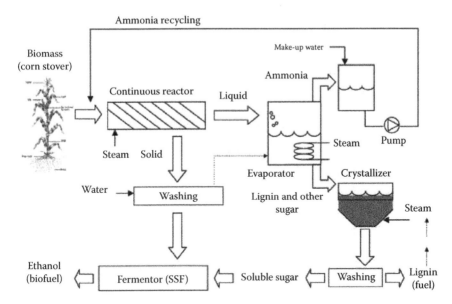

**FIGURE 8.18** Ammonia recycle percolation (ARP) process diagram. (From Kim, T.H. and Lee, Y.Y., *Bioresour. Technol.*, 96, 2007, http://www.sciencedirect.com/science/article/pii/S0960852405000684, 2005.)

Under these conditions, aqueous ammonia reacts primarily with lignin and causes its depolymerization and cleavage of lignin-carbohydrate linkages.[1,2] The ammonia pretreatment does not produce inhibitors for the downstream biological processes.

### 8.2.10.3 Specific Effects on Lignocellulosic Biomass

In general, AFEX and ARP processes are not differentiated in the literature, although AFEX is carried out with liquid ammonia and ARP with an aqueous ammonia solution.[1] The AFEX pretreatment simultaneously reduces lignin content and removes some hemicelluloses while decrystallizing cellulose. It can have a profound effect on the rate of cellulose hydrolysis. The cost of ammonia, and especially of ammonia recovery, drives the cost of the AFEX pretreatment.

Recently, a significant amount of research has been carried out to determine the optimum conditions for ammonia pretreatment of lignocellulosic materials.[1] Teymouri et al.[112] evaluated the optimum process conditions such as ammonia loading, moisture content of biomass, temperature, and residence time necessary for maximum effectiveness of the AFEX process on corn stover (~17% lignin). A comparison of enzymatic digestibility of corn stover with AFEX at different ratios of ammonia to biomass is shown in Figure 8.19.

The optimum pretreatment conditions were found to be a temperature of 90°C, an ammonia/dry corn stover mass ration of 1:1, a moisture content of 60% (dry weight basis), and a residence time of 5 min.

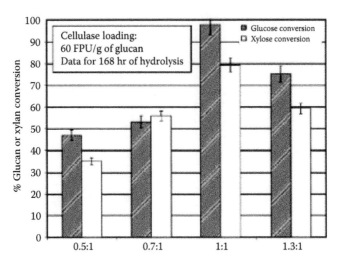

**FIGURE 8.19** Effect of ammonia loading (grams of $NH_3$/gram of dry biomass) on the subsequent enzymatic conversion of glucans and xylans for APEX pretreatment of corn stover at 90°C and 60% moisture content. (From Dale, B.E. and Moreira, M.J., *Biotechnol. Bioeng. Symp.*, 12, 31, http://www.osti.gov/energycitations/product.biblio.jsp?osti_id=5893198, 1982.)

### 8.2.11 CARBON DIOXIDE EXPLOSION

Another pretreatment that utilizes fiber separation via a rapid pressure drop is $CO_2$ explosion and particularly supercritical $CO_2$ explosion.[113] The method is similar to steam explosion and ammonia fiber explosion; high pressure carbon dioxide is injected into the batch reactor to impregnate the biomass and then liberated by an explosive decompression.[5] It was hoped that, on account of $CO_2$ forms carbonic acid when dissolved in water, the acid increases the hydrolysis rate.[1,114] Carbonic acid may offer the benefits of acid catalysts without the use of an acid like sulfuric acid.[5] The pH of carbonic acid is determined by the partial pressure of $CO_2$ in the reactor and can be neutralized by releasing the reactor pressure. Carbon dioxide molecules are comparable in size with water and ammonia and should be able to penetrate small pores accessible to water and ammonia molecules. Compared with steam explosion, $CO_2$ explosion offers the benefit of lower operating temperatures (35°C–80°C) which reduces sugar decomposition.[118] Dale et al.[115] used the method to pretreat alfalfa (4 kg of $CO_2$/kg of fiber at a pressure of 5.62 MPa) and obtained 75% of the theoretical glucose released during 24 h of the subsequent enzymatic hydrolysis.[1] The yields were relatively low compared with those of steam or ammonia explosion pretreatments but high compared with that of enzymatic hydrolysis without pretreatment.

An alternative use of $CO_2$ in pretreatment is the extraction with *supercritical* $CO_2$ (SC-$CO_2$). $CO_2$ becomes supercritical under relatively mild conditions (critical temperature = 31°C and critical pressure = 7.38 MPa). SC-$CO_2$ has been widely used as an extraction solvent. For several years, it has been considered as possible pretreatment process for lignocellulosic biomass. Sahle Demessie et al.[116] reported the increased permeability of Douglas-fir heartwood by SC-$CO_2$. Zheng et al.[117] compared SC-$CO_2$ explosion with steam and ammonia explosion for pretreatment of cellulosic materials such as recycled paper mix, sugarcane bagasse, and repulping waste of recycled paper and found that $CO_2$ explosion was more cost efficient than ammonia explosion.

### 8.2.12 MECHANICAL/ALKALINE PRETREATMENT

Combined mechanical/alkaline pretreatment consists of a continuous mechanical pretreatment (such as milling, extrusion, and refining) of lignocellulosic materials with the aid of an alkali.[5] The resulting fractions include a soluble fraction (containing lignin, hemicelluloses, and inorganic components) and a cellulose-enriched solid fraction.

As opposed to the acid-catalyzed pretreatment methods, the general principle behind alkaline pretreatment methods is the removal of lignin, whereas cellulose and a portion of the hemicelluloses remain in the solid fraction.[5] The solid fraction is subsequently subjected to enzyme hydrolysis for the production of C6- and C5-sugars. Therefore, this pretreatment process is especially suitable in combination with fermentation routes in which both C6- and C5-sugars can be converted to valuable products.

By performing mechanical and chemical pretreatments in one step, the accessibility of cellulose to enzymes is improved, resulting in higher delignification and improved enzymatic hydrolysis.[5] Furthermore, the moderate operating temperatures of this process prevent the formation of degradation and oxidation products.

### 8.2.13 BIOLOGICAL PRETREATMENT

In biological pretreatments, microorganisms such as wood brown-, white-, and soft-rot fungi are employed to degrade components of biomass based on their enzymatic systems.[1] Brown-rot and soft-rot fungi attack polysaccharides predominantly and do not degrade lignin extensively, whereas white-rot fungi are the most efficient in causing lignin degradation. Lignin degradation by white-rot fungi takes place through the action of lignin-degrading enzymes such as peroxidases and laccases (oxidases). These enzymes are regulated by carbon and nitrogen sources. White-rot fungi are the most effective for biological pretreatment of lignocellulosic biomass (Figure 8.20).

The advantages of biological pretreatments include low energy requirement and mild environmental conditions.[1,5] Nevertheless, the rate of biological hydrolysis is usually low, resulting in long residence times. Recently, Tanjore[119] studied biological pretreatments of corn stover through aerobic and anaerobic solid substrate fermentation. Shi et al.[120] investigated cotton stalks degradation by solid state cultivation using *Phanerochaete chrysosporium*. They showed that moisture content significantly affected lignin degradation.

**FIGURE 8.20**   White-rot fungi in a Birch tree. (From http://en.wikipedia.org/wiki/Wood-decay_fungus.)

## 8.3  SUMMARY

The mode of action of various important pretreatments on the structure of ligno-cellulosic biomass and the generation of fermentation inhibitors is summarized in Table 8.2.[1,2,5]

Table 8.2 leads to the following conclusions:

- Wet oxidation, AFEX, ARP, and $CO_2$ explosion reduce cellulose crystallinity.
- All pretreatments partially or totally remove hemicelluloses.
- Alkaline, organosolv, wet oxidation, AFEX, ARP, mechanical/alkaline, and biological partially or totally remove lignin.
- Fermentation inhibitors are formed with liquid hot water, acid, and steam explosion.
- No or small amounts of fermentation inhibitors are formed with alkali, organosolv, wet oxidation, AFEX, ARP, $CO_2$ explosion, mechanical/ alkaline, and biological.[121]

Specific advantages and disadvantages for various pretreatment methods are summarized in the following[1]:

1. Liquid hot water may generate fermentation inhibitors, but using water as a solvent is an economic and environmental advantage.[2,5,122] Chemical pretreatments that use alkalis, organic solvents, or oxidants are largely focused on lignin removal, thereby enhancing enzymatic degradability of cellulose.[5] Alkaline hydrolysis requires long residence times and leads to the formation of irrecoverable salts and their incorporation into biomass. Organosolv requires solvents to be drained from the reactor, evaporated, condensed,

## TABLE 8.2
## Effect of Various Pretreatments on the Structure of Lignocellulosic Biomass (In Addition to Increasing Surface Area) and the Formation of Fermentation Inhibitors

| Pretreatment | Effect on Cellulose | Effect on Hemicelluloses | Effect on Lignin | Inhibitor Formation |
|---|---|---|---|---|
| Liquid hot water | | Removal[a] (*major*) | Alteration | Yes |
| Dilute acid | | Removal[a] (*major*) | Alteration (minor) | Yes |
| Strong acid | Total hydrolysis | Hydrolysis | Alteration | Yes |
| Alkaline | | Removal (minor)[a] | Removal[b] (*major*) | |
| Organosolv | | Removal (minor)[a] | Removal[b] (*major*) | |
| Wet oxidation | Decrystallization | Removal (minor)[a] | Removal[b] (*major*) | |
| Ionic liquids | Dissolution | Dissolution | Dissolution | |
| Steam explosion | | Removal[a] (*major*) | Alteration (minor) | Yes |
| AFEX | Decrystallization | Removal[a] | Removal[b] | No |
| ARP | Decrystallization | Removal (minor)[a] | Removal[b] (*major*) | No |
| CO₂ explosion | Decrystallization | Removal[a] | | No |
| Mechanical/alkaline | | Removal (minor)[a] | Removal[b] (*major*) | |
| Biological | | Degradation | Degradation | |

*Source:* Kumar, P. et al., *Ind. Eng. Chem. Res.*, 48, 3713, http://pubs.acs.org/doi/abs/10.1021/ie801542g, 2009; Mosier, N. et al., *Biores. Technol.*, 96, 673, http://www.sciencedirect.com/science/article/pii/S0960852404002536, 2005; Harmsen, P. et al., *Literature Review of Physical and Chemical Pretreatment Processes for Lignocellulosic Biomass*, Wageningen UR, http://www.biobased.nl/everyone/1426/5/0/30/, 2010.

[a] Hydrolysis, solubilization, and extraction from the solid phase.

[b] Solubilization and extraction from the solid phase.

and recycled, resulting in high cost.[1] Acid hydrolysis generates fermentation inhibitors and requires corrosion-resistant equipment, resulting in high cost.[1]

2. Steam explosion is an attractive, potentially cost-effective process because it makes limited use of chemicals, it does not result in excessive dilution of the resulting sugars, and it requires low energy input with no recycling or environmental cost.[122] Limitations of steam explosion include destruction of a portion of the xylan fraction, incomplete disruption of the lignin-carbohydrate matrix, and generation of fermentation inhibitors. AFEX/ARP do not produce fermentation inhibitors but are somewhat not efficient for biomass with high lignin content.[1,122] The low cost of $CO_2$ as a pretreatment solvent, no generation of toxins, the use of low temperatures, and high solids capacity are attractive features of $CO_2$ explosion.[122] However, the high cost of equipment that can withstand high-pressure conditions of the treatment is a limitation.

3. The main disadvantage of biological pretreatment is that the rate of hydrolysis is very low.[1,5]

Biomass pretreatment remains a key bottleneck in the processing of lignocellulosics for biofuels and other biobased products.[122] Although some pretreatment methods show apparent advantages, it is unlikely that one method will become the method of choice for the diversity of biomass feedstocks, which can react differently to a given technology.

## REFERENCES

1. P. Kumar, D.M. Barrett, M.J. Delwiche and P. Stroeve, *Ind. Eng. Chem. Res.* 48, 3713, 2009. http://pubs.acs.org/doi/abs/10.1021/ie801542g.
2. N. Mosier, C. Wyman, B. Dale, R. Elander, Y.Y. Lee, M. Holtzapple and M. Ladisch, *Biores. Technol.* 96, 673, 2005. http://www.sciencedirect.com/science/article/pii/S0960852404002536.
3. G. Brodeur, E. Yau, K. Badal, J. Collier, K.B. Ramachandran and S. Ramakrishnan, *Enzyme Res.* 2011, 787532, 2011. http://www.ncbi.nlm.nih.gov/pmc/articles/PMC3112529/#!po=3.12500.
4. A.M. Raspolli Galletti and C. Antonetti, Eurobioref, Castro Marina, 2011. http://www.eurobioref.org/Summer_School/Lectures_Slides/day2/Lectures/L04_AG%20Raspolli.pdf.
5. P. Harmsen, W. Huijgen, L. Bermudez and R. Bakker, *Literature Review of Physical and Chemical Pretreatment Processes for Lignocellulosic Biomass*, Wageningen UR, 2010. http://www.biobased.nl/everyone/1426/5/0/30/.
6. Y. Zheng, Z. Pan and R. Zhang, *Int. J. Agric. Biol. Eng.* 2, 51, 2009. http://ijabe.org/index.php/ijabe/article/download/168/83.
7. V. Chaturvedi and P. Verma, *Biotech* 3, 415, 2013. http://www.ncbi.nlm.nih.gov/pmc/articles/PMC3781263/#!po=84.3750.
8. K.W. Waldron, *Bioalcohol Production: Biochemical Conversion of Lignocellulosic Biomass*, Elsevier, 2010. http://www.woodheadpublishing.com/en/book.aspx?bookID=1533.
9. G. Moxley, *Autohydrolysis and Steam Explosion*, Novozymes, Bagsværd, Denmark.
10. Y. Zhu, *Overview of Biomass Pretreatment Technologies*, Novozymes, Bagsværd, Denmark, 2011.
11. W.S.L. Mok and M.J. Antal, *Ind. Eng. Chem. Res.* 31, 1157, 1992. http://pubs.acs.org/doi/abs/10.1021/ie00004a026.
12. http://www.inbicon.com/en/biomass-refinery/core-technology
13. Inbicon, *Inbicon Biomass Refinery*, Kalundborg. http://www.inbicon.com/Biomass%20Refinery/Pages/Inbicon_Biomass_Refinery_at_Kalundborg.aspx and http://www.inbicon.com/en/global-solutions/danish-projects.
14. Y.Y. Lee, P. Iyer and R.W. Torget, *Adv. Biochem. Eng. Biotechnol.* 65, 93, 1999. http://www.springerlink.com/content/tf1ttedvjwqlm9df/.
15. K.J. Zeitsch, *The Chemistry and Technology of Furfural and its Many By-Products*, Sugar Series 13, Elsevier, New York, 2000. http://www.sciencedirect.com/science/bookseries/01677675.
16. J.F. Saeman, *Ind. Eng. Chem.* 37, 43, 1945. http://pubs.acs.org/doi/abs/10.1021/ie50421a009.
17. Y. Chen, R.R. Sharma-Shivappa, D. Keshwani and C. Chen, *Appl. Biochem. Biotechnol. Part A* 142, 276, 2007. http://www.springerlink.com/content/a2542710357w3u12.
18. S.A. Rhydolm, *Pulping Processes*, Wiley Interscience, New York, 1966.
19. D.N.S. Hon and N. Shiraishi, *Wood and Cellulosic Chemistry*, Marcel Dekker, New-York, 1991.

20. Institute of Paper Science and Technology, Basics of kraft pulping. http://www.ipst.gatech. edu/faculty/ragauskas_art/technical_reviews/Basics%20of%20Kraft%20Pulping.pdf.
21. H. Xu, Biobleaching of kraft pulp with recombinant manganese peroxidase, Proquest Information and Learning Company, 2007.
22. J.L. Wertz and O. Bedue, *Lignocellulosic Biorefineries*, EPFL Press, Lausanne, Switzerland, 2013.
23. J.L. Wertz, O. Bedue and J.P. Mercier, *Cellulose Science and Technology*, EPFL Press, Lausanne, Switzerland, 2010.
24. M. McLeod, *Paperi ja Puu-Paper Timber* 89, 1, 2007. http://kraftpulpingcourse. knowledgefirstwebsites.com/f/Top_Ten.pdf.
25. http://en.wikipedia.org/wiki/Kraft_process
26. H. Tran and E. K. Vakkilainnen, *The Kraft Chemical Recovery System*, Tappi, Peachtree Corners, GA.
27. M. Moshkelani, M. Marinova and J. Paris, The forest biorefinery and its implementation in the pulp&paper industry, in *2nd European Conference on Polygeneration*, Taragona, Spain, March 30–April 1, 2011. http://six6.region-stuttgart.de/sixcms/media. php/773/ABS-03-Jean-Paris.pdf.
28. A. van Heiningen, Forest Biorefinery Research at the University of Maine, in *NRBP Steering Committee Meeting*, Latham, NY, May 25, 2006. http://www.nrbp.org/events/ meetings/060524/presentations/heiningen.pdf.
29. H. Mao, J.M. Genco, A. van Heiningen and H. Pendse, *BioResources* 5, 525, 2010. http:// www.ncsu.edu/bioresources/BioRes_05/BioRes_05_2_0525_Mao_GVP_Kraft_Mill_ Biorefinery_Acetic_Acid_Ethanol_723.pdf.
30. Valmet. http://www.valmet.com/products/pulping-and-fiber/chemical-recovery/lignin-separation/lignoboost-process/.
31. http://www.biofuelsdigest.com/bdigest/2013/11/03/metso-is-supplying-a-lignoboost-plant-to-stora-ensos-new-biorefinery/
32. S. Mcintosh and T. Vancov, *Bioresour. Technol.* 101, 6718, 2010. http://www. sciencedirect.com/science/article/pii/S0960852410005936.
33. K.C. Nlewem, M.E. Thrash, *Bioresour. Technol.* 101, 5426, 2010. http://www. sciencedirect.com/science/article/pii/S0960852410003020.
34. Y. Li, R. Ruan, P.L. Chen, Z. Liu, X. Pan, X. Lin, Y. Liu, C.K. Mok and T. Yang, *Trans. ASAE* 47, 821, 2004. http://asae.frymulti.com/abstract.asp?aid=16078&t=2.
35. D.R. Keshwani, J.J. Cheng, J.C. Burns, L. LI and V. Chiang, Microwave Pretreatment of switchgrass to enhance enzymatic hydrolysis, in *ASABE Annual International Meeting*, 2007. http://asae.frymulti.com/abstract.asp?aid=23472&t=2.
36. Z.H. Hu and Z.Y. Wen, *Biochem. Eng. J.* 38, 369, 2008. http://www.sciencedirect.com/ science/article/pii/S1369703x07002756.
37. M. Pedersen, K.S. Johansen and A.S. Meyer, *Biotechnol. Biofuels* 4, 11, 2011. http:// www.ncbi.nlm.nih.gov/pubmed/21569460.
38. G. Zacchi, Pretreatment of biomass for ethanol production, in *International Conference on Lignocellulosic Ethanol*, Copenhagen, Denmark, October 13–15, 2010. http:// ec.europa.eu/energy/renewables/events/doc/2010_10_13/1_pretreatment_guido_zacchi. pdf.
39. R.T. Elander, B.E. Dale, M. Holtzapple, M.R. Ladisch, Y.Y. Lee, C. Mitchinson, J.N. Saddler and C.E. Wyman, *Cellulose* 16, 649, 2009. http://www.springerlink.com/ content/d47121610u58n877.
40. B.C. Saha and M.A. Cotta, *Biomass Bioenerg.* 32, 971, 2008. http://www.sciencedirect. com/science/article/pii/S0961953408000184.
41. S.C. Rabelo, R.M. Filho and A.C. Costa, *Appl. Biochem. Biotechnol.* 153, 139, 2009. http://www.springerlink.com/content/m746r7x236707242.

42. R. Sierra, C. Granda and M.T. Holzapple, *Biotechnol. Prog.* 25, 323, 2009. http://onlinelibrary.wiley.com/doi/10.1002/btpr.83/abstract.
43. P. Kumar, D.M. Barett, M.J. Delwiche and P. Stroeve, *Ind. Eng. Chem. Res.* 48, 3713, 2009. http://www.che.ncsu.edu/ILEET/CHE596web_Spr2010/resources/biomass-biofuels/Methods-for-Pretreatment.pdf.
44. S. Kim and M.T. Holtzapple, *Bioresour. Technol.* 97, 778, 2006. http://www.ncbi.nlm.nih.gov/pubmed/15961306.
45. F.K. Agbogbo and M.T Holzapple, Recent advances in the MixAlco process for the production of mixed alcohol fuel. http://www.eri.ucr.edu/ISAFXVCD/ISAFXVAF/RcAMAP.pdf.
46. V.S. Chang, W.E. Kaar, B. Burr and M.T. Holtzapple, *Biotechnol. Lett.* 23, 1327, 2001. http://www.springerlink.com/content/pg75137v4k838580/.
47. Terrabon, MixAlco—Overview, 2008. http://www.terrabon.com/mixalco_overview.html.
48. http://www.huffingtonpost.com/2012/10/23/terrabon-inc-bioenergy-bankruptcy-texas_n_2006313.html
49. T.J. Mcdonough, The chemistry of organosolv delignification, IPST Technical Paper Series Number 455, 1992. http://smartech.gatech.edu/bitstream/handle/1853/2069/tps-455.pdf?sequence=1.
50. E. Muurinen, Organosolv pulping, a review and distillation study related to peroxyacid pulping, Doctoral thesis, Oulo University Library, Finland, 2000. http://herkules.oulu.fi/isbn9514256611/isbn9514256611.pdf.
51. X. Zhao, K. Cheng and D. Liu, *Appl. Microbiol. Biotechnol.* 82, 815, 2009. http://www.ncbi.nlm.nih.gov/pubmed/19214499.
52. P. Alvira, E. Tomas-Pejo, M. Ballesteros and M.J. Negro, *Bioresour. Technol.* 101, 4851, 2010. http://www.scribd.com/doc/53790135/Pre-Treatment-Technologies-for-an-Efficient-Bio-Ethanol-Production-Process.
53. H. Chen and W. Qiu, *Biotechnol. Adv.* 28, 556, 2010. http://www.ncbi.nlm.nih.gov/pubmed/20546879.
54. J.J. Villaverde, P. Ligero and A. de Vega, *Open Agric. J.* 4, 102, 2010. http://www.benthamscience.com/open/toasj/articles/V004/SI0085TOASJ/102TOASJ.pdf.
55. F.B. Sun and H.Z. Chen, *Bioresour. Technol.* 99, 5474, 2008. http://www.sciencedirect.com/science/article/pii/S0960852407009285.
56. R. El Hage, Prétraitement du Miscanthus x giganteus. Vers une valorisation optimale de la biomasse cellulosique, Thèse, Université de Nancy, France, 2010. http://www.scd.uhp-nancy.fr/docnum/SCD_T_2010_0063_EL_HAGE.pdf.
57. N. Brosse, R. El Hage, P. Sannigrahi and A. Ragauska, *Cell. Chem. Technol.* 44, 71, 2010. http://www.cellulosechemtechnol.ro/pdf/CCT44,1-3%20(2010)/P.71-78.pdf.
58. X. Pan, D. Xie, R. Yu and J. Saddler, *Biotechnol. Bioeng.* 101, 39, 2008. http://onlinelibrary.wiley.com/doi/10.1002/bit.21883/pdf.
59. http://www.lignol.ca/
60. http://www.greencarcongress.com/2007/08/canadian-wood-c.html
61. European Biocore Project. http://www.biocore-europe.org/page.php?optim=An-innovative-fractionation and http://cordis.europa.eu/fetch?CALLER=FP7_PROJ_EN&ACTION=D&DOC=1&CAT=PROJ&RCN=94207.
62. G. Avignon, M. Delmas, Procede de production de pate a papier, lignines, sucres et acide acetique par fractionnement de matiere vegetale lignocellulosique en milieu acide formique/acide acetique, EP 1180171, 2002 and WO 0068494, 2000. https://www.google.com/patents/EP1180171B1.
63. H. Quoc Lam, Y. Le Bigot, M. Delmas and G. Avignon, *Ind. Crops Prod.* 14, 139, 2001. http://www.cimv.fr/uploads/new-procedure-for-the-destructuring-vegetable-matter-industrial-crops-2001.pdf.

64. R. Bakker, Advanced chemical/physical fractionation, in *Workshop on the BioSynergy Project*, Reims, France, 2010. http://www.biosynergy.eu/fileadmin/biosynergy/user/docs/WP1-Advanced_Physical__chemical_fractionation-RBakker-BIOSYNERGY_workshop_Reims__FR__17-11-2010.pdf.

65. CIMV. http://www.cimv.fr/technologie/technologie-cimv/technologie-cimv.html.

66. Biocore project: "An innovative concept to valorize the biomass." http://www.biocore-europe.org/page.php?optim=An-innovative-concept.

67. http://www.cimv.fr/recherche-developpement/partenariat-biocore.html

68. M.J. Selig, T.B. Vinzant, M.E. Himmel and S.R. Decker, *Appl. Biochem. Biotechnol.* 155, 397, 2009. http://www.springerlink.com/content/b7157368g1021435/.

69. B.C. Saha and M.A. Cotta, *Biotechnol. Prog.* 22, 449, 2006. http://ddr.nal.usda.gov/bitstream/10113/25800/1/IND43819470.pdf.

70. B. Qi, X. Chen, Y. Su and Y. Wan, *Ind. Eng. Chem. Res.* 48, 7346, 2009. http://pubs.acs.org/doi/abs/10.1021/ie8016863.

71. B. Wang, X. Wang and H. Feng, *Bioresour. Technol.* 101, 752, 2010. http://www.ncbi.nlm.nih.gov/pubmed/19762230.

72. B.C. Saha and M.A. Cotta, *New. Biotechnol.* 27, 10, 2010. http://www.sciencedirect.com/science/article/pii/S1871678409012540.

73. G. Banerjee, S. Car, J.S. Scott-Craig, D.B. Hodge and J.D. Walton, *Biotechnol. Biofuels* 4, 16, 2011. http://www.biotechnologyforbiofuels.com/content/4/1/16/abstract.

74. H.B. Klinke, B.K. Ahring, A.S. Schimdt and A.B. Thomsen, *Bioresour. Technol.* 82, 15, 2002. http://www.sciencedirect.com/science/article/pii/S0960852401001523.

75. A.B. Bjerre, A.B. Olesen, T. Fernqvist, A. Ploger and A.S. Schmidt, *Biotechnol. Bioeng.* 49, 568, 1996. http://www.ncbi.nlm.nih.gov/pubmed/18623619.

76. C. Martin, M. Marcet and A.B. Thomsen, *Bioresources* 3, 670, 2008. http://www.ncsu.edu/bioresources/BioRes_03/BioRes_03_3_0670_Martin_MT_Bagasse_Pretreatment_WetOx_SteamExp.pdf.

77. H. Palonen, A.B. Thomsen, M. Tenkanen, A.S. Schmidt and L. Viikari, *Appl. Biochem. Biotechnol.* 117, 1, 2004. http://www.springerlink.com/content/q3v545878317m682.

78. G. Lissens, H. Klinke, W. Verstraete, B. Ahring and A.B. Thomsen, *J. Chem. Technol. Biotechnol.* 79, 889, 2004. http://onlinelibrary.wiley.com/doi/10.1002/jctb.1068/abstract.

79. T. Leskinen, A.W.T. King and D.S. Argylopoulos, Chapter 6, Fractionation of lignocellulosic materials with ionic liquids. http://www4.ncsu.edu/~dsargyro/Publications/documents/ArgyropoulosChaptrer6FINAL.pdf.

80. A.M. da Costa Lopes, K.G. Joao, A.R.C. Morais, E. Bogel-Lukasik and R. Bogel-Lukasik, *Sustain. Chem. Process.* 1, 3, 2013. http://www.sustainablechemicalprocesses.com/content/1/1/3/abstract.

81. H. Olivier-Bourbigou and Y. Chauvin, *J. Mol. Catal. A: Chem.* 214, 9, 2004. http://www.sciencedirect.com/science/article/pii/S1381116904001402.

82. H. Olivier-Bourbigou, L. Magna and D. Morvan, *Appl. Catal. A: Gen.* 373, 1, 2010. http://www.sciencedirect.com/science/article/pii/S0926860x09007030.

83. J. Lu, J. Yan and J. Texter, *Prog. Polym. Sci.* 34, 431, 2009. http://www.sciencedirect.com/science/article/pii/S0079670008001226.

84. N.V. Plechkova and K.R. Seddon, *Chem. Soc. Rev.* 37, 123, 2008. http://pubs.rsc.org/en/Content/ArticleLanding/2008/CS/b006677j.

85. M.J. Earle and K.R. Seddon, *Pure Appl. Chem.* 72, 1391, 2000. http://www.iupac.org/publications/pac/72/7/1391/.

86. Katholieke Universiteit Leuven, Ionic Liquids Info, 2011. http://www.kuleuven.be/ionic-liquids/ionic_liquid.php.

87. R.P. Swatlowski, S.K. Spear, D. John, J.D. Holbrey and R.D. Rogers, *J. Am. Chem. Soc.* 124, 4974, 2002. http://pubs.acs.org/doi/abs/10.1021/ja025790m.

88. M.B. Turner, S.K. Spear, J.G. Huddleston, J.D. Holbrey and R.D. Rogers, *Green Chem.* 5, 443, 2003. http://pubs.rsc.org/en/content/articlelanding/2003/gc/b302570e.

89. H. Zhao, *J. Chem. Technol. Biotechnol.* 101, 891, 2010. http://onlinelibrary.wiley.com/doi/10.1002/jctb.2375/full.

90. S. Bose, D.W. Armstrong and J.W. Petrich, *J. Phys. Chem. B* 114, 8221, 2010. http://pubs.acs.org/doi/abs/10.1021/jp9120518.

91. http://en.wikipedia.org/wiki/Steam_explosion

92. E. Chornet and R.P. Overend, *Proceedings of the International Workshop on Steam Explosion Techniques: Fundamentals and Industrial Applications*, 21, 1988. http://www.amazon.fr/Steam-Explosion-Techniques-Fundamentals-International/dp/2881244572.

93. M. Ibrahim, Thesis, Virginia Tech, 1998. http://scholar.lib.vt.edu/theses/available/etd-2998-114756/unrestricted/e-body1.pdf.

94. T. Jeoh, Thesis, Virginia Polytechnic Institute and State University, 1998. http://scholar.lib.vt.edu/theses/available/etd-011499-120138/unrestricted/ETD.PDF.

95. N. Jacquet, C. Vanderghem, C. Blecker and M. Paquot, *Biotechnol. Agron. Soc. Environ.* 14, 561, 2010. http://popups.ulg.ac.be/Base/document.php?id=6226.

96. M. Tanahashi, S. Takada, T. Aoki, T. Goto, T. Higuchi and S. Hanai, *Wood Res.* 69, 36, 1983. http://ci.nii.ac.jp/els/110000012836.pdf?id=ART0000336830&type=pdf&lang=en&host=cinii&order_no=&ppv_type=0&lang_sw=&no=1310116852&cp=.

97. M. Tanahashi, Characterization and degradation mechanisms of wood components by steam explosion and utilization of exploded wood, abstract of the PhD Thesis by the author, Kyoto University, 1989. http://repository.kulib.kyoto-u.ac.jp/dspace/bitstream/2433/53271/1/KJ00000017898.pdf.

98. M. Tanahashi, K. Tamabuchi, T. Goto, T. Aoki, M. Karina and T. Higuchi, *Wood Res.* 75, 1, 1998. http://ci.nii.ac.jp/els/110000012875.pdf?id=ART0000336908&type=pdf&lang=en&host=cinii&order_no=&ppv_type=0&lang_sw=&no=1310116757&cp=.

99. G. Excoffier, A. Peguy, M. Rinaudo and M.R. Vignon, Evolution of lignocellulosic components during steam explosion. Potential applications, in B. Focher, A. Marzetti and V. Crescenzi (Eds.), *Proceedings of the International Workshop on Steam Explosion Techniques: Fundamentals and Industrial Applications*, 83, 1998. http://www.amazon.fr/Steam-Explosion-Techniques-Fundamentals-International/dp/2881244572.

100. B. Focher, A. Marzetti, P.L. Beltrame, P. Carniti and A. Visciglio, Steam explosion of wheat straw. Product fractionation and enzymatic hydrolysis of the cellulosic component, in B. Focher, A. Marzetti and V. Crescenzi (Eds.), *Proceedings of the International Workshop on Steam Explosion Techniques: Fundamentals and Industrial Applications*, 331, 1998. http://www.amazon.fr/Steam-Explosion-Techniques-Fundamentals-International/dp/2881244572.

101. R.H. Marchessault and J.M. St-Pierre, A new understanding of the carbohydrate system, in I. Chemrawn, L.E. St-Pierre and G.R. Brown (Eds.), *Future Sources of Organic Raw Materials*, 613, Pergamon Press, Oxford, 1980.

102. P.J. Morjanoff and P.P. Gray, *Biotechnol. Bioeng.* 29, 733, 1987. http://onlinelibrary.wiley.com/doi/10.1002/bit.260290610/abstract.

103. M.R. Vignon, C. Garcia-Jaldon and D. Dupeyre, *Int. J. Biol. Macromol.* 17, 395, 1995. http://www.ncbi.nlm.nih.gov/pubmed/8789346.

104. E. Viola, M. Cardinale, R. Santarcangelo, A. Villone and F. Zimbardi, *Biomass Bioenerg.* 32, 613, 2008. http://www.sciencedirect.com/science/article/pii/S0961953407002346.

105. M. Mes-Hartree, B.E Dale and W.K. Craig, *Appl. Microbiol. Biotechnol.* 29, 462, 1988. http://www.springerlink.com/content/m653710m565251gt.

106. L.E. Gollapalli, B.E. Dale and D.M. Rivers, *Appl. Biochem. Biotechnol.* 98, 23, 2002. http://www.ncbi.nlm.nih.gov/pubmed/12018251.

107. M.T. Holtzapple, J.H. Jun, G. Ashok, S.L. Patibandla and B.E. Dale, *Appl. Biochem. Biotechnol.* 28/29, 59, 1991. http://www.springerlink.com/content/x462833787910938/.

108. Y. Sun and J. Cheng, *Bioresour. Technol.* 83, 1, 2002. http://www.sciencedirect.com/science/article/pii/S0960852401002127.
109. J.D. Mcmillan, Pretreatment of lignocellulosic biomass, in M.E. Himmel, J.O. Baker and R.P. Overend (Eds.), *Enzymatic Conversion of Biomass For Fuels Production*, American Chemical Society, Washington, DC, 292, 1994.
110. H.H. Yoon, Z.W. Wu and Y.Y. Lee, *Appl. Biochem. Biotechnol.* 51/52, 5, 1995. http://cat.inist.fr/?aModele=afficheN&cpsidt=3554674.
111. T.H. Kim and Y.Y. Lee, *Bioresour. Technol.* 96, 2007, 2005. http://www.sciencedirect.com/science/article/pii/S0960852405000684.
112. F. Teymouri, L. Laureano-Perez, H. Alizadeh and B.E. Dale, *Appl. Biochem. Biotechnol.* 115, 951, 2004. http://www.ncbi.nlm.nih.gov/pubmed/15054244.
113. J. Ruffell, Pretreatment and hydrolysis of recovered fibre for ethanol production, Thesis, The University of British Columbia, 2008. http://opac.tistr.or.th/Multimedia/Web/0051/wb0051907.pdf.
114. P.G. Walsum and H. Shi, *Bioresour. Technol.* 93, 217, 2004. http://www.sciencedirect.com/science/article/pii/S0960852403003353.
115. B.E. Dale and M.J. Moreira, *Biotechnol. Bioeng. Symp.* 12, 31, 1982. http://www.osti.gov/energycitations/product.biblio.jsp?osti_id=5893198.
116. E. Sahle Demessie, A. Hassan, K.L. Levien, S. Kumar and J.J. Morrell, *Wood Fiber Sci.* 27, 296, 1995. http://swst.metapress.com/content/bg25qk0780343206.
117. Y. Zheng, H. Lin and G.T. Tsao, *Biotechnol. Prog.* 14, 890, 1998. http://www.ncbi.nlm.nih.gov/pubmed/9841652.
118. http://en.wikipedia.org/wiki/Wood-decay_fungus
119. D. Tanjore, Biological pretreatments of corn stover biomass through aerobic and anaerobic solid substrate fermentation, Ph.D., The Pennsylvania State University, 2009. http://gradworks.umi.com/33/99/3399715.html.
120. J. Shi, M.S. Chinn and R.R. Sharma-Shivappa, *Bioresour. Technol.* 99, 6556, 2008. http://www.ncbi.nlm.nih.gov/pubmed/18242083.
121. M.J. Ray, D.J. Leak, P.D. Spanu and R.J. Murphy, *Biomass Bioener.* 34, 1257, 2010. http://www.sciencedirect.com/science/article/pii/S0961953410001108.
122. V.B. Agbor, N. Cicek, R. Sparling, A. Berlin and D.B. Levin, *Biotechnol. Adv.* 29, 675, 2011. http://www.ncbi.nlm.nih.gov/pubmed/21624451.

# 9 Valorization of Hemicelluloses

## 9.1 INTRODUCTION

As a valuable renewable resource composed of monomeric C5 and C6 sugars, hemicelluloses are promised to a bright industrial future due to the value chains that can be initiated from these sugars.[1,2]

Hemicelluloses in the cell wall have the primary role of interacting with other polymers to ensure the adequate properties of the wall.[3] The most important biological function of hemicelluloses is indeed their contribution to strengthen the cell wall by interaction with cellulose and, in some walls, with lignin. However, in a large number of cases, hemicelluloses can also function as seed storage carbohydrates (reserve materials). This has happened independently many times in evolution, and it has been suggested that from a taxonomic viewpoint, hemicelluloses are as important as starch is in the role of storing carbohydrate in seeds. Much of our knowledge of hemicelluloses comes from the study of seed polysaccharides rather than the polymers in vegetative tissues (i.e., the leaves, stems, and roots).[4] The major storage hemicelluloses in seeds are polymers of the sugar mannose, particularly as mannans, glucomannans, or galactomannans, which are reserve of carbohydrate present in the *endosperm* and *perisperm* cell walls of some species.[5] More rarely, seeds store xyloglucans in the cotyledon cell wall, for example, tamarind, nasturtium, and mustard, or arabinogalactans, for example, lupins. Hemicellulose-rich seeds do not usually store starch.

In this chapter, we will successively cover the valorization of:

- Seed storage hemicelluloses and other hemicelluloses.
- Hemicelluloses extracted from papermaking processes.
- Hemicelluloses extracted from second-generation biorefineries focusing on biochemical conversion.

## 9.2  SEED STORAGE HEMICELLULOSES AND OTHER HEMICELLULOSES

### 9.2.1  Main Types and Sources of Storage Hemicelluloses

Cell wall storage hemicelluloses are found as the principal storage compounds in seeds of many taxonomically important groups of plants (Table 9.1).[6]

The *galactomannans* are matrix polysaccharides present in large amounts in the endosperm cell walls of leguminous seeds.[7] Leguminous endosperm cell walls are specialized in both structure and composition. A thin primary cell wall, of apparently normal composition, is present plus a very thick inner wall layer, composed almost exclusively of galactomannans (no cellulose, apparently no other polysaccharide components). The galactomannans are laid down during the development of the seed and are mobilized after germination, acting as a carbohydrate reserve for the developing seedling. Galactomannans are an example of a cell wall storage polysaccharide (a seed cell wall polysaccharide which serves as a storage molecule). Cell wall storage polysaccharides are all similar in structure to hemicelluloses or pectins present in the primary or secondary cell wall, from which they almost certainly arose in the course of evolution. Galactomannans are known from many economically important plants such as coconut, guar, and locust bean.[3] Not only they are especially abundant in legumes, but also occur in other seeds such as tobacco and coffee.

Other examples of cell wall storage hemicelluloses include *xyloglucans*, which are important storage polymers in the seeds of nasturtium and tamarind[8]; *galactans*, which are present in lupins; *glucomannans*, which are present in the konjac plant (in this case, the storage organ is a corm and not the seed)[3]; *arabinoxylans*, which are present in seeds of dicots such as flax and psyllium and also in cereal endosperm; and *mixed-linked glucans*, which are present in cereal endosperm.

---

**TABLE 9.1**

**Examples of Cell Wall Storage Hemicelluloses in Seeds**

| Hemicelluloses | Typical Source |
|---|---|
| Galactomannans | Endosperms, mainly of leguminous seeds (e.g., of guar) |
| Glucomannans | Endosperms, e.g., of *Asparagus*, *Endymion*, *Iris* |
| Mannans | Endosperms of palm seeds (e.g., dates); Endosperms of Umbelliferae |
| Xyloglucans (XyGs) | Cotyledons of several species, e.g., tamarind, nasturtium |
| Galactans | Cotyledons of some lupins |
| β-1,3;1,4-glucans | Endosperms of cereal grasses |

## 9.2.2   Uses of Hemicelluloses in General

Seed storage hemicelluloses are used directly as products in the food industry, for example, guar and locust bean gums (galactomannans), konjac gum (glucomannans), and tamarind gum (XyGs).[3] Furthermore, the hemicelluloses give important properties to many food and feed products. In the baking industry, the arabinoxylans affect baking quality. Mixed-linkage glucans and arabinoxylans are well-known antinutritional compounds in animal feed, and they can cause filtering and haze problems in the brewery industry due to their viscosity. To correct these problems, hemicellulases are added to feed and are used in the baking and brewery industries. Mixed-linkage glucans have a documented cholesterol-lowering effect in hypercholesterolemic humans and daily intake of mixed-linkage glucans is recommended by the U.S. Food and Drug Administration. Furthermore, xylooligosaccharides (XOS) and arabinoxylooligosaccharides (AXOS) have *prebiotic* properties.[9–11]

Fragments of cell wall polysaccharides such as xyloglucan, xylan, galactoglucomannan, cellulose, and rhamnogalacturonan have also been shown to act as elicitors that trigger the defense mechanism of plants.[12] Among studied oligosaccharidic elicitors, COS-OGA is currently undergoing evaluation at the European level for registration as a plant protection product. The COS-OGA active substance consists of a complex of chitosan fragments (chitooligosaccharides, COS), which are fond in fungal cell walls and crustacean exoskeletons, that are associated with pectin fragments (oligogalacturonides, OGA) originating from plant cell walls.[13]

Hemicelluloses find also applications as sustainable films and coatings (Figure 9.1).[14]

The targeted uses of hemicelluloses have primarily been packaging films for foodstuffs and biomedical applications. Oxygen permeability was typically comparable with values found for other biopolymer films. The modification of hemicelluloses to create more hydrophobic films reduced the water vapor permeability. However, modified hemicellulose coatings intended for food still exhibited water vapor permeabilities,

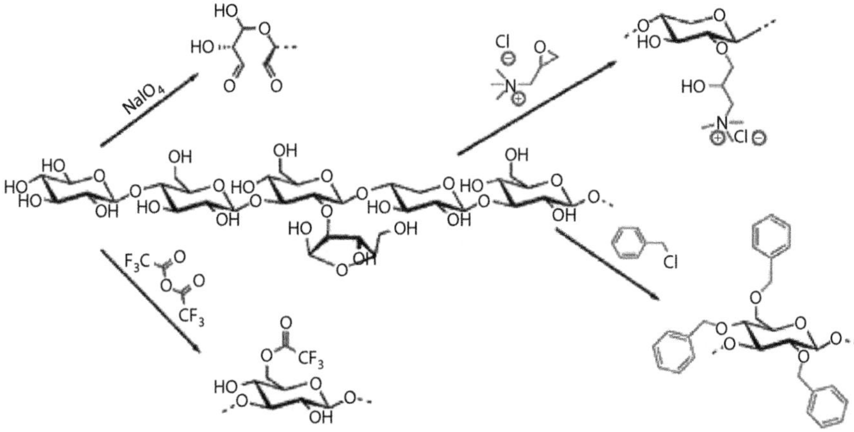

**FIGURE 9.1**   Films and coatings from hemicelluloses. (Reprinted with permission from Hansen, N.M.L. and Plackett, D., *Biomacromolecules.* 9(6), 1493–1505, 2008. Copyright 2008, American Chemical Society.)

several magnitudes higher than those of other polymers currently used for this purpose. Research on hemicelluloses for biomedical applications has included biocompatible hydrogels and coatings with enhanced cell affinity. Several possibilities exist for chemically modifying hemicelluloses, and studies of films from modified hemicelluloses have identified other potential applications such as selective membranes.

Last but not least, as will be shown in Section 9.4 and was reviewed in Chapter 5, hemicellulose hydrolysis results in sugars, largely pentoses, which can be chemically or biochemically converted into ethanol or a variety of chemicals. Platform molecules have recently been intensively investigated as intermediates for transformation of lignocellulosic biomass into fuels and chemicals.[15,16] Progress is underway on strategies for process integration of chemocatalytic biomass valorization.

### 9.2.3 Uses of Some Specific Hemicelluloses

#### 9.2.3.1 Galactomannans

The largest market for guar galactomannans is in the food industry. Applications include[17]:

- *Baked goods*: Increases dough yield, gives greater resiliency, and improves texture and shelf life; in pastry fillings, it prevents *weeping* (syneresis) of the water in the filling, keeping the pastry crust crisp.
- *Dairy*: Thickens milk, yogurt, kefir, and liquid cheese products; helps maintain homogeneity and texture of ice creams and sherbets.
- *Meat*: Functions as lubricant and binder.
- *Dressings and sauces*: Improves the stability and appearance of salad dressings, barbecue sauces, relishes, ketchups, and so on.
- *Miscellaneous*: Dry soups, instant oatmeal, sweet desserts, canned fish in sauce, frozen food items, and animal feed.

Nonfood industrial applications for guar galactomannans include[17]:

- *Textile industry*: Sizing, finishing, and printing.
- *Paper industry*: Improved sheet formation.
- Explosives industry, as waterproofing agent mixed with ammonium nitrate, nitroglycerin, and so on.
- Pharmaceutical industry, as binder or as disintegrator in tablets.
- *Cosmetics and toiletries industries*: Thickener in toothpastes, conditioners in shampoos (usually in chemically modified version).
- Oil and gas drilling, hydraulic fracturing.
- Mining.
- *Hydroseeding*: Formation of seed bearing *guar tack*.

Guar galactomannans are also known for their nutritional and medicinal effects.[17]

Locust bean galactomannans are used as a thickening agent and gelling agent in food technology.[18,19] They are soluble in hot water.

**TABLE 9.2**
**Applications of Hemicelluloses**

| Type | Application |
|---|---|
| Hemicelluloses in general | Films and coatings; biomedical |
|  | Fuels and chemicals (Sections 9.3 and 9.4) |
| Seed storage hemicelluloses in general | Food industry |
| Seed storage galactomannans | Food industry, textiles, paper, explosives, pharmaceuticals, cosmetics, oil and gas drilling, mining, hydroseeding |
| Konjac root glucomannans | Water-soluble dietary fibers |
| Tamarind seed XyGs | Paper, food, textiles, pharmaceuticals, cosmetics |

### 9.2.3.2 Glucomannans

Konjac glucomannans are water-soluble dietary fibers and can be used as a gelling agent, thickener, stabilizer, emulsifier, and film former.[20]

### 9.2.3.3 Xyloglucans

Tamarind XyGs are having applications in paper, food, textile industry, and so on.[21] In recent years, research has been initiated on the use of tamarind xyloglucans in pharmaceutical and cosmetic applications. Pharmaceutical applications include binder in tablet dosage form, ophthalmic drug delivery, sustained drug delivery, ocular drug delivery, and controlled release of spheroids.

### 9.2.4 SUMMARY OF HEMICELLULOSE APPLICATIONS

A summary of hemicellulose applications is presented in Table 9.2.

## 9.3 PAPERMAKING: VALUE PRIOR TO PULPING

The concept of value prior to pulping has been proposed, where the hemicelluloses are either partially or completely extracted for the production of biofuels and chemicals.[22] The remaining solids (mainly cellulose and lignin) can be further delignified for wood pulp or fiber production. Such preextraction of hemicelluloses would mesh with efforts to convert current pulp mills into biomass-integrated biorefineries. The extracted hemicellulosic fraction could be transformed into biofuels, biochemicals, and biomaterials such as bioethanol, biohydrogen, furfural, xylitol, bifunctional organic molecules, barrier films, and hydrogels.

The hemicelluloses released during the pulping process accumulate and increase the effluent load, contributing to the posttreatment cost in a papermaking process.[22] Therefore, preextraction of hemicelluloses from wood chips prior to pulping not only can offer new feedstocks for the production of chemicals and fuels but also would remedy some of the operational problems in pulp and paper mills. In addition, extracting hemicelluloses prior to pulping can also benefit the energy recovery from chemical pulping liquors, because the heating value of hemicelluloses is about the half of that of lignin.

So far, all the research devoted to the value prior to pulping concept has been focusing on chemical pulping processes, and few attempts have been applied to mechanical pulping processes.[22] The implementation of hemicellulose extraction prior to pulping (prehydrolysis) is expected to affect the operation of a conventional kraft pulp mill.[23] The magnitude of impacts will depend especially on the extraction conditions.

The pulp and paper industry provides an opportunity to expand the range of the products that are manufactured to an integrated forest biorefinery, thus improving the utilization of the woody biomass.[24] During kraft pulping, hemicelluloses are degraded into low molecular weight saccharinic acids and end up in the black liquor together with degraded lignin. To recover energy, black liquors are concentrated and burned. As the heating value of hemicelluloses is considerably lower than that of lignin, extracting the hemicelluloses before the pulping stage for generation of high value products has the potential to improve overall economics. The combination of hemicelluloses extraction with chemical pulping processes is one approach to generate a sugar feedstock amenable to biochemical transformation to fuels and chemicals.

## 9.4 BIOREFINERIES

Biorefining is the sustainable processing of biomass into a spectrum of biobased products, and bioenergy. Biorefineries are already applied for ages in, for example, the food industry. Large-scale implementation of biorefineries for nonfood applications, however, is still lacking except for ethanol production.

The world is investing deeply in long-term solutions to ensure that the most can be made from what the planet produces. Biorefineries combine technologies and processes that convert various types of biomass into biofuels, chemicals, polymers, and other high-value products.

### 9.4.1 PENTOSE FERMENTATION

A major challenge in bioethanol production using lignocellulosic feedstock is inefficient utilization of hemicelluloses, which account for ~25% of lignocellulosic biomass.[25] Xylose, comprising >60% of recoverable sugars from hemicelluloses, is a major product of the hemicellulose hydrolysis. Utilization of this carbon source would significantly increase the ethanol yield from an estimated 60 gal/dry ton to 90 gal/dry ton.

Due to the very attractive properties of the yeast *Saccharomyces cerevisiae* in industrial fermentation (winemaking, baking, and brewing), there have been considerable efforts to design recombinant pentose and arabinose fermenting strains of this yeast.[26] On account of *S. cerevisiae* cannot utilize xylose but does utilize and ferment its isomer D-xylulose (a five-carbon sugar including a ketone group), the obvious way to allow xylose metabolism is to introduce a heterogeneous pathway converting xylose to xylulose.[27] Several approaches have been explored to express a pentose utilization pathway from naturally pentose-utilizing bacteria and fungi in *S. cerevisiae*. Figure 9.2 summarizes the initial pathways for D-xylose utilization in bacteria and fungi.

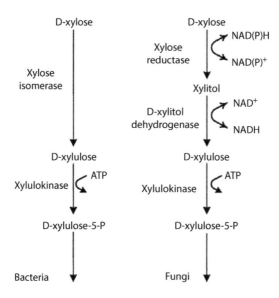

**FIGURE 9.2**   The initial xylose utilization pathways in bacteria and fungi. (Adapted from Hahn-Hagerdal, B. et al., *Appl. Microbiol. Biotechnol.*, 74, 937, http://www.ncbi.nlm.nih.gov/pubmed/17294186, 2007.)

In naturally D-xylose-utilizing bacteria, xylose is isomerized to D-xylulose by xylose-isomerase. Xylulose is then phosphorylated to xylulose 5-phosphate, which is an intermediate of the pentose phosphate pathway (PPP, secondary pathway for glucose metabolism). A similar pathway has been found in an anaerobic fungus; however, most naturally xylose-utilizing fungi contain a more complex pathway consisting of reduction–oxidation reactions involving the cofactors NAD(P)H and NAD(P)+. Xylose is reduced to xylitol by a xylose reductase (XR), and xylitol is then oxidized to D-xylulose by a xylitol dehydrogenase (XDH). As in bacteria, xylulose is phosphorylated to D-xylulose 5-phosphate by a xylulokinase (XK). The conversion of L-arabinose into intermediates of the PPP requires more enzymatic reactions than the conversion of xylose (Figure 9.3).

Xylose-fermenting strains of *S. cerevisiae* can be constructed by introducing genes encoding xylose isomerase from bacteria and fungi, or genes encoding xylose reductase and xylilol dehydrogenase from fungi.[26] Also the endogeneous gene encoding xylulokinase has to be overexpressed to obtain significant xylose fermentation. Transport proteins are needed for uptake of xylose, and other sugars in yeast. In *S. cerevisiae*, it has been found that xylose is transported by the hexose transporters, but the affinity for xylose is ~200-fold lower than for glucose (Figure 9.4). Hence, xylose uptake is competitively inhibited by glucose. It has been suggested that xylose is taken up by both high- and low-affinity systems of glucose transporters, but the uptake is increased in the presence of low glucose concentrations. Consequently, to obtain efficient cofermentation of xylose and glucose in SSCF (simultaneous saccharification and cofermentation) with recombinant *S. cerevisiae*, it is necessary to keep the glucose concentration low.

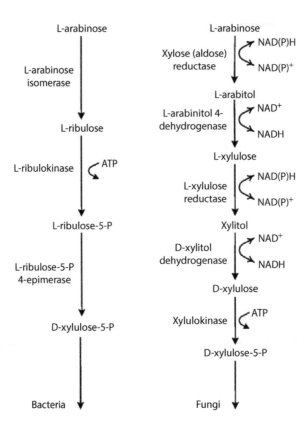

**FIGURE 9.3**   The initial L-arabinose pathways in bacteria and fungi. (Adapted from Hahn-Hagerdal, B. et al., *Appl. Microbiol. Biotechnol.,* 74, 937, http://www.ncbi.nlm.nih.gov/pubmed/17294186, 2007.)

### 9.4.2   BUILDING BLOCKS FROM SUGARS

Hemicelluloses contain a number of C5 and C6 sugars in their backbone and side chains. In this sense, they are unique polysaccharides that have the potential to generate a wide variety of building blocks and added value products. Sources of glucose from starch and cellulose are also mentioned in this section to give a complete view of sugar-derived molecules.

In 2004, the DOE (U.S. Department of Energy) identified twelve building block chemicals that can be produced from sugars (carbohydrates including hemicelluloses and cellulose) via biological or chemical conversion.[28] The twelve building blocks can be subsequently converted to a number of high-value biobased chemicals or materials. Building block chemicals, as considered for this analysis, are molecules with multiple functional groups that possess the potential to be transformed into new families of useful molecules. The twelve sugar-based building blocks are as follows:

- 1,4-diacids (succinic, fumaric, and malic)
- 2,5-furan dicarboxylic acid

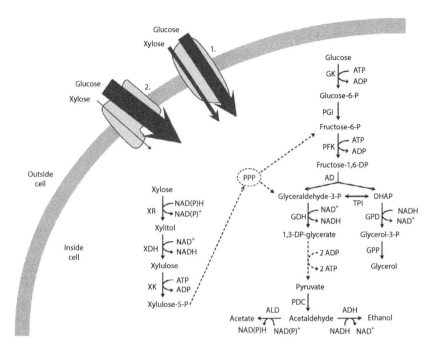

**FIGURE 9.4** Simplified scheme of sugar transport and metabolism in *S. cerevisiae*. (1) Low- and intermediate-affinity hexose transporters. (2) High-affinity hexose transporters. PPP, pentose phosphate pathway; XR, xylose reductase; XDH, xylitol dehydrogenase; XK, xylulokinase, GK, Glucokinase; PGI, phosphoglucose isomerase; PFK, phosphofructokinase; AD, aldolase; TPI, triose phosphateisomerase; GDH, glyceraldehyde-3-P dehydrogenase; GPD, glycerol-3-P dehydrogenase; GPP, glycerol-3-phosphotase; PDC, pyruvate decarboxylase; ALD, acetaldehyde dehydrogenase; ADH, alcohol dehydrogenase. (From Olofsson, K. et al., *Biotechnol. Biofuels*, 1, 7, http://www.biotechnologyforbiofuels.com/content/1/1/7, 2008.)

- 3-hydroxy propionic acid
- Aspartic acid
- Glucaric acid
- Glutamic acid
- Itaconic acid
- Levulinic acid
- 3-hydroxybutyrolactone
- Glycerol
- Sorbitol
- Xylitol/arabinitol

These building blocks include 1,4-acids (succinic, fumaric, and malic), two acid–alcohols, one ketoacid, two aminoacids, one lactone, one furanic derivative, and polyols, that is, mainly multifunctional molecules.

In the 6 years since the original DOE report, considerable progress in biobased products development has been made. In 2010, Bozell and Peterson presented an updated group of candidate structures based on advances since 2004.[29] Some of the compounds described by Bozell are members of DOE's original list. Several new compounds also appear, and represent advances in technology development.

The list presented by Bozell included 10 compounds:

- Ethanol.
- Furans.
- Glycerol and derivatives: glycerol is not a carbohydrate, but structurally it can be considered as a *mini-sugar*, in that transformations appropriate to glycerol may be applied to carbohydrates.
- Biohydrocarbons.
- Lactic acid.
- Succinic acid.
- Hydroxypropionic acid/aldehyde.
- Levulinic acid.
- Sorbitol.
- Xylitol.

Several organic acids (fumaric, malic, aspartic, glucaric, glutamic, and iataconic), 3-hydroxylactone, and arabinitol all from the 2004 list were not included in the 2010 list. Ethanol, furans, glycerol and derivatives, biohydrocarbons, lactic acid, and hydroxypropionic aldehyde were added to the 2004 list.

In the next sections, biobased building blocks are classified according to the number of carbon atoms in the compound (from C2 to C6 and C6+) and include organic acids, alcohols, furans, aldehydes, and lactams.

### 9.4.2.1   C2 Molecules

Recent technology developments and strategic commercial partnerships have positioned ethanol as a feedstock for chemical production, improving its platform potential (Table 9.3). In particular, it can be converted by dehydration into ethylene, a key C2 monomer in the synthesis of many polymers including polyethylene, polypropylene, PVC, and PET.[30,31]

---

### TABLE 9.3
### C2 Platform Molecules

| Molecule Molecular Formula Structure | Manufacturing Process (Producers) | Main Application(s) (Producers) |
|---|---|---|
| Ethanol $C_2H_6O$ OH | Glucose fermentation (*many producers in the word*) | Biofuel Important building block in the chemical industry Ethylene glycol and PET (*Coca-Cola*) Polyethylene (*Braskem*) Polypropylene (*Braskem*) |

Coca-Cola's PlantBottle was showcased as part of The Coca-Cola Company's pavilion at Expo Milano 2015 a global showcase for sustainable innovation.[32] Coca-Cola has displayed 100% biobased content PlantBottles produced using paraxylene produced at Virent's Madison, the first demonstration scale production of a PET plastic bottle made entirely from plant-based materials.[33]

### 9.4.2.2 C3 Molecules

The C3 building blocks include lactic acid, acrylic acid, 3-hydroxypropionic acid, 3-hydroxypropionic aldehyde, 1,3-propanediol, and glycerol (Table 9.4).

1. *Lactic acid* is commercially produced by glucose fermentation. The primary use for lactic acid is the production of polylactic acid. Although lactic acid cab undergoes direct polymerization, the process is more effective if lactic acid is first converted to a low molecular prepolymer (MW ~5000) and then polymerized to the lactide. Lactic acid has been suggested as a platform material for the production of several downstream materials. Lactic acid undergoes ready esterification to give lactate esters, of interest as new green solvents. Catalytic reduction of lactic acid leads to propylene glycol, which can be further dehydrated to give propylene oxide. Alternatively, lactic acid can be dehydrated to give acrylic acid and esters, but in practice, this conversion proceeds in low yield.
2. Biobased *acrylic acid* is produced from either PHA (polyhydroxyalkanoates) or 3-hydroxypropiocic acid. Acrylic acid leads to acrylates, superabsorbent polymers, water treatment chemicals, coatings, and adhesives.[34]
3. *Glycerol* is a byproduct of the transesterification reaction of triglycerides (vegetable oil) with ethanol in the manufacture of biodiesel. It can lead to
   a. Products resulting from glycerol reduction such as ethylene glycol.
   b. Products obtained by glycerol dehydration via loss of a primary hydroxyl group (hydroxypropionaldehyde and acrolein) or at the secondary hydroxyl group (hydroxyacetone).
   c. *1,3-Propanediol* (1,3-PDO), which can result from the biochemical conversion of glycerol. 1,3-PDO is currently produced mainly from glucose fermentation using a transgenic *Escherichia coli* developed by Genencor and DuPont. The dialcohol is one of the components of DuPont's Sorona (1,3-PDO and terephthalic acid), a polymer used in textiles and carpeting.[29] The polymer, an aromatic polyester, is called polytrimethylene terephthalate (PTT).
   d. *Epichlorhydrin* by chlorination. Solvay has developed a new process for the production of biobased epichlorhydrin from glycerol, called Epicerol.

Glycerol is somehow the vegetable equivalent of propylene in petrochemistry.[35] Biotransformation and catalytic chemistry are two pillars for the development of glycerol chemistry.

**TABLE 9.4**

**C3 Platform Molecules**

| Molecule<br>Molecular Formula<br>Structure | Manufacturing Process<br>(Producers) | Main Applications |
|---|---|---|
| Lactic acid<br>$C_3H_6O_3$ 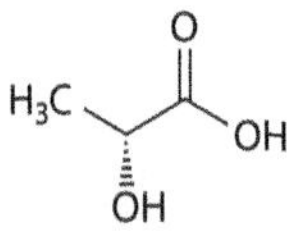<br><br>H₃C ⋮ OH structure (O double bond, OH) | Glucose fermentation<br>(*Galactic/Total*) | Lactate esters (*Galactic*)<br>Propylene glycol and propylene oxide<br>Acrylic acid and esters<br>PLA (*NatureWorks*) |
| Acrylic acid<br>$C_3H_4O_2$ 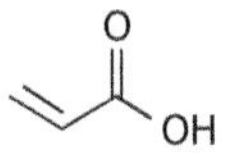<br><br>structure (O double bond, OH) | From polyhydroxy-<br>alkanoates (PHA)<br>(*Metabolix*)<br>From 3-hydroxypropionic<br>acid (*BASF, Cargill,<br>Novozymes*) | Acrylates and acrylic polymers<br>Superabsorbent polymers[34]<br>Products for water treatment<br>Coatings and adhesives |
| 3-hydroxypropionic acid<br>$C_3H_6O_3$ 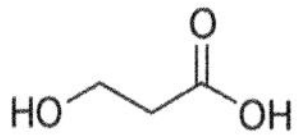<br><br>HO structure O OH | Glycerol fermentation<br>(*DuPont*) | Acrylates[36]<br>1,3-Propanediol |
| 3-hydroxypropionic aldehyde<br>(HPA)<br>$C_3H_6O_2$<br>Structure: Aldehyde instead of<br>carboxylic in the earlier<br>structure | Glycerol fermentation<br>(*Dupont*) | Acrylates[28]<br>1,3-Propanediol<br>3-hydroxypropionic acid<br>Acrolein |
| 1,3-Propanediol<br>$C_3H_8O_2$ <br><br>HO OH | • Glucose fermentation<br>(*DuPont*)[37]<br>• Glycerol fermentation | Polytrimethylene terephthalate<br>(PTT)[38]<br>Cosmétiques, alimentation et<br>médicaments |
| Glycerol<br>$C_3H_8O_3$ 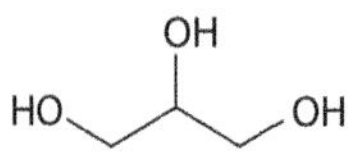<br><br>OH<br>HO OH | Byproduct in the<br>production of biodiesel<br>from vegetable oil<br>(triglycerides) | Ethylene glycol, propylene glycol,<br>acetol and lactic acid obtained by<br>glycerol reduction<br>Hydroxypropionaldehyde and<br>acrolein (a precursor of acrylic<br>acid), and hydroxyacetone obtained<br>by glycerol dehydratation l[39,40]<br>1,3-Propanediol (*DuPont*) obtained<br>(a) by glycerol fermentation;[41–43]<br>(b) by glucose fermentation<br>Epichlorhydrin (*Solvay*)[44] obtained by<br>reaction with HCl, leading to epoxy<br>resins |

### 9.4.2.3 C4 Molecules

Table 9.5 presents the major C4 building blocks:

1. The C4 platform molecules include the succinic, maleic, and fumaric acids. Succinic acid is a widely investigate building block available from bio-chemical transformation of sugars. A process based on recombinant *E. coli* has been licensed by Roquette and is part of a Roquette/DSM joint venture. An alternative *E. coli* strain originally developed by the U.S. Department

**TABLE 9.5**
**C4 Platform Molecules**

| Molecule<br>Chemical Formula<br>Structure | Manufacturing Process (Producers) | Main Application(s)<br>(Producers) |
|---|---|---|
| Succinic acid<br>$C_4H_6O_4$<br> | Glucose fermentation (*BioAmber, Roquette/DSM*) | Monomer in nylon and polyester polymerization<br>1,4-butanediol, tetrahydro-furan and γ-butyrolactone<br>Fumaric acid and maleic acid by dehydrogenation |
| Malic acid<br>$C_4H_6O_5$<br> | Glucose fermentation | See succinic acid |
| Fumaric acid<br>$C_4H_4O_4$<br> | Glucose fermentation[48] | See succinic acid |
| Aspartic acid<br>$C_4H_7NO_4$<br> | Chemical synthesis<br>Protein extraction<br>Fermentation<br>Enzymatic conversion | Aspartame (a sweetener)<br>2-Amino-1,4-butanediol<br>Amino-2-pyrrolidone<br>Substituted amino-diacids<br>Aspartic anhydride<br>Amino-γ-butyrolactone<br>3-Aminotetrahydrofuran |
| Isobutanol<br>$C_4H_{10}O$<br> | Sugar fermentation by yeasts and separation of isobutanol (*Gevo*) | Chemicals and fuels (*Gevo*)<br>Partnership with *Coca-Cola* on 100% biobased PET for bottles (*Gevo*) |

(*Continued*)

**TABLE 9.5 (Continued)**
**C4 Platform Molecules**

| Molecule Chemical Formula Structure | Manufacturing Process (Producers) | Main Application(s) (Producers) |
|---|---|---|
| 1,4-Butanediol $C_4H_{10}O_2$  | Catalytic conversion of biobased succinic acid (*BioAmber*)[49] • Sugar fermentation (*Genomatica* and its licensees such as *BASF*[50] et *Novamont*[51]) • Catalytic conversion of γ-butyrolactone (*Metabolix*)[52] | Tetrahydrofuran Polybutylene terephthalate (PBT) (*Lanxess, DSM*) |
| Ethyl acetate $C_4H_8O_2$ | Acetogenic process for the fermentation of sugars to acetic acid (*Zeachem*)[53] | Compounds in C2 (ethanol), C3, C4 et C6 by a biochemical and thermochemical process (*Zeachem*) |
| 3-hydroxybutyrolactone $C_4H_6O_3$  (above: γ-butyrolactone) | Production via chemical transformations[28] Oxidative degradation of starch[28] | Acrylate-lactone[28] 2-Amino-3-hydroxy tetrahydrofuran 3-Aminotetrahydrofuran 3-Hydroxytetrahydrofuran Epoxy-lactone γ-Butenyl-lactone |
| Furan $C_4H_4O$ | Catalytic decarboxylation of furfural | Precursor in fine chemistry Tetrahydrofuran ($C_4H_8O$) |

of Energy has been licensed by Bioamber, which recently commissioned a production facility. Maleic and fumaric acids, which are available from dehydrogenation/cyclization of succinic acid, have a chemistry similar to that of succinic acid. Succinate esters are precursors for petrochemical products such as 1,4-butanediol, tetrahydrofuran, γ-butyrolactone, or various pyrrolidone derivatives. Succinic is a component of biobased polymers such as nylons and polyesters.

2. Aspartic acid is one of the 22 genetically coded 22 amino acids.[45] It is an essential part of metabolism among many species, including humans, for protein production.[28] There are several configurations of aspartic acid produced; however, L-aspartic is by far the most common. L-aspartic acid is mainly used for the production of aspartame, a synthetic sweetener.

3. Isobutanol is one of the isomers of butanol. It is very used not only as solvent in chemical reactions but also as reactive in organic synthesis.[46] It is

naturally produced during sugar fermentation. It is also a byproduct of the decomposition of organic matter. Isobutanol is an important building block with applications in several chemicals and fuels markets. The company Gevo has a partnership with Coca-Cola to produce paraxylene (a precursor of terephthalic acid, representing 70% of PET[47]) from biobased isobutanol, for the production of 100% biobased PET bottles.

4. 1,4-butanediol results from the catalytic conversion of succinic acid, from sugar fermentation or from catalytic conversion of γ-butyrolactone. It leads to THF (tetrahydrofuran) and to PBT (polybutylene terephthalate). 3-hydroxybutyrolactone is obtained chemically in particular by oxidative degradation of starch.[28] Hydroxybutyrolactone can lead to various lactone and tetrahydrofuran derivatives.

5. Furan is mainly obtained by decarboxylation of furfural. It is a precursor in fine chemistry. It leads to tetrahydrofuran by hydrogenation.

### 9.4.2.4 C5 Molecules

C5 building blocks are shown in Table 9.6. They include glutaric, levulinic, itconic and glutamic acids, isoprene, furfural, and C5 polyols.

1. Glutaric acid occurs in plant and animal tissues, in particular in the juice of immature sugar beets. It is used in organic synthesis and as an intermediate for the manufacture of polymers such as polyamides and polyesters, ester plasticizers, and corrosion inhibitors.[54] 1,5-Pentanediol is manufactured by hydrogenation of glutaric acid and its derivatives. Glutaric acid can be prepared by fermentation with recombinant *E. coli*.[55]

2. Levulinic acid is produced from acid treatment of C6 sugars. Its formation proceeds by initial loss of water to form hydroxymethyfurfural (HMF) as an intermediate. Readdition of water to HMF induces ring cleavage to form levulinic acid and an equivalent of formic acid. Levulinic acid is catalytically transformed into substituted pyrrolidines, lactones, levulinate esters, and diphenolic acid, a potential green replacement for bisphenol A in the production of polycarbonates.

3. Itaconic acid is a C5 dicarboxylic acid, also known as methyl succinic acid. Itaconic acid is currently produced via fungal fermentation and is used primarily as a specialty monomer. The major applications include the use as a copolymer with acrylic acid and in styrene–butadiene systems.

4. Isoprene is a high value hydrocarbon with a world market of $1–2 billion. The immediate precursor to isoprene and naturally occurring polyisoprenoids is isopentenyl diphosphate (IPP). Isoprene can be obtained by fermentation of glucose. It is mainly used to the production of cis-1,4-polyisoprene, a rubber used in the manufacture of tyres and surgical gloves.

5. Glutamic acid is classified, together with aspartic acid, as an acidic amino acid. It has one additional methylene group in its side chain than does aspartic acid. In neuroscience, glutamate is an important neurotransmitter that plays the principal role in neural activation. Monosodium glutamate (MSG)

## TABLE 9.6
## C5 Platform Molecules

| Molecule Chemical Formula Structure | Manufacturing Process (Producers) | Main Application(s) |
|---|---|---|
| Glutaric acid $C_5H_8O_4$ 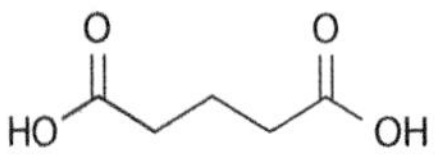 | Fermentation with recombinant *E. coli* | Monomer for the production of polyesters and polyamides 1,5-Pentanediol (a plasticizer and precursor to polyesters) |
| Levulinic acid $C_5H_8O_3$ 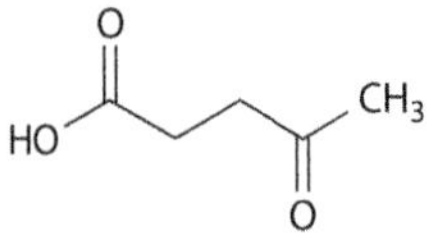 | Acid treatment of C6 sugars to form hydroxymethylfurfural as an intermediate (HMF; $C_6H_6O_3$; see structure below)  | Substituted pyrrolidones, lactones, levulinate esters Diphenolic acid |
| Itaconic acid $C_5H_6O_4$ 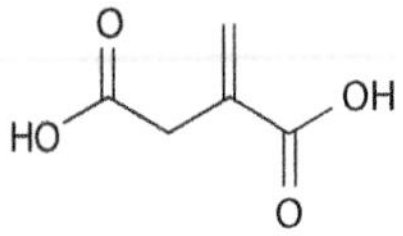 | Fungal fermentation | Comonomer in the production of copolymers with acrylic acid and in styrene-butadiene systems |
| Isoprene $C_5H_8$ 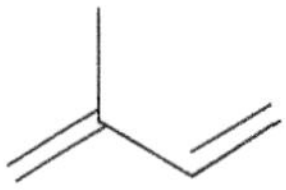 | Glucose fermentation (*Ajinomoto* in collaboration with *Bridgestone*; *DuPont* in collaboration with *Goodyear*; *Amyris* in collaboration with *Michelin*; *Elevance* in collaboration with *Hutchinson*; *Aemetis*)[57] | Polyisoprene |
| Glutamic acid $C_5H_9NO_4$ 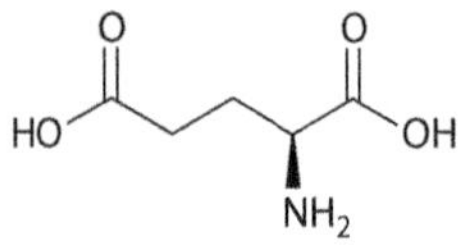 | Fermentation | Food additive |
| Xylitol and arabitol $C_5H_{12}O_5$ 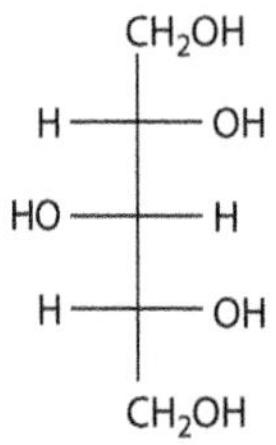 | Catalytic hydrogenation of xylose[29] Microbial conversion of xylose to xylitol Extraction from biomass pretreatments | Sweetener Ethylene glycol, propylene glycol, glycerol, and lactic acid[28] Xylaric acid Mixture of hydroxy furan Polyesters Antifreeze |

*(Continued)*

**TABLE 9.6 (*Continued*)**
**C5 Platform Molecules**

| Molecule Chemical Formula Structure | Manufacturing Process (Producers) | Main Application(s) |
|---|---|---|
| Furfural $C_5H_4O_2$ | Xylose dehydration | Furfuryl alcohol and tetrahydrofurfuryl alcohol (THFA) Phenol-furfural type resins Solvent |
| Alkyl polypentosides (APP) | Fischer's glycosylation | Dishwashing and laundry detergents Cosmetics cleaning products |

is the sodium salt of glutamic acid, one of the most abundant naturally occurring nonessential amino acids. MSG is found in many vegetables and fruits. It is used in the food industry as a flavor enhancer.

6. Xylitol and arabitol have the same chemical formula $C_5H_{12}O_5$. The main chain of the two polyols includes five alcohol groups. They are produced by hydrogenation of sugars or by extraction from biomass pretreatment processes.[28] The two polyols are sweeteners and are used for the manufacture of ethylene glycol, propylene glycol, glycerol, lactic acid, xylaric acid, and a mixture of hydroxy furans. They are also starting materials for polyesters.

7. Furfural results from the dehydration of xylose. It leads to the manufacture of furfuryl alcohol and tetrahydrofurfuryl alcohol, and phenol-furfural type resins. It can also serve as a solvent.

8. Pentose-derived surfactants (APP for Alkyl PolyPentosides) are gaining interest as efficient ingredients based on agriculture waste valorization.[56] They can be obtained on an industrial scale by Fisher-Type glycosylation. Natural fatty alcohol from coconut or palm kernel oil can be used to build up the hydrophobic part of the surfactant. They have uses in dishwashing and laundry detergents, cosmetics, and cleaning products.

### 9.4.2.5 C6 Molecules

C6 building blocks (Table 9.7) mainly include adipic, saccharic, 2,5-furan dicarboxylic acid, caprolactam, and sorbitol.

1. Adipic acid is known to be a monomer of nylon 6,6. Biobased adipic acid developed by Verdezyne is produced by fermentation. It is also a monomer of polyurethane, and its esters are PVC plasticizers.

**TABLE 9.7**
**C6 Platform Molecules**

| Molecule Chemical Formula Structure | Manufacturing Process (Producers) | Applications (Producers) |
|---|---|---|
| Adipic acid $C_6H_{10}O_4$ | Fermentation of vegetable oils, alkanes, or sugars (*Verdezyne*)[57] | Monomer of nylon 6,6 Monomer of polyurethane PVC plasticizers (adipic acid esters) |
| Saccharic acid (or glucaric) $C_6H_{10}O_8$ | Oxidation of glucose by nitric acid Glucose catalytic oxidation with a bleach | Esters and sels Polyhydroxypolyamides New polyesters Laundry surfactants |
| 2,5-furan dicarboxylic acid(FDCA) $C_6H_4O_5$ | Selective oxidation of hydroxymethylfurfural (*Avantium*) | Substitute of terephthalic acid in the synthesis of polyesters such as PET (*Avantium*) Partnership of *Avantium* with *Coca-Cola* on 100% biobased PEF (polyethylene furanoate) as a substitute to 100% biobased PET[58] |
| Caprolactam $C_6H_{11}NO$ | Production from lysine, from biobutadiene, from bioacrylonitrile and from biobenzene[59] | Monomer of nylon-6 |
| Sorbitol $C_6H_{14}O_6$ | Chemical reduction of glucose Production from sucrose and mixtures of fructose and glucose | Intermediate in the production of hydrocarbons Sweetener Ethylene glycol, propylene glycol, lactic acid, glycerol Isosorbide 1,4-Sorbitan and 2,5-anhydrosugars; antifreeze Polyesters |

2. Saccharic (also called glucaric) acid is obtained by oxidation of glucose with nitric acid. It leads to new nylons (polyhydroxypolyamides). Saccharic acid (and its esters) leads also to the production of new polyester types. In addition, it could find applications in the market of laundry surfactants.

3. 2,5-Furan dicarboxylic acid is obtained by selective oxidation of hydroxy-methylfurfural. It is a potential substitute of terephthalic in the manufacture of polyesters such as PET. The Dutch company Avantium works in partnership with Coca-Cola to produce a 100% biobased PEF (polyethylene furanoate). Virent and Gevo also work in partnership with Coca-Cola to produce a 100% biobased PET.
4. Polycaprolactam can be produced from lysine, biobutadiene, bioacrylonitrile, and biobenzene. It is a monomer of nylon 6.
5. Sorbitol, a sweetener, is obtained by chemical reduction of glucose. It is also produced from sucrose or from mixtures of fructose and glucose. It is an intermediate in the production of hydrocarbons and is a starting material for many products going from ethylene glycol to propylene glycol, lactic acid, glycerol, isosorbide, 1,4-sorbitan, and polyesters.

### 9.4.2.6 C6+ Molecules

Examples of platform compounds with more than six carbon atoms are farnesene (a $C_{15}H_{24}$ hydrocarbon) and generally long-chain hydrocarbons ($C_nH_m$) (Table 9.8).

1. The U.S. Company Amyris produces farnesene by fermentation of sugars. Farnesene is a starting material for biobased fuels such as diesel.
2. The U.S. Company Virent produces hydrocarbons by catalytic transformation of soluble sugars (« aqueous phase reforming »), which are the starting materials for biobased fuels such as gasoline, kerosene, and diesel.

**TABLE 9.8**
**C6+ Platform Molecules**

| Molecule Chemical Formula Structure | Manufacturing Processes (Producers) | Applications (Producers) |
|---|---|---|
| Farnesene $C_{15}H_{24}$ | Sugar fermentation (*Amyris*) | Diesel and other biobased fuels (*Amyris*) |
| Long chain hydrocarbons $C_nH_m$ such as octane $C_8H_{18}$: | Catalytic transformation of soluble sugars into hydrocarbons (*Virent*) | Gasoline, diesel, and kerosene (*Virent*) |

### 9.4.3  Case Study: EU Biocore Project—An Innovative Fractionation

The first step of the EU Biocore project (2010–2014) was the fractionation that aims to extract the three major components of biomass.[60,61] This step is critical because it governs subsequent steps and largely determines how fractions can be valorized. Lignocellulosic biomass is particularly difficult to fractionate because it is a composite material that is chemically and structurally complex, being held together by covalent and noncovalent bonds. Hemicelluloses are the first extracted fraction in numerous fractionation processes of lignocellulosic materials.[62]

The composition of biomass, in terms of the proportion of its three major components, varies according to botanical origin. Moreover, the chemical structure of the three components also varies from one species to another. In particular, the hemicellulose component is mainly composed of xylans in grasses and deciduous wood, but in resinous species, it is composed of glucomannans.

Biocore used a CIMV organosolv technology to extract the fractions. The CIMV Company, a Biocore partner, has developed an acetic/formic acid-based organosolv process. The CIMV process cogenerates partially depolymerized hemicelluloses and sulfur-free, low-molecular-weight lignins. Besides, the derived cellulose pulp has very low lignin content.[63] Biocore developed various valorization routes for the primary biorefinery products.[64] In addition to the use of cellulose pulp for paper applications, Biocore had as objective to demonstrate technoeconomic feasibility for the use of lignins as bioadhesives and hemicelluloses as feed/food additives (*prebiotics*).

Once biomass has been properly fractionated into cellulose, hemicelluloses, and lignin, it is possible to develop a whole family of biorefinery products within appropriate valorization chains.[65] In Biocore, both white biotechnology, which uses enzymes and microbes as catalysts, and chemistry were deployed to develop a whole range of molecules that will target different market sectors. A key feature of Biocore was to provide several polymer types. This is because forecasts indicate that biobased polymers will constitute one of the most dynamic future markets for biobased products.

## REFERENCES

1. ACS Symposium #864, *Hemicelluloses: Science and Technology*, P. Gatenholm and M. Tenkanen, (Eds.). Washington, DC, 2003. http://pubs.acs.org/isbn/9780841238428.
2. L.J. Gibson, *J. R. Soc. Interface*. http://rsif.royalsocietypublishing.org/content/early/2012/08/07/rsif.2012.0341.full.
3. H.V. Scheller and P. Ulvskov, *Ann. Rev. Plant Biol.* 61, 263, 2010. http://arjournals.annualreviews.org/doi/full/10.1146/annurev-arplant-042809-112315?amp;searchHistoryKey=%24%7BsearchHistoryKey%7D.
4. E.L. Hirst, D.A. Ries and N.G. Richardson, *Biochem. J.* 95, 453, 1965.
5. J.D. Bewley, M. Black and P. Halmer, *The Encyclopedia of Seeds: Science, Technology and Uses*, CABI, Wallingford, 2006.
6. M.S. Buckeridge, *Plant Physiol.* 154, 1017, 2010. http://www.plantphysiol.org/content/154/3/1017.
7. P.M. Dey and J.B. Harbone, *Plant Biochemistry*, Academic Press, San Diego, CA, 1997.
8. S.E. Marcus, Y. Verhertbruggen, C. Herve, J.J. Ordaz-Ortiz, V. Farkas, H.L. Pedersen, W.G.T. Willats and J.P. Knox, *BMC Plant Biol.* 8, 60, 2008. http://www.biomedcentral.com/1471-2229/8/60.

9. E. Escarnot, R. Agneesens, B. Wathelet and M. Paquot, *Food Chem.* 122, 857, 2010.
10. E. Escarnot, M. Aguedo, R. Agnessens, B. Wathelet and M. Paquot, *J. Cereal Sci.* 53, 45, 2011.
11. E. Escarnot, M. Aguedo and M. Paquot, *Carbohydr. Polym.* 85, 419, 2011.
12. I.A. Larskaya and T.A. Gorshkova, *Biochemistry (Moscow)* 80, 881, 2015.
13. G. van Aubel, R. Buonatesta and P. van Cutsem, *Crop Prot.* 65, 129, 2014.
14. N.M.L. Hansen and D. Plackett, *Biomacromolecules* 9, 1493, 2008. http://pubs.acs.org/doi/abs/10.1021/bm800053z.
15. J.L. Wertz, Molécules plateformes biobasées issues de glucides, ValBiom, 2013. www.valbiom.be.
16. I. Delidovitch, K. Leonhard and R. Palkovits, *Energy Environ. Sci.* 7, 2803, 2014. http://pubs.rsc.org/en/content/articlelanding/2014/ee/c4ee01067a#!divAbstract.
17. http://en.wikipedia.org/wiki/Guar_gum
18. http://en.wikipedia.org/wiki/Locust_bean_gum
19. S. Gillet, M. Simon, M. Paquot and A. Richel, *Biotechnol. Agron. Soc. Environ.* 18, 97, 2014.
20. http://www.glucomannan.com/gum.htm
21. S.V. Patil, D.R. Jadge and S.C. Dhawale, Pharmainfo.net, Tamarind Gum: A Pharmaceutical Overview, Latest reviews 6, Issue 4, 2008. http://www.pharmainfo.net/reviews/tamarind-gum-pharmaceutical-overview.
22. W. Liu, Z. Yuan, C. Mao, Q. Hou and K. Li, *BioResources* 6, 3469, 2011.
23. M. Hamaguchi, J. Kautto and E. Vakkilainen, *Chem. Eng. Res. Des.* 91, 1284, 2013. http://www.sciencedirect.com/science/article/pii/S0263876213000488.
24. P. Reyes, C. Parra, R.T. Mendonca and J. Rodriguez, SBFC, Clearwater Beach, FL, 2014. http://sim.confex.com/sim/36th/webprogram/Paper26538.html.
25. R. Avanasi Narasimhan, ScholarsArchive@OSU, 2010. https://ir.library.oregonstate.edu/xmlui/handle/1957/19569?show=full.
26. K. Olofsson, M. Bertilsson and G. Liden, *Biotechnol. Biofuels* 1, 7, 2008. http://www.biotechnologyforbiofuels.com/content/1/1/7.
27. B. Hahn-Hagerdal, K. Karhumaa, C. Fonseca, I. Spencer-Martins and M.F. Gorwa-Grauslund, *Appl. Microbiol. Biotechnol.* 74, 937, 2007. http://www.ncbi.nlm.nih.gov/pubmed/17294186.
28. DOE, EERE, Biomass, *Top Value Added Chemicals from Biomass*, 2004. http://www.nrel.gov/docs/fy04osti/35523.pdf.
29. J.J. Bozell and G.R. Peterson, *Green Chem.* 12, 539, 2010. http://pubs.rsc.org/en/content/articlelanding/2010/gc/b922014c#!divAbstract.
30. http://fr.wikipedia.org/wiki/%C3%89thanol#Mati. C3.A8re_premi.C3.A8re
31. http://cenblog.org/the-chemical-notebook/2010/10/brakem-to-make-propylene-from-ethanol/
32. http://www.biofuelsdigest.com/bdigest/2015/06/04/virent-coca-cola-hit-key-production-milestone-with-the-100-biobased-plant-bottle/
33. http://formule-verte.com/coca-colavirent-une-bouteille-en-pet-a-100-biosourcee-presentee-a-milan/
34. http://www.metabolix.com/Products/Biobased-Chemicals/Chemical-Products/Bio-Based-Acrylic-Acid
35. http://www.jle.com/e-docs/00/04/74/F7/article.phtml
36. S.M. Raj, C. Rathnasingh, J.E. Jo and S. Park, *Process Biochem.* 43, 1440, 2008.
37. http://duponttateandlyle.com/about_us
38. http://www.ncbi.nlm.nih.gov/pubmed/10531640
39. http://www.ncbi.nlm.nih.gov/pmc/articles/PMC239268/pdf/aem00164-0072.pdf
40. http://www.ncbi.nlm.nih.gov/pmc/articles/PMC183339/
41. http://www2.dupont.com/Renewably_Sourced_Materials/en_US/proc-buildingblocks.html

42. http://aem.asm.org/content/72/1/96
43. http://www.pfb.info.pl/files/kwartalnik2/1_2011/13.Drozdzynska.pdf
44. http://www.npt.nl/cms/images/stories/Verslagen/Presentatie_Klumpe_Solvay_15042010.pdf
45. http://fr.wikipedia.org/wiki/Acide_aspartique
46. http://fr.wikipedia.org/wiki/2-m%C3%A9thylpropan-1-ol
47. http://www.icis.com/blogs/green-chemicals/2011/06/bioplastic-bottle-update-viren
48. http://www.additifs-alimentaires.net/E297.php?src_ponct
49. http://www.bio-amber.com/products/en/products/bdo_1_4_butanediol
50. http://formule-verte.com/basf-va-produire-du-butanediol-biosource-grace-a-genomatica
51. http://formule-verte.com/novamont-lance-sa-ive-generation-de-bioplastiques
52. http://www.metabolix.com/Products/Biobased-Chemicals/Chemical-Products/BIO-GBL-BDO/
53. http://www.zeachem.com/
54. http://www.reciprocalnet.org/recipnet/showsamplebasic.jsp?sampleId=27344563
55. http://www.biofuelsdigest.com/biobased/tag/acid/
56. F. Martel, B. Estrine, R. Plantier-Royon, N. Hoffman and C. Portella, *Top Curr. Chem.* 294, 79, 2010.
57. http://www.chemicals-technology.com/projects/verdezyne-adipic-acid-plant-california/
58. http://polymerinnovationblog.com/polyethylene-furanoate-pef-100-biobased-polymer-to-compete-with-pet
59. http://greenchemicalsblog.com/2012/09/01/invista-seeks-more-bio-based-nylon-feedstock/
60. Project Biocore. http://www.biocore-europe.org/page.php?optim=An-innovative-fractionation.
61. http://www.biocore-europe.org/page3681.html?optim=biocore-in-brief
62. S. Raharja, L. Rigal, P.F. Vidal, A.V. Bridgwater et al., eds., *Developments in Thermochemical Biomass Conversion*, Springer Science+Business Media, Dordrecht, the Netherlands, 1997.
63. Frost & Sullivan, London, 2013. http://www.frost.com/prod/servlet/press-release.pag?docid=288351416.
64. European Commission, Strategic energy technologies information system, 2014. http://setis.ec.europa.eu/setis-magazine/bioenergy/biocore-mixing-feedstock.
65. Project Biocore. http://www.biocore-europe.org/page.php?optim=Complementarities-between-Biotechnologies-and-Chemistry.

# 10 Valorization of Lignin

## 10.1 INTRODUCTION

### 10.1.1 BACKGROUND

Lignin valorization, in the era of lignocellulosic biorefineries, is a particularly promising thematic. As a biobased polymer, lignin is unusual because of its heterogeneity and lack of a defined primary structure. Lignin is the glue that holds plant cell walls together and is the main aromatic carbon source generated in nature.[1] Traditionally, most large-scale industrial processes that use plant polysaccharides have burned lignin to generate the power needed to productively transform biomass.[2] Lignin is most often valorized by combustion in paper mills for energy production. The advent of biorefineries that convert cellulosic biomass into biobased products and bioenergy will generate substantially more lignin than necessary to power the operation, and therefore considerable efforts are underway to transform it to value-added products. Its valorization in low-molecular weight compounds (such as benzene, toluene, xylenes—BTX) would generate a market of 100 billion dollars as building blocks of aromatic compounds.[3]

It is generally estimated that the research on cellulose to produce ethanol started 20 years ago, research on hemicelluloses including C5 sugar valorization started 10 years ago, and lignin valorization starts approximately now.

## 10.1.2    Biorefining and Sustainable Development

As already mentioned, biorefining is a process that converts biomass into biobased products and bioenergy.

When seeking for sustainability, future biorefineries will have to optimally valorize the whole plant while minimizing environmental footprint. First-generation biorefineries convert edible biomass into energy and nonenergy products, while second-generation biorefineries convert lignocellulosic biomass into energy and products. Glucose is present in both raw materials but it is less accessible in lignocellulose, making a pretreatment necessary.[4]

Biorefining, at the heart of the bioeconomy of the future, should allow the birth of a more innovative, more resource-efficient, and more competitive society.[5] In the same way, bioeconomy will optimize land use and food security through a sustainable, resource-efficient and largely waste-free utilization of renewable materials while protecting environment.

Potential high-value products from isolated lignin include the following:

- Low-cost carbon fibers.
- Engineering plastics and thermoplastic elastomers, polymeric foams and membranes.
- A variety of fuels and chemicals notably sourced from petroleum.

These lignin coproducts must be low cost and function as well as oil-based counterparts.[2] Each product stream has its own distinct challenges:

- Development of renewable lignin-based polymers requires improved processing technologies which can be coupled to tailored bioenergy crops incorporating lignin with desired properties.
- For fuels and chemicals, multiple strategies have emerged for lignin depolymerization and upgrading, including biochemical, chemical, thermochemical, and catalytic treatments.
- The multifunctional nature of lignin has historically yielded multiple product streams that require extensive separation and purification procedures, but engineering plant feedstocks reduce this challenge.[2]

## 10.1.3    Integration into Lignocellulosic Biomass

Lignin is mainly deposited in terminally differentiated cells of supportive and water-conducting tissues. Lignin is one of the major obstructions in the conversion of plant biomass to pulps, papers, biobased products, or biofuels.[6] It protects polysaccharides in plant cell walls from microbial degradation. As lignin is strongly bound by ether and ester linkages to polysaccharides, cellulose and hemicellulose polymers are able to get only a limited access to microbial hydrolytic enzymes. For the efficient use of polysaccharides in pulps or biofuels, lignin should be reduced or removed. In addition, the residual products generated during the removal of lignin can inhibit the subsequent processes of saccharification and fermentation. Work has been reported on the liberation of lignin and soluble lignin–carbohydrate complexes (LCC) by chemical attack.[7]

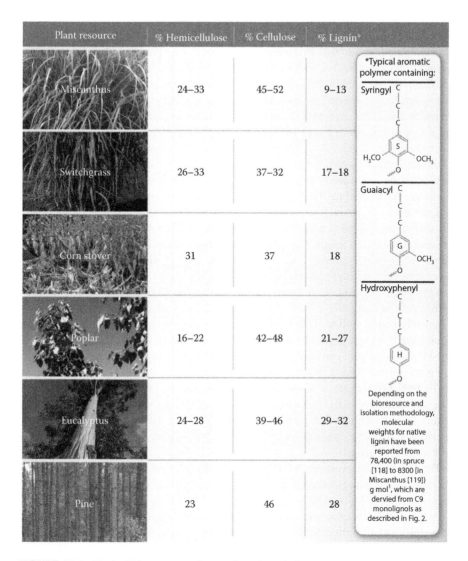

| Plant resource | % Hemicellulose | % Cellulose | % Lignin* | |
|---|---|---|---|---|
| Miscanthus | 24–33 | 45–52 | 9–13 | *Typical aromatic polymer containing: Syringyl |
| Switchgrass | 26–33 | 37–32 | 17–18 | |
| Corn stover | 31 | 37 | 18 | Guaiacyl |
| Poplar | 16–22 | 42–48 | 21–27 | Hydroxyphenyl |
| Eucalyptus | 24–28 | 39–46 | 29–32 | Depending on the bioresource and isolation methodology, molecular weights for native lignin have been reported from 78,400 (in spruce [118] to 8300 [in Miscanthus [119]) g mol¹, which are dervied from C9 monolignols as described in Fig. 2. |
| Pine | 23 | 46 | 28 | |

**FIGURE 10.1**   Typical biomass constituents for selected plant resources.

Figure 10.1 highlights the relative amounts of lignin and polysaccharides in several key plants and woody resources.[2]

## 10.2   MAIN LIGNIN EXTRACTION PROCESSES

Paper chemical processes and some other related processes include two steps for the separation of lignocellulosic polymers:

- Removal and solubilization of hemicelluloses and lignin
- Filtration of insoluble cellulose and filtrate concentration

### 10.2.1 KRAFT PROCESS

Kraft pulping (Section 9.3) is the dominant chemical pulping process in the world.[8] It uses strong alkali with a sodium sulfide catalyst to separate lignin from the cellulose fibers. After pulping, the cellulose fibers go through several bleaching stages to remove residual lignin and produce a strong, white, stable papermaking pulp. The lignin and hemicelluloses that are dissolved in the pulping stage is known as black liquor and is sent to a recovery system where it is burned. Typically, delignification requires several hours at 170°C–176°C (Figure 8.6). Under these conditions, lignin and hemicelluloses degrade to give fragments that are soluble in the strongly basic liquid. The recovery stage is crucial to the functioning of a kraft mill—it supplies much of the energy needed to operate the mill and regenerates the inorganic pulping chemicals.

Kraft lignins have different chemical properties than lignosulfonates. There are sulfur-containing groups present, so kraft lignin is only soluble in alkaline solution (pH above 10). Kraft lignin can thus be precipitated from black liquor by lowering the pH to 10 with a suitable acid. This lignin is used in niche applications such as dispersants for dyes and pesticides.

Kraft lignin contains about 1.5%–3.0% of sulfur.[9] The sulfur is believed to be present in lignin as inorganic sulfur, as elemental sulfur, as adsorbed polysulfide, and/or organically bound sulfide. The possibilities of minimizing the sulfur content in kraft lignin have been investigated extensively. By acidification of the black liquor, most of the dissolved lignin can be isolated through precipitation, but low-molecular mass lignin fragments remain in the solution. The yield from the precipitation depends on the final pH and of the ion strength of the solution. The use of carbon dioxide is beneficial for technical purposes, but the pH cannot be lowered below 8.5. More lignin can be precipitated if a strong mineral acid is used. By using carbon dioxide, the weakly acidic phenolic hydroxyl groups of lignin are protonated, but not the carboxylic acid groups.

Thiolignins have been made for studies on the intermediates formed in the lignin reactions taking place during kraft pulping.[9] Thiolignins have been prepared with sulfur contents as high as 12%–18%. By treating wood with sodium hydrogen sulfide at pH 8 to 9 at 100°C, the thiolignins are formed. When heated with sodium hydroxide at 160°C, the thiolignins lose a great part of their sulfur and the product is very similar to kraft lignins.

### 10.2.2 SULFITE PROCESS

As a source of papermaking-grade bleached pulp, the sulfite process has been largely displaced by kraft (alkaline) pulping.[7] Nonetheless, lignosulfonates isolated from spent sulfite pulping liquors are the most important commercial source of lignin today with global production being about 1 million metric tons per year. Lignosulfonates contain sulfonate ($RSO_3^-$) groups bonded to the polymer and are therefore soluble in water at a wide range of pH. The common applications of lignosulfonates are as dispersants, binders, complexing agents, and emulsifying agents.

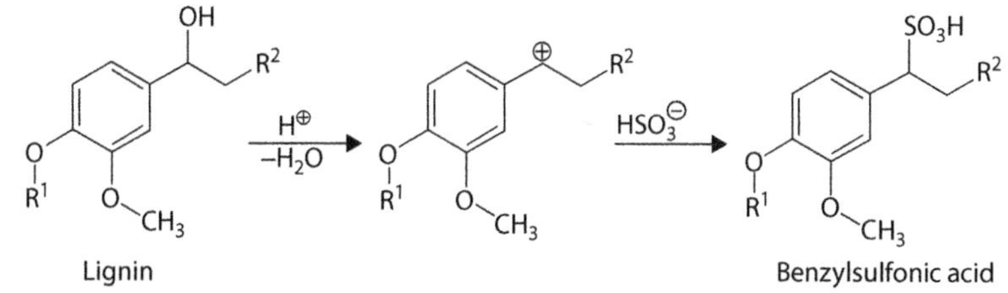

**FIGURE 10.2** Formation of lignosulfonates; Q = common group in lignin. (Courtesy of http://en.wikipedia.org/wiki/Lignosulfonates.)

The major part of delignification during sulfite process involves acid scission of ether linkages. The primary site for scission of ether linkages is the $\alpha$ carbon of the propyl side chain. The electrophilic resonance-stabilized carbocations produced during the scission of the ether linkage react with the bisulfite ions ($HSO_3^-$) to give sulfonates (Figure 10.2).

### 10.2.3 ORGANOSOLV PROCESS

Organosolv pretreatment process uses an organic solvent or mixtures of organic solvents with water to remove lignin before enzymatic hydrolysis of the cellulosic fraction.[2] In addition to lignin removal, hemicellulose hydrolysis occurs resulting in improved enzymatic digestibility of cellulose. Common solvents for the processes include ethanol, methanol, acetone, and ethylene glycol. Temperatures used for the processes can be as high as 200°C, but lower temperatures can be sufficient depending on, for example, the type of biomass and the use of a catalyst. Possible catalysts include inorganic and organic acids.

Originally developed in the years 1980–1990 as an *environmentally friendly* alternative to kraft and sulfite pulping, this process received regain of interest recently with the development of biorefineries. Main advantages are the absence of odoriferous sulfur compounds, increase of yield compared with Kraft pulp process, and possibility to apply to all feedstocks; main drawbacks are energy consumption and complicated recovery systems for solvent.

Main reactions during organosolv cooking are hydrolysis of ether linkages in lignin and hemicellulose–lignin bonds.[11] Easily hydrolyzable $\alpha$-ether bonds are more readily broken, but it is likely that $\beta$-aryl ether bonds are also broken under the conditions of many processes. Most of the organosolv processes use either a neutral solvent, with or without an acid added catalyst, or an acidic solvent. In cases where no acid is added, the liquor becomes acidic as a result of release of acetic acid from the wood. Under these conditions, neutral or acidic solvolysis reactions of lignin may be expected to occur.

Model compounds have shown that $\alpha$-aryl ether linkages are more easily split than $\beta$-aryl ether linkages, especially when they occur in a lignin structural unit containing a free hydroxyl group in the *para* position.[9]

The likelihood of β-ether cleavage is greater in more acidic systems.[9] Formation of either a resonance-stabilized benzyl carbocation or a similarly stabilized transition state is followed by elimination of water to form the readily hydrolyzed enol ether. This releases guaiacol and forms β-hydroxyconiferyl alcohol, which exists in equilibrium with its keto form. In general, β-ether cleavage is likely to be more important in the pulping of hardwoods than in the pulping of softwoods. Hardwoods are delignified more easily than softwoods, mainly due to differences in β-ether reactivity, α-ether concentration, lignin concentration, and propensity to undergo condensation reactions.

## 10.3 PROPERTIES AND APPLICATIONS OF EXTRACTED LIGNINS

### 10.3.1 LIGNIN PROPERTIES

Lignin is a three-dimensional amorphous polymer mainly consisting of methoxylated phenylpropane structures.[12–14] It is a thermoplastic polymer and can be compounded after compatibilization with most thermoplastics for making composites.[15] In general, softwood lignins have a slightly higher $T_g$ (glass transition temperature) than hardwood lignins, which is probably due to a more cross-linked structure in softwoods. Unmodified lignin in wood has the lowest $T_g$ (65°C–105°C), and for modified lignins, $T_g$ is reported to be higher: milled wood lignin (lignin extracted from finely ground wood) 110°C–160°C and kraft lignin (lignin found in black liquor at kraft pulp mills) 124°C–174°C.[16]

### 10.3.2 SHORT-, MEDIUM-, AND LONG-TERM APPLICATIONS

Lignin is one of the most abundant renewable resources. Worldwide, about 50 million tons of lignin are produced annually as residue in paper production processes.[17] Most lignin waste is burned to generate energy for the pulp mills. However, based on its interesting functions and properties, lignin offers perspective for higher added value applications in renewable products. World production of lignosulfonates, which are water-soluble anionic polyelectrolyte polymers derived from the production of wood pulp using sulfite pulping, is estimated to be close to one million tons.[18] World production of kraft lignins not used as fuel is about 0.1 million tons, whereas that of sulfur-free lignins (organosolv) is limited. Most lignin waste is burned to generate energy for the pulp mills.

Lignocellulosic biorefineries dedicated to biofuels production are expected to produce huge quantities of lignin. Consequently, the optimal lignin use for these biorefineries should be determined. An approach taking into account biofuels and related biobased products should be considered.

Lignin valorization shows various challenges for this biorefining approach. Chemically, it differs from sugars by a complex aromatic structure. Instead of cellulose which has a linear substructure of glucose units, lignin has a high degree of structural variability, which differs from biomass source and the lignin recovery process.[19] Lignin applications fall into three categories.

**TABLE 10.1**
**Main Short-, Mid-, and Long-Term Applications of Lignin**

| | Category | Examples |
|---|---|---|
| Short-term (<5 years)/low added value | Bioenergy, synthesis gas | Heat, power |
| Mid-term (~2019)/mid to high added value | Macromolecules | Carbon fibers<br>Polymer modifiers<br>Resins/adhesives/binders |
| Long-term (~2025)/low to high added value | Aromatic monomers and oligomers | Benzene, toluene, xylenes<br>Vanillin<br>Lignin monomers<br>Building blocks (phenolics, styrene) |

*Source:* Wertz, J.L., *Molecules from Lignin Valorization*, ValBiom, 2015; van Dam, J., Gosselink, R., and de Jong, E., *Lignin Applications*, Wageningen UR, Agrotechnology & Food Innovations. http://www.biomassandbioenergy.nl/infoflyers/LigninApplications.pdf; Gosselink, R. et al., *Valorization of Lignin Resulting from Biorefineries*, Agrotechnology & Food Sciences Group, Wageningen UR, 2008. http://www.rrbconference.com/bestanden/downloads/145.pdf.

In the category of established applications (within 5 years), lignin is only used as carbon source, and drastic means are used to break its macromolecular structure. In mid-term applications (about 2019), the macromolecular structure is advantageously maintained. The third application category (long-term: about 2025) uses technologies that degrade the macromolecular structure of lignin but maintain the aromatic character of the produced molecules.

A summary of the present and potential applications of lignin is given in Table 10.1.

### 10.3.3 PRESENT APPLICATIONS

Lignin has several relatively low-value applications such as follows[21]:

- Fuel, yielding more energy when burned than cellulose
- Cement additive, especially as cement set retarder
- Asphalt additive, especially due to its antioxidant characteristics
- Feed binder to plasticize and hold the pellet together
- Additive in biomass-based fuel pellets[22]

The production of vanillin is a current application with high added value.

The application as macromolecular binder for wood and wood-based panels is already current and with medium added value (MDF, medium density fiberboard, and plywood).[23]

### 10.3.4 Medium-Term Applications

The medium-term (about 2019) applications with medium to high added value include the following:

- Synthesis of macromolecules
- Lignin-based composites, especially lignin matrix composites and component for polymeric materials such as starch film, conductive polymers, polyesters, and polyurethanes[23,24]
- Precursors to make carbon fibers, soil conditioners, nitrogenous fertilizers, and pulping catalysts[16,25]

#### 10.3.4.1 Carbon Fibers

The high costs of conventional carbon fiber, which starts with expensive polyacrylonitrile (PAN) polymer precursor, have limited their widespread use in wind turbine blades.[26] Therefore, there is a huge potential to reduce the production cost of carbon fibers by replacing PAN with low-cost materials from natural sources.

Lignin as a renewable resource can be used as a precursor for the production of carbon fibers (Figure 10.3). The low cost of lignin has been projected to result in savings of 37%–49% in the final production cost of carbon fiber. Replacing PAN by lignin for wind turbine blades has a triple pay-off: it uses renewable resources, it optimizes energy and material costs for producing carbon fibers, and the fibers themselves will be used for wind turbines to produce renewable energy. However, lignin is a very brittle biopolymer that cannot be spun, stretched/aligned, and spooled into fibers without modification.

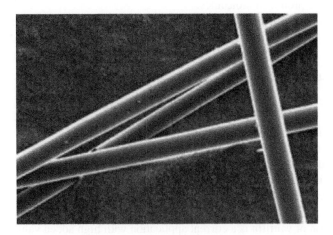

**FIGURE 10.3** Photomicrograph of a carbon fiber precursor produced by firing a fiber that is 99% lignin. (Courtesy of ORNL, U.S. Government; From Oak Ridge National Laboratory, Carbon-fiber composites for cars, http://web.ornl.gov/info/ornlreview/v33_3_00/carbon.htm.)

The aim at Iowa State University is to devise methods to improve the processability of lignin-based precursor so that small diameter lignin fiber can be produced, spun, and stretched prior to pyrolysis into carbon fiber.[26] Its state-of-the-art polymer processing equipment includes a lab-scale production unit consisting of a twin-screw microcompounder, a monofilament spin unit, and a fiber stretching and conditioning unit.

Oak Ridge National Laboratory (ORNL) and GrafTech have collaborated to develop the performance of high-temperature thermal insulation prototypes made from lignin-based carbon fiber.[29] A process flowchart developed by ORNL and GrafTech includes the following[29]:

- Lignin compounding and pelletizing
- Lignin fiber melt blowing
- Fiber stabilization
- Fiber carbonization

ORNL has estimated the production cost of lignin fibers produced by melt blowing to form a mat which is then stabilized and carbonized to $4–5/lb.[30]

To obtain lignin-derived carbon fiber, isolated lignin is typically first processed into fibers by extruding filaments from a melt or solvent swollen gel.[2] Then, the spun fibers are thermally stabilized in air where the lignin fiber is oxidized. At this stage, the filaments become pyrolyzable without melting or fusion. During pyrolysis under nitrogen or inert atmosphere, the fibers become carbonized through the elimination of hydrocarbon volatiles, their oxidized derivatives, carbon monoxide, carbon dioxide, and moisture.

### 10.3.4.2 Plastics and Composites

Another lignin application is plant-derived plastics and composites.[2] The synthesis of engineering plastics and thermoplastic elastomers, polymeric foams, and membranes from lignin with unique properties or comparable or higher properties to those from petroleum has been reported for some time. The most frequent lignin source for these studies has come from chemical pulping operations that are directed primarily at lignin removal from cellulosic fibers through a series of alkaline depolymerization or lignin sulfonation reactions. Although the lignin structure from such operations may be far removed from that needed for most high-value material applications, these sources have found commercial markets such as additive for cement, dust suppression, and drilling fluids for oil recovery.

Future development of green lignin-based polymers pivots on new processing technologies coupled to tailor-made bioenergy crops containing lignin with desired chemical and physical properties for a variety of lignin-based material applications.[2]

Lignin has been studied as an adhesive for more than 100 years but there are only a few industrial applications.[31] Being the natural glue in plants and having a phenolic nature makes lignins an attractive replacement for wood adhesives. The reason for the current interest is the high availability and low price of lignin. An adhesive system for wood composites consisting mainly of lignin has yet to be developed. Lignin has less reactive sites in the aromatic ring than phenols, and the steric effects caused by the macromolecular structure further hinder its reactivity. The low reactivity leads to slow curing and causes problems where curing speed is a key parameter. Modifications such as phenolation, methylolation, and demethylation have been shown to have a positive effect on the reactivity of lignin.

Phenol-formaldehyde (PF) resins are formed by polycondensation reaction of phenol and formaldehyde.[32] Phenol hydroxyl groups at *para* and *ortho* positions primarily react to formaldehyde. Phenol-formaldehyde adhesives are usually used to bond exterior grade plywood for high bonding strength and water resistance.[33] However, all components are based on petrochemicals and nonrenewable materials. An investigated area has been the utilization of lignin, although other studies have been carried out on the use of lignin in other adhesives such as urea-formaldehyde, resorcinol-formaldehyde, or melamine-formaldehyde.[34]

Danielson et al. have investigated the potential for partially replacing phenol with Kraft lignin in the PF resin designed for application as an adhesive in the production of plywood.[35] The kraft lignin, considered to be an environmentally friendly alternative to phenol, was precipitated from black liquor recovered from kraft pulping of softwood. Kraft lignin phenol-formaldehyde (KLPF) resin was prepared in a one-step preparation with different additions of lignin. The mechanical properties of test samples made from KLPF resins were equal to or better than those of test samples made from PF resin only.

## 10.3.5  LONG-TERM APPLICATIONS

Long-term (about 2025) applications for lignin and modified lignin include the following[23,24]:

- Depolymerized lignin for aromatic base chemicals (e.g., phenols), the main routes for lignin depolymerization being base-catalyzed depolymerization followed by hydrodeoxygenation, pyrolysis and fractionation, catalytic liquefaction, hydrocracking, or supercritical depolymerization
- Photostabilization and upgrading of lignin-rich mechanical wood pulp and paper
- UV stabilizer and coloring agent
- Surfactant
- Raw material for aromatic synthon production
- Carrier for the control release of bioactive compounds especially in the agriculture area (pesticides, herbicides, insecticides, and fertilizers)[36]

### 10.3.5.1  Aromatic Catabolism

So far, lignin has not been cost-effectively converted into fuels and chemicals.[37] With the development of lignocellulosic biorefineries around the world to produce fuels and chemicals from biomass-derived carbohydrates, the amount of waste lignin will strongly increase, warranting new lignin upgrading strategies. In nature, some microorganisms have evolved pathways to *catabolize* (break down) lignin-derived aromatics (Figure 10.4). The utilization of these natural aromatic pathways may enable new

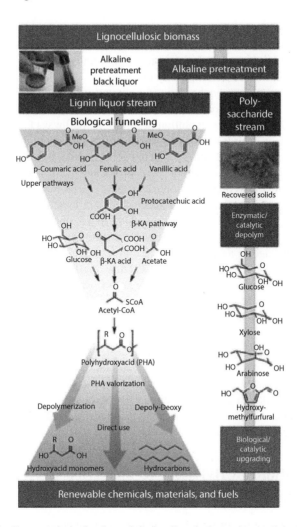

**FIGURE 10.4** Integrated production of fuels, chemicals, and materials from biomass-derived lignin via natural aromatic catabolic pathways and chemical catalysis. Biomass fractionation can yield streams enriched in lignin and polysaccharides, which can be converted along parallel processes. The challenges associated with lignin's heterogeneity are overcome in a *biological funneling* process through upper pathways that produce central intermediates such as protocatechuic acid. The aromatic rings of these intermediates are metabolized to acetyl-CoA. Residual glucose and acetate present will also be metabolized to acetyl-CoA, the primary entry point to PHA production. PHAs are converted to alkenoic acids, and further depolymerized and deoxygenated (depoly-deoxy) into hydrocarbons, thus demonstrating valuable products from lignin. (Reproduced from Borregaard LignoTech, http://www.lignotech.no. With permission of PNAS.)

routes to overcome the lignin utilization barrier that, in turn, may enable a broader slate of molecules derived from lignocellulosic biomass. In other words, developing biological conversion processes for one such lignin-utilizing organism may enable new routes to overcome the heterogeneity of lignin.

In nature, some fungi and bacteria depolymerize lignin by using powerful oxidative enzymes. This pool of aromatic compounds present during biomass decomposition likely triggered evolution of microbial pathways for using aromatic molecules as carbon sources. Many aromatic-catabolizing organisms use *upper pathways*, wherein a diverse battery of enzymes funnels aromatic molecules to central intermediates such as catechol and protocatechuate (Figure 10.4). From these intermediates, dioxygenase enzymes cleave carbon–carbon bonds in the aromatic rings to produce ring-opened species that are metabolized to central carbon metabolism, thus enabling microorganisms to metabolize a broad range of aromatic species. These pathways have long been studied for their catalytic novelty in C–C bond cleavage in aromatic rings. From a biomass conversion standpoint, these upper pathways offer an approach to funnel the heterogeneous portfolio of molecules produced from lignin depolymerization to targeted intermediates for upgrading to fuels, chemicals, and materials.

### 10.3.5.2  Fuels

Despite the anticipated improvements in engineered lignin structure and tailored pretreatments, some lignin fractions from a biorefinery are not suitable for conversion into materials but can still be valuable for conversion into fuels.[2] Lignin depolymerization is challenging given the broad distribution of bond strengths in the various C–O and C–C linkages in lignin and the tendency for low-molecular weight species to undergo recondensation.

## 10.4  INDUSTRIAL ACTIVITIES AND BIOREFINING PROJECTS

### 10.4.1  BORREGAARD: A LIGNOCELLULOSIC BIOREFINERY

Since the beginning of the twentieth century, Norwegian company Borregaard is in the business of lignocellulosic products. Now part of Orkla Group (32,000 people and EUR 7.5 billion turnover), Borregaard is divided into five divisions according to the market and product type: Borregaard ChemCell is responsible for specialty cellulose and bioethanol; Borregaard LignoTech for lignin-based products and trading activities; Borregaard Synthesis for fine chemicals/pharmaceuticals; Borregaard Ingredients for vanillin and omega-3 for foodstuffs; and Borregaard Energy for electrical power.[38]

In Europe, Borregaard LignoTech is the world's leading producer of lignin and lignosulfonate-based products, with over 60 years supplying binding and dispersing agents to the chemical industry.[39,40] Applications and functionalities of LignoTech's lignin-based products include the following:

- Binding agents
- Crystal growth modifier

- Dispersing agents
- Dust suppressant
- Emulsion stabilizers
- Extrusion aids
- Retarders
- Rheology control

These products are used in areas such as concrete additives, textile dyes, agrochemicals, batteries, oil drilling, ceramic materials, briquettes, and animal feed. Potential for commercial lignin applications is estimated to be enormous in the following decades. Carbon fibers and composite materials are among the most attractive examples of exploitable markets.

Borregaard aims to use as much as possible of their raw materials (wood) in the production process (Figure 10.5).[41,42] The company utilizes the different component parts of timber in a range of different products, hemicelluloses being used for the ethanol plant.

The wood pulp production plant in Sarpsborg (Norway) is a good example of the biorefinery concept developed by Borregaard.[43] The plant consumes spruce wood and transforms it into specialty cellulosic pulp and chemicals. The biorefinery in Sarpsborg uses every year 650,000 m³ of solid wood and produces 20 million liters of bioethanol, 160,000 tons of specialty cellulose (for production of cellulose acetate, nitrocellulose, viscose...), ~150,000 tons of lignosulfonates, and 1000 tons of vanillin.

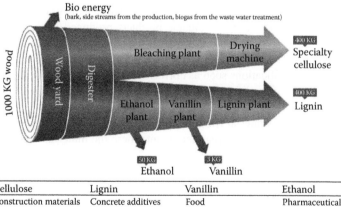

| Cellulose | Lignin | Vanillin | Ethanol |
|---|---|---|---|
| Construction materials | Concrete additives | Food | Pharmaceutical industry |
| Cosmetics | Animal feed | Perfumes | Bio fuel |
| Food | Mining | Pharmaceuticals | Paint/varnish |
| Tablets | Batteries | | Car care |
| Paint/varnish | Briquetting | | |
| Filters | Soil conditioner | | |
| Textiles | | | |

**FIGURE 10.5** Borregaard's biorefinery concept. (From Chowdhury, M.A., *Int. J. Biol. Macromol.*, 65, 136, 2014.)

The process includes several steps:

- The bark is removed and the log is cut into chips. The bark is sent to incineration.
- The chips are cooked in acidic conditions (also called sulfite pulping process). In these conditions, lignin residues are oxidized and transformed almost completely in lignosulfonates (also known as lignin sulfonates or sulfonated lignins).[10] Lignosulfonates are water-soluble anionic polyelectrolyte polymers. Most delignification in sulfite pulping involves acidic cleavage of ether bonds. The electrophilic carbocations produced during ether cleavage react with bisulfite ions ($HSO_3^-$) to give sulfonates (Equations 10.1 and 10.2).

$$R\text{-}O\text{-}R' + H^+ \rightarrow R^+ + R'OH \qquad (10.1)$$

$$R^+ + HSO_3^- \rightarrow R\text{-}SO_3H \qquad (10.2)$$

The primary site for ether cleavage is the $\alpha$-carbon of the propyl side chain. While lignosulfonates are formed, considerable hydrolysis of hemicelluloses occurs, which make the process particularly valuable to obtain high-purity pulps. The cellulose is then separated and goes through a range of bleaching and final purification stages.[42]

- The remaining parts of the timber, which includes binding agents and sugar compounds from hemicelluloses, are separated in the pulping process. This flow of raw materials (waste liquor) is sent to the ethanol factory, where the sugar is converted into ethanol through fermentation.
- The sugar-free waste liquor contains lignosulfonates, which is the starting point for the production of valuable products produced in the lignin and vanillin (Figure 10.6) factories. Lignosulfonates are polydisperse (20–80 kDa) and contain 0.6–1.2 sulfonate groups per monomer.[42] They can be used for various applications such as corrosion inhibition, concrete formulation, and rheology modification. Vanillin (used in food, perfumes…) is obtained by catalytic oxidation of lignosulfonates.
- The residual organic material from the timber which is not included in any products goes to a biological treatment plant where the material is broken

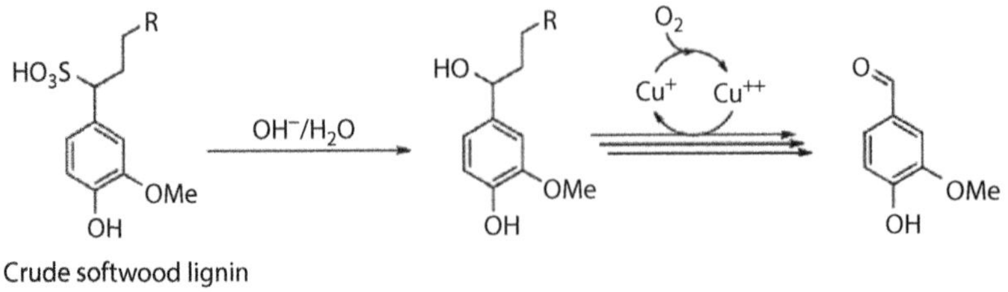

**FIGURE 10.6** Oxidation of lignosulfonate to vanillin. (From Sorlie, P.A., Borregaard, Corporate presentation, http://hugin.info/111/R/1565694/486007.pdf, 2011.)

down with the help of bacteria. In this process, biogas (methane) is produced, which is used as an energy source. The biological mud from the breakdown process is burned in the bark incineration plant.

Borregaard has also participated to several research and development projects on biorefineries, including two important projects financed by European Union Seventh Framework Programme, namely EuroBioRef and Suprabio.

## 10.4.2 BURGO: A LIGNOSULFONATE PRODUCER

Burgo's lignin sulfonates are obtained by a process that separates lignin from cellulose by means of sulfonation.[44,45] Lignin is then solubilized as a calcium salt of lignin sulfonic acid. During the sulfonation reaction, certain molecules of cellulose and lignin hydrolyze to form wood sugars, aldonic acids, and other water-soluble substances, which are all recovered in production of lignin sulfonates.

Burgo Group lignin sulfonates are produced entirely from Norway spruce to ensure that the quality level is constant, a quality that is difficult to achieve if various woods are used.

Burgo Group lignin sulfonates can be used in several fields as follows[43]:

* Concrete, cement mixtures
* Surface-active agents
* Carbon black, mineral granulation
* Emulsions
* Colorants, bonders, and resins
* Soil stabilization
* Tanning
* Ceramics and firebricks
* Industrial detergents
* Antiparasite products
* Slurry conditioning
* Water treatment and conditioning
* Mineral flotation

The Burgo lignin sulfonate range is produced at the Tolmezzo (Italy) mill that has a capacity of 45,000 tons per year of sulfite pulp.[46] The lignin sulfonate range is available in powder, pallet, or liquid form.[47]

## 10.4.3 VALMET: LIGNOBOOST

LignoBoost (Section 8.2.5.2 LignoBoost) is a unique technology for extracting high-quality lignin from a kraft pulp mill.[48,49]

Since 2008, the LignoBoost technology is owned and commercialized by Valmet (previously Metso's Pulp, Paper, and Power businesses). Two commercial plants (Domtar, USA and Stora Enso, Finland) are presently running.

### 10.4.4 LIGNOL

The Canadian Company Lignol Energy Corporation (LEC) produces high-purity lignin as an ethanol coproduct through an *organosolv* process (Section 8.2.6). A clear benefit of any organosolv-based process is that it fractionates biomass into its three components at relatively high purity. Organosolv lignin from Lignol is suitable for the following demonstrated applications[50]:

- Phenol-formaldehyde resin and wood adhesive substrate
- Printed circuit board encapsulated resins
- Foundry resins and molding compounds
- Friction materials, green strength binders, organic particles
- Antioxidants in rubber, lubricants, feed additives
- Rubber tackifiers
- Renewable surfactants, concrete admixtures, air entrainers, superplasticizers
- Carbon fiber and activated carbon production
- Animal feed applications

LEC's wholly own subsidiary, Lignol Innovations Ltd. (LIL), has biorefinering at post-pilot plant stage, with outputs including cellulosic ethanol, high-value cellulose and high-purity lignin.[51] LIL's biorefining platform represents a transformative technology for the forest products industry worldwide. LIL's modified solvent-based pretreatment technology facilitates the rapid, high-yield conversion of cellulose to ethanol and the production of value-added biochemical coproducts, including high-purity HP-L™ lignin.

HP-L lignin represents a new class of high-purity lignin extractives (and their subsequent derivatives) which can be engineered to meet the chemical properties and functional requirements of a range of industrial applications that until now has not been possible with traditional lignin by-products generated from other processes. LIL has developed two distinct but related technologies, each offering its own value proposition and investment opportunity.[52] One covers biorefining technology and the other covers lignin IP and lignin applications. LIL believes that longer term significant revenues could result from licensing or selling its lignin IP portfolio and is preparing to pursue strategies along these lines.

In August 2014, Lignol announced that its largest shareholder, Difference Capital Financial Inc., has placed the Company in receivership.[53]

### 10.4.5 CIMV

In France, the CIMV organosolv process has been developed with the objective of isolating sulfur-free linear lignin for applications such as binding agent for phenolic powder resins, polyol component in polyurethane formulations, and phenolic component of epoxy resins.[54,55] The process has been designed for the manufacture of whitened pulp, nondegraded linear lignin, and C5 sugar syrup from annual fiber crops and hardwood.[56]

In August 2014, CIMV announced that it has signed a collaboration agreement to commercialize second-generation biofuel and biobased chemical technology with Dyadic, a global biotechnology company based in Florida, United States whose technologies are used to develop, manufacture, and sell enzymes and other proteins.[57] CIMV and Dyadic will work together to develop more efficient, fully integrated processes to produce environmentally low impact biofuels and biobased chemicals.

CIMV is the only biorefinery company currently in operation, which can recycle lignins and supply completely pure glucose to the market.[58] In 2015, the Horizon 2020 project 2G Biopic structured around CIMV's process was launched.[59]

## 10.4.6 WAGENINGEN

Wageningen UR (university and research center) Food & Biobased Research Unit is both founder and coordinator of the Wageningen UR Lignin Platform launched in June 2011.[60]

The objective of the Wageningen UR Lignin Platform is to promote interdisciplinary, scientifically challenging and precompetitive research and to create a network on the valorization of lignin for the industrial production of lignin-derived chemicals.[61] Lignin valorization will be an essential part of the integral biorefinery concepts and will be an important driver for development of economic and sustainable biorefineries. The Wageningen UR Lignin Platform brings together unique competences covering the whole chain from cultivating lignocellulosic biomass to lignin-derived biobased products.

Wageningen UR Food & Biobased Research has been the coordinator of the project LignoValue devoted to value-added valorization of lignin for optimal biorefinery of lignocellulose toward energy carriers and products.[62]

This project focuses on:

- Primary biorefinery involving pretreatment and fractionation for production of streams of cellulose, hemicelluloses, and lignin.
- Secondary biorefinery in which lignin is converted into phenols, performance products such as resins, fuel additives, electricity, and heat via thermochemical routes such as pyrolysis and depolymerization under supercritical conditions.
- Development of an integral biorefinery in which the technologies have been optimally integrated and all biomass fractions consist of an optimal quality for further refining and valorization.

## 10.4.7 IOWA STATE UNIVERSITY

During fast pyrolysis, the polyaromatic structure of lignin is depolymerized to smaller phenolic compounds.[63] Phenolic compounds found in bio-oil are both monomers and oligomers. As the water-insoluble fraction of bio-oil mainly consists of phenolic oligomers, it is also called *pyrolytic lignin*. While phenolic monomers are valuable platform chemicals, pyrolytic lignin is less preferred. Pyrolytic lignin is not only known to attribute to thermal instability and high viscosity of bio-oil, but it also causes deactivation of catalyst during catalytic upgrading. For these reasons, lignin-derived compounds would be less problematic if they were present as monomers instead of oligomers. However, yields of phenolic monomers in bio-oil are usually very low (<10 wt%), whereas pyrolytic lignin is the majority of lignin derivatives, accounting for 25–30 wt% of bio-oil. Thus, understanding the formation mechanism of pyrolytic lignin during pyrolysis is very important in promoting lignin depoly-merization to phenolic monomers. However, the origin of pyrolytic lignin is not yet very clear. Two primary hypotheses have been discussed in the literature: the thermal ejection as lignin fragments and the repolymerization of primary phenolic monomers in the vapor phase.

The research goal of Iowa State University was to understand the formation path-way of pyrolytic lignin.[63] During experiments, lignin and phenolic model compounds were pyrolyzed in a micropyrolyzer system. The study revealed that the primary pyrolysis products of lignin are mostly phenolic monomers and dimers with molecu-lar weight below 400 Daltons (Da). Primary phenolic monomers with unsaturated C=C bonds or methoxy groups further react in the vapor phase to form phenolic oligomers or phenolic monomers with higher molecular weights. Phenolic dimers with biphenol, phenyl coumaran, and diaryl structures are also the primary pyrolysis products because these structures have relatively high thermal stability in conven-tional pyrolysis temperature. The study further indicated that these primary pheno-lic dimers with m/z (mass-to-charge ratio) <400 Da are likely evaporated instead of thermally ejected.

## 10.5  CONCLUSIONS

The effective utilization of lignin is a key for the accelerated development of lig-nocellulosic biorefineries.[2] However, it has been long said in the pulp industry that *one can make anything from lignin except money*. Lignin can be a viable, commercially relevant sustainable feedstock for a new range of materials and uses.[2]

First, the advent of new lignocellulosic biorefineries will introduce an excess supply of different nonsulfonated, native, and transgenically modified lignins into the process streams. Secondly, future research will continue to establish to what extent the lignin structure in plants can be altered to yield a product that can be readily recovered via pretreatment and has the appropriate tailored structures to be valorized for materials, chemicals, and fuels. Thirdly, although lignin sequencing remains a vision, approaches based largely on new techniques have greatly improved our knowledge of the structures of lignin and its products. These results need to be further integrated into computational modeling to provide a predictive tool of lignin's chemical and physical properties in multiple environments. Such insights may help redesign lignin within its cross-linked complex biological matrix to meet subsequent processes and end product goals.

Overall, the need to understand and manipulate lignin from its assembly within plant cell walls to its extraction and processing into value-added products aligns with the potential to obtain a deeper understanding of complex biological structures. This is particularly true because the valorization of lignin cannot come at the expense of the effective utilization of cellulose and hemicelluloses.

The conversion of biomass-derived polysaccharides has provided mankind with renewable fuels and chemicals for more than a century.[36] Conversely, the only use for lignin to date on a scale concomitant with polysaccharide-derived fuels and chemicals is heat and power. It has been shown that an aromatic catabolism overcomes the inherent challenges in lignin valorization, thus enabling comprehensive utilization of biomass polymers to produce renewable fuels, chemicals, and materials for a sustainable energy economy.

## REFERENCES

1. D. Meier, Catalytic hydrocracking of lignins to useful aromatic feedstocks, in *DGMK Conference*, Berlin, Germany, 2008. http://www.dgmk.de/petrochemistry/abstracts_content16/Meier.pdf.
2. A.J. Ragauskas, G.T. Beckham, M.J. Biddy, R. Chandra, F. Chen, M.F. Davis, B.H. Davison et al., *Science* 344, 1246843, 2014. doi:10.1126/science.1246843.
3. http://anellotech.com/sites/default/files/Anellotech-release05-07-2014.pdf.
4. http://www.ethanolrfa.org/bio-refinery-locations/.
5. http://ec.europa.eu/research/participants/data/ref/h2020/other/legal/jtis/bbi-sira_en.pdf.
6. J. Kim, B. Choi, Y.-H. Park, B.-K. Cho, H.-S. Lim, S. Natarajan, S.-U. Park, and H. Bae, *ScientificWorldJournal* 2013, 2013, 421578. http://www.ncbi.nlm.nih.gov/pmc/articles/PMC3800569/.
7. F.E. Brauns and D.A. Brauns, *The Chemistry of Lignin: Covering the Literature for the Years 1949–1958*, Academic Press, New York, 1960.
8. Lignoworks, 2014. http://www.lignoworks.ca/content/what-lignin.
9. S. Svensson, Minimizing the sulphur content in Kraft lignin, STFI-PACKFORSK, 2008. http://www.diva-portal.org/smash/get/diva2:1676/FULLTEXT01.pdf.
10. http://en.wikipedia.org/wiki/Lignosulfonates.
11. T.J. Mcdonough, The chemistry of organosolv delignification, IPST Technical Paper Series Number 455, 1992. http://smartech.gatech.edu/bitstream/handle/1853/2069/tps-455.pdf?sequence=1.
12. J. Zakzeski, P.C.A. Bruijnincx, A.L. Jongerius and B.M. Weckhuysen, *Chem. Rev.* 110, 3552, 2010. http://pubs.acs.org/doi/abs/10.1021/cr900354u.
13. P. Dole and F. Bouxin, *Macromolecular and Molecular Uses of Lignin*, GFP 2008, Lyon, France, 2008. http://gfp2008.ccsd.cnrs.fr/docs/00/32/05/99/PDF/abstract_gfp2008_DOLE.pdf.
14. L. Jouanin, Lignines, Académie d'Agriculture de France, 2010. http://www.academie-agriculture.fr/mediatheque/seances/2010/20100217resume2.pdf.
15. T. Liu and Q.W. Wang, *Adv. Mater. Res.* 113–114, 606, 2010. http://www.scientific.net/AMR.113-116.606.
16. I. Brodin, *Chemical Properties and Thermal Behaviour of Kraft Lignins*, KTH Royal Institute of Technology, Stockholm, Sweden, 2009. http://kth.diva-portal.org/smash/get/diva2:234300/FULLTEXT01.
17. J. van Dam, R. Gosselink and E. de Jong, *Lignin Applications*, Wageningen UR, Agrotechnology & Food Innovations. http://www.biomassandbioenergy.nl/infoflyers/LigninApplications.pdf.

18. R.J.A. Gosselink, J.E.G. van Dam, P. de Wild, W. Huijgen, T. Bridgwater, H.J. Heeres, A. Kloekhorst, E.L. Scott and J.P.M. Sanders, *Valorisation of lignin—Achievements of the lignovalue project 3rd Nordic Wood Biorefinery Conference*, Stockholm, Sweden, 2011. http://library.wur.nl/WebQuery/wurpubs/419444.

19. J.E. Holladay, J.F. White, J.J. Bozell and D. Johnson, *Top Value Added Chemicals from Biomass—Volume II: Results of screening for Potential Candidates from Biorefinery Lignin*, Pacific Northwest National Laboratory, 2007. http://www1.eere.energy.gov/bioenergy/pdfs/pnnl-16983.pdf.

20. J.L. Wertz, *Molecules from Lignin Valorization*, ValBiom, Gembloux, Belgium, 2015.

21. J.L. Wertz and O. Bedue, *Lignocellulosic Biorefineries*, EPFL Press, Lausanne, Switzerland, 2014.

22. M. Kluko, Densified fuel pellets, United States Patent Application 20090205546, 2009. http://www.freepatentsonline.com/y2009/0205546.html.

23. Wageningen UR, 2010. http://www.biobasedproducts.wur.nl/UK/projects/fibres.

24. N. Eisenreich, EU project: BIOCOMP, new classes of engineering composites materials from renewable resources, 2008. http://www.biocomp.eu.com/uploads/Final_Summary_Report.pdf.

25. H. Hatakeyama, *Chemical Modification, Properties and Usage of Lignin*, T.Q. Hu, Ed., 2002. http://www.springer.com/life+sciences/forestry/book/978-0-306-46769-1.

26. Iowa State University, Advanced carbon fibers from lignin, 2014. http://polycomp.mse.iastate.edu/advanced-carbon-fibers-from-lignin.

27. https://commons.wikimedia.org/wiki/File:Windturbine_HamburgWasser_Steinwerder_02.jpg.

28. Oak Ridge National Laboratory, Carbon-fiber composites for cars. http://web.ornl.gov/info/ornlreview/v33_3_00/carbon.htm.

29. Oak Ridge National Laboratory, Commercialization of new carbon fiber materials based on sustainable resources for energy applications, 2013. http://info.ornl.gov/sites/publications/files/Pub41318.pdf.

30. http://www.inda.org/BIO/rise2012_526_PPT.pdf.

31. V. Hemmila, J. Trischler and D. Sandberg, in C. Brischke and L. Meyer, Eds., *Proceedings of the 9th Meeting of the Northern European Network for Wood Science and Engineering*, Hannover, Germany, 2013. http://lnu.diva-portal.org/smash/get/diva2:661146/FULLTEXT01.pdf.

32. B. Klasnja and S. Kopitovic, *Holz als Roh-und Werkstoff* 50, 282, 1992. http://link.springer.com/article/10.1007/BF02615352#page-1.

33. W. Zhang, Y. Ma, Y. Xu, C. Wang and F. Chu, *Int. J. Adhes. Adhes.* 40, 11, 2013.

34. T.Q. Hu, *Chemical Modification, Properties and Usages of Lignin*, Springer, New York, 2002.

35. B. Danielson, http://www.researchgate.net/publication/249574372_Kraft_lignin_in_phenol_formaldehyde_resin._Part_1._Partial_replacement_of_phenol_by_kraft_lignin_in_phenol_formaldehyde_adhesives_for_plywood.

36. M.A. Chowdhury, *Int. J. Biol. Macromol.* 65, 136, 2014.

37. J.G. Linger, D.R. Vardon, M.T. Guarnieri, E.M. Karp, G.B. Hunsinger, M.A. Franden, C.W. Johnson et al., *Proc. Natl. Acad. Sci. USA*, 111, 12013, 2014. http://www.pnas.org/content/111/33/12013.full.

38. http://www.borregaard.com/About-us/Organisation.

39. Borregaard LignoTech. http://www.lignotech.no.

40. http://www.borregaard.com/Business-Areas/Borregaard-LignoTech.

41. http://www.borregaard.com/content/view/full/10227.

42. P.A. Sorlie, Borregaard, Corporate presentation, 2011. http://hugin.info/111/R/1565694/486007.pdf.

43. M. Lersch, Creating value from wood. The Borregaard biorefinery. http://www.bioref-integ.eu/fileadmin/bioref-integ/user/documents/Martin_Lersch__Borregaard_-_Creating_value_from_wood_-_The_Borregaard_biorefinery.pdf.
44. http://www.burgo.com/en/group/figures/ls.
45. http://www.burgo.com/sites/default/files/files/brochure_bretax.pdf.
46. M. Payne, Pulp & Paper International, 1993. http://www.risiinfo.com/db_area/archive/ppi_mag/1993/9308/93080107.htm.
47. http://www.burgo.com/en/group/paper-mills/tolmezzo.
48. Innventia. http://www.innventia.com/en/Our-Expertise/Chemical-Pulping-and-bleaching/Biorefinery-products/Biorefinery-separation-processes/LignoBoost/.
49. Innventia, Press release: EU approves Swedish support to LignoBoost demonstration. http://www.innventia.com/templates/STFIPage____9421.aspx.
50. Lignol Energy, Cellulosic ethanol—The sustainable fuel, in *TAPPI, International Conference on Renewable Energy*, 2007. http://www.tappi.org/content/Events/07renew/07ren06.pdf.
51. Lignol, 2013. http://www.lignol.ca/LIL.html.
52. Lignol, 2014. http://www.newswire.ca/fr/story/1332839/lec-provides-corporate-update-reports-fiscal-2014-third-quarter-results.
53. Lignol, 2014. http://www.lignol.ca/news/News-2014/NewsRelease20140827.pdf.
54. CIMV, The biorefinery concept, CIMV Technology. http://www.cimv.fr/cimv-technology/cimv-technology/5-.html?lang=en.
55. Biocore Project, Chemical and thermochemical transformations. http://www.biocore-europe.org/page.php?optim=Chemical-and-thermochemical-transformations.
56. M. Delmas, *Chem. Eng. Technol.* 31, 792, 2008. http://www.cimv.fr/uploads/vegetal-refining-and-agrichemistry-chemical-engineering-and-technology-2008.pdf.
57. Dyadic. http://dyadic.com.
58. http://www.cimv.fr/research-development/cimv-research-development.html.
59. http://www.greencarcongress.com/2015/06/20150604-biopic.html.
60. Wageningen UR, 2011. http://www.wur.nl/UK/about.
61. Food & Biobased Research, Wageningen UR, 2011. http://www.fbr.wur.nl/UK/newsagenda/archive/agenda/2011/Workshop_Wageningen_UR_Lignin_Platform.htm.
62. http://library.wur.nl/WebQuery/wurpubs/419444.
63. Iowa State University, Understanding formation of pyrolytic lignin, 2014. https://www.cset.iastate.edu/research/current-research/pyroliticlignin/.

# 11 Perspectives

## 11.1 INTRODUCTION

The economy of tomorrow will be more biobased than today. Simultaneously resolving food supply, energy security, oil-based product supply, and environmental concerns is a critical challenge for the humanity today. Only the use of new technologies will allow us to progressively bridge the gap between economic growth and environmental sustainability. The bioeconomy uses not only renewable resources such as lignocellulose but also municipal solid waste and algae, instead of fossil resources.

Lignocellulosic biomass, composed primarily of cellulose, hemicelluloses, and lignin, can be used to produce bioenergy and biobased products. The operation occurs in a biorefinery, which is analogous to an oil refinery. The overall objective of a lignocellulosic biorefinery is to maximize the added value generated from the entire biomass feedstock.

The economic viability of a lignocellulosic biorefinery is highly depending on how hemicelluloses and lignin present in biomass are valorized. Hemicelluloses are present in all green plants, and lignin is present only in vascular plants. Hemicelluloses and lignin represent more than 40% of total biomass.

In a bioeconomy, by definition, biomass is sustainably produced and used for a range of applications.[1,2] Such an economy offers great opportunities for the development of a circular economy, the optimal use of raw materials, and economic growth. The importance of the bioeconomy in the world will only increase in the coming years. The biobased economy, which is not explicitly defined by the European Commission, is embedded in the bioeconomy. It is that part of the bioeconomy in which biobased products and bioenergy are made using biomass as feedstock.

This chapter will start with a presentation of the U.S. Department of Energy Biomass Program and the European Bio-Based Industries (BBI) Joint Technology Initiative (JTI). Promising building blocks from biomass will be reviewed. Biobased economy in the United States and in Europe will be analyzed. Finally, the sustainability challenge will be discussed.

## 11.2   LIGNOCELLULOSIC BIOREFINERIES

Most lignocellulosic biorefineries are expected to be ready for large-scale production in the coming years. However, there are more and more exceptions:

- Poet-DSM opened first commercial cellulosic ethanol plant in Iowa, U.S. in September 2014.[3] The plant, named Project Liberty, converts baled corn cobs, leaves, husk, and stalk into renewable fuel. At full capacity, it will convert 770 tons of biomass per day to produce ethanol at a rate of 25 million gallons (95 million liters) per year. Total investment is $250 million.
- UPM has started commercial production of wood-based renewable diesel from forestry residue in January 2015; the company has invested EUR 175 million in the new Lappeenranta biorefinery producing 120 million liters (100,000 tonnes) of renewable diesel annually.[4]
- Abengoa's Hugoton (Kansas) plant has the capacity to convert 300,000 dry tons of agricultural residues such as corn stover and wheat straw into 25 million gallons (95 million liters) of ethanol and 21 megawatts of renewable electricity.[5] Total investment is $500 million. In December 2015, Abengoa decided to suspend production at its new plant.
- DuPont's commercial-scale cellulosic ethanol facility in Nevada, Iowa, United States was completed in 2015; the facility will produce 30 million gallons per year (110 million liters per year).[6] DuPont is investing more than $200 million to construct its cellulosic biorefinery in Nevada that will be fueled by agricultural residue harvested from a 30 mile radius around the facility.
- Stora Enso is a Finnish pulp and paper manufacturer formed by the merger of Swedish Stora and Finnish Enso.[7] Stora Enso's Sunila Pulp Mill in Finland extracts lignin from pine and spruce. This first dedicated biorefinery investment was completed in early 2015. The new machinery was installed in the architectural milieu that dates back to 1938. Prior to making the 32-million-euro investment decision, Stora Enso carefully looked into areas where lignin extraction would be profitable. The initial markets are anticipated in the construction and automotive industries, where lignin offers a sustainable alternative to the phenols used in plywood and wood-paneling glues and the polyols used in foams.

**FIGURE 11.1**   Thistle as raw material at Matrìca in Sardinia. (From http://www.matrica.it/; Courtesy of Roberta Barbieri, Versalis s.p.a, San Donato Milanese, Italy.)

- Matrìca is a 50:50 joint venture between Versalis, the leading Italian manufacturer of petrochemical products, and Novamont, a global force in the bioplastics sector[8] Matrìca's renewable products are derived from raw materials of vegetable origin and come into by means of an innovative integration of agriculture and industry. Starting from selected raw materials including thistle (Figure 11.1) with low levels of environmental impact, Matrìca produces a series of innovative building blocks that are deployed in various different industries: bioplastics, biolubricants, home and personal care products, plant protection, additives for the rubber and plastics industry, and food fragrances. Matrìca contributes to the development and integration of agriculture, the local area and industry, affording tremendous advantages on all fronts.

## 11.3   BIOBASED ECONOMY IN THE UNITED STATES

### 11.3.1   Opportunities in the Emerging Biobased Economy

A recent report submitted to U.S. Department of Agriculture has explored the opportunities associated with the biobased economy (excluding fuel, food, and feed).[9] Some of the key findings included:

- Government policies and industry business-to-business sustainability programs are driving the biobased economy.
- Across the globe, nations are investing in public/private partnerships to expand their biobased economy for domestic and international customers.

- In the U.S., the USDA BioPreferred program and federally supported research continue to drive investment in R&D and make available broader sets of biobased consumer products.
- Although there is wealth of data regarding the economic impact of the bio-economy in Europe and various nations, there is a lack of understanding and quantification of the economic benefits of the bioeconomy and specifically the nonfuel bioeconomy in the U.S.
- There are challenges facing the continued expansion of the bioeconomy. These include reliable availability of raw materials with increased climate and severe weather impacts, water availability, and stability of the markets.
- The biobased economy is, in fact, growing, and it offers great potential for increased job creation in the U.S.
- Continued investments are needed to establish a biobased infrastructure while ensuring that the economics of biobased feedstocks are competitive with existing, petroleum-based feedstocks.
- An economic impact model is required to study the potential impacts of the bioeconomy and policies that can encourage investment.[9]

The market for biobased products is growing because of the efforts of manufacturers, consumers, and government officials to promote the development and acceptance of these products as they become commercially viable. One of the main objectives of the efforts to increase the market share for biobased products is to reduce the U.S.'s dependence on foreign oil. However, the widespread use of such products also would help rejuvenate the rural economy so that it is less dependent on government subsidies and also would help create a self-sustaining sector that relies on domestic renewable resources. Some of the many biobased products that are currently produced are bioplastics, lubricants, solvents, surfactants, and other biosynthetics. In addition, many biofuel coproducts are emerging that can be produced from a variety of different sources of biomass.

Purchasing of biobased products continues to be supported through government mandates on procurement policies.[9] However, sales of biobased products also are driven increasingly by U.S. customers' preferences based on their growing awareness of the environmental impacts of their purchasing decisions. Surveys suggest that customers want to buy *green*, biobased products, but they continue to be more sensitive to prices than their European counterparts. As the bioeconomy expands, challenges arise in the development of biobased products. One of these challenges is the uncertainty created by policy changes impacting current farming practices.

## 11.3.2 MULTI-YEAR PROGRAM PLAN

The Bioenergy Technologies Office is one of the 11 technology development offices within the Office of Energy Efficiency and Renewable Energy at the U.S. Department of Energy.[10] The Multi-Year Program Plan (MYPP) sets forth the goals and structure of the Bioenergy Technologies Office. It identifies the research, development, demonstration, and deployment (RDD&D) activities that the Office will focus on more than the next five years and outlines why these activities are important for meeting the energy and sustainability challenges facing the nation.

The mission of the Office is to develop and transform the U.S. renewable biomass resources into commercially viable, high-performance bioproducts, and biopower through targeted research, development, demonstration, and deployment supported through public and private partnerships.

The goal of the Office is to develop commercially viable bioenergy and bioproducts technologies

- To enable sustainable, nationwide production of biofuels that are compatible with today's transportation infrastructure, can reduce greenhouse gas emissions relative to petroleum-derived fuels, and can displace a share of petroleum-derived fuels to reduce U.S. dependence on foreign oil.
- To encourage the creation of a new domestic bioenergy and bioproducts industry.

The Office portfolio is organized according the biomass-to-bioenergy supply chain—from the feedstock source to the end user (Figure 11.2).

The Office will continue to support basic science and RDD&D of advanced biomass utilization technologies. Detailed life-cycle analysis of environmental, economic, and social impacts will continue to inform decisions regarding Office activities.

This MYPP is designed to allow the Office to progressively enable deployment of increasing amounts of biofuels, bioproducts, and bioenergy across the nation from a widening array of feedstocks. This approach will have a significant near-term impact on offsetting petroleum consumption and facilitate the shift to renewable, sustainable bioenergy technologies in the long term, while allowing the market to determine the ultimate implementation across diverse U.S. resources.

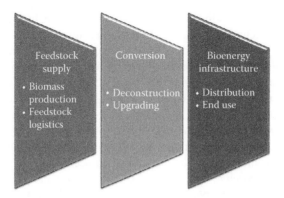

**FIGURE 11.2** Biomass to bioenergy supply chain. (Adapted from Bioenergy Technologies Office, Energy Efficiency & Renewable Energy, DOE, Multi-year program plan, http://www1. eere.energy.gov/bioenergy/pdfs/mypp_may_2013.pdf, 2013.)

## 11.4 REPRESENTATIVE PROGRAMS IN THE UNITED STATES

### 11.4.1 DEPARTMENT OF ENERGY'S GENOMIC SCIENCE PROGRAM

Natural systems, from simple microbes to complex, highly diverse ecosystems, hold secrets of life that fascinate curious minds, drive scientific inquiry, and offer biological solutions to many energy and environmental challenges facing us today.[11] Even the simplest of these systems, the single microbe is so complex that we do not yet possess a full understanding of how a living system works.

The Department of Energy (DOE)'s Genomic Science Program is driven by a grand challenge in biology: understanding biological systems so well that we can develop predictive, computational models of these systems. The DNA code—the genome—is the starting point for understanding any biological system.[11]

The Genomic Science Program aims to develop a predictive understanding of biological systems relevant to energy production, environmental remediation, and climate change mitigation.[12] The genome sequences of organisms studied in these projects are provided largely by the DOE Joint Genome Institute, an important user facility and a world leader in generating sequences of organisms. Synthetic biology is the design and construction of biological systems that do not exist in the natural world.

To focus the most advanced biotechnology-based resources on the challenges of biofuel production, the DOE established three Bioenergy Research Centers (BRCs).[13]

1. The *Joint Bioenergy Institute* (JBEI) is one of the three DOE Bioenergy Research Centers (BRCs) established by DOE's Office of Science in 2007 on the basis of a nationwide competition to accelerate fundamental research breakthroughs for the development of advanced, next-generation biofuels. JBEI is a partnership led by Lawrence Berkeley National Laboratory (LBNL).[14]

   JBEI brings the sunlight-to-biofuels pipeline under one roof in four interdependent research divisions that focus on
   - Designing and developing new bioenergy crops.
   - Enhancing biomass deconstruction.
   - Developing routes to new biofuels through synthetic biology.
   - Creating technologies that advance biofuels research.[14]

   Researchers in JBEI's Feedstocks Division are investigating metabolic pathways involved in the biosynthesis of lignin. This unique basic research program could help transform lignin into a valuable source of chemicals and polymers while improving the economics of converting cellulosic biomass into fuels.[14]

2. The *Bioenergy Science Center* (BESC) is a multi-institutional (18 partners), multidisciplinary research organization focused on the fundamental understanding and elimination of biomass recalcitrance.[15]

   BESC's approach to improve accessibility to the sugars within biomass involves (1) designing plant cell walls for rapid deconstruction and (2) developing multitalented microbes for converting plant biomass into biofuels in a single step (consolidated bioprocessing).[15] Addressing the obstacle of biomass recalcitrance will require a multiscale understanding of plant cell walls from biosynthesis to deconstruction pathways. This integrated understanding would generate models, theories, and finally processes that will be used to understand and overcome biomass recalcitrance.

BESC is organized into three research focus areas:[15]

- Biomass formation and modification
- Biomass deconstruction and conversion[15]
- Characterization and modeling

3. The only DOE BRC based at an academic institution, the GLBRC (*Great Lakes Bioenergy Research Center*) is guided by an educational philosophy that emphasizes understanding the complex relationships among energy production, technology, economics, society, and the environment.[16]

Research areas include sustainability, plants, deconstruction, and conversion.[17] At the GLBRC, sustainability researchers are exploring complex issues in agricultural and industrial systems. Research focuses on understanding the attributes and mechanisms responsible for the environmental sustainability of biofuel production systems, such as environmental impacts—many of which may be positive—and socioeconomic factors including incentives and policy options.

At the GLBRC, plants researchers are developing the next generation of improved crops. On account of crops will continue to be grown for food and feed in the future, research focused on enhancing plants with desirable energy traits must be pursued without sacrificing grain yield and quality.

Plants research projects fall under three general categories:

- Reducing lignocellulosic biomass recalcitrance through plant cell wall modification
- Improving the value of the biomass grown for bioenergy production
- Integrating these and other beneficial traits into bioenergy crops that exhibit improved nutrient use and stress tolerance for sustainable production[18]

Current *BRC strategies* can be summarized as follows:

- Develop next-generation bioenergy crops by unraveling the biology of plant development
- Discover and design enzymes and microbes with novel biomass-degrading capabilities
- Develop transformational microbe-mediated strategies for advanced biofuels production[18]

## 11.4.2 New U.S. Research Areas

Recent key U.S. articles dealing with hemicelluloses and lignin are numerous:

- Lignin valorization through integrated biological funneling and chemical catalysis. The utilization of these natural aromatic catabolic pathways may enable a broader slate of molecules derived from lignocellulose.[19]

- Lignin valorization: improving lignin processing in the biorefinery.[20] Advances in biotechnology and chemistry hold promise for greatly expanding the scope of products derived from lignin. Refinement of biomass pretreatment technologies facilitates lignin recovery and enables catalytic modifications for desired chemical and physical properties.
- Efficient biomass pretreatment using ionic liquids (ILs) derived from lignin and hemicelluloses.[21–23] *Bionic liquids*—ILs derived from lignin and hemicelluloses—show great promise for liberating fermentable sugars from lignocellulose and improving the economics of biofuels refineries. Due to their potential for large-scale deployment, ILs derived from inexpensive, renewable reagents are highly desirable.

With respect to overall sugar yield, these bionic liquids show that they perform comparably with traditional ILs in biomass pretreatment. The concept of deriving ILs from lignocellulosic biomass shows significant potential for the realization of a *closed-loop* process for lignocellulosic biorefineries (Figure 11.3).

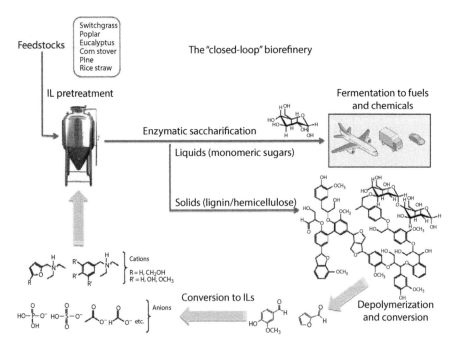

**FIGURE 11.3** Hypothetical process flow for a closed-loop biorefinery using ionic liquids derived from lignocellulosic biomass. (From European Commission, Bio-based economy in Europe: State of play and future potential Part 2, http://ec.europa.eu/research/bioeconomy/pdf/biobasedeconomyforeuropepart2_allbrochure_web.pdf, 2011.)

## 11.5 BIOBASED ECONOMY IN EUROPE

### 11.5.1 BIOBASED ECONOMY IN EUROPE

A socially and environmentally beneficial bioeconomy already exists to some extent in Europe.[24] The European forest-based sector is a particularly good example in which public and private organizations have successively worked together in solving complex challenges.

A number of sectorial policies have been put in place at European level to support the development of a biobased economy, but further integration between policies is needed to avoid contradictions in policy goals and ensure a level playing field for all actors. Further obstacles are identified in the impacts of climate change, which are likely to increase existing problems, and societal concerns regarding biotechnologies. Missed opportunities arise also from the insufficient implementation of existing legislation and difficult access to public money for public-private partnerships and demonstration projects.

The three pillars of the EU bioeconomy strategy and action plan are as follows:

- Investing in research, innovation, and skills
- Reinforced policy interaction and stakeholder engagement
- Enhancement of markets and competitiveness in bioeconomy sectors[25]

### 11.5.2 BIOECONOMY OBSERVATORY

The EU bioeconomy observatory was launched at the bioeconomy stakeholders' conference (Turin, October 8–9, 2014). It was set up by the European Commission's Joint Research Centre (JRC).[26]

The bioeconomy observatory website includes statistics on investments in research, innovation and skills, mapping of policy initiatives at EU and national levels, bioeconomy profiles for EU, Member States and EU regions, socio-economic analysis of biobased value chains, and environmental sustainability assessment of biobased products. Data and information are grouped along the three key pillars highlighted in the EU bioeconomy strategy: research, policy, and markets.

The creation of the bioeconomy observatory was foreseen in the EU's bioeconomy strategy.[27] It will allow a regular assessing on the progress and impact of the bioeconomy. It will supply policy makers and stakeholders with reference data and analyses, providing a solid basis for policy development and decision making. Information and data are currently collected at EU, Member States, and regional levels. Eventually, they will also be available for key international partners such as Brazil, Malaysia, or the United States. The bioeconomy observatory website is the first step of a continuous upgrading process, based on comments from stakeholders and new available data.[28]

## 11.6 REPRESENTATIVE PROGRAMS IN THE EUROPE

### 11.6.1 Bio-Based Industries

The Bio-Based Industries (BBI) Joint Undertaking is a Public-Private Partnership (PPP) between the European Commission and the Bio-based Industries Consortium (BIC).[29]

In July 2013, the European Commission proposed its Innovation and Investment Package containing five JTIs, (Joint Technology Initiatives) including the newcomer on biobased industries.[30]

The PPP is an instrument to support industrial research and innovation, to overcome the innovation *valley of death*, the path from research to the marketplace. It encourages partnership with the private sector to fund and bring together the resources needed to address the challenges involved in commercializing major society-changing new technologies.

The industry is organized in a Bio-based Industries Consortium. The Consortium currently brings together in February 2015 about 80 European large and small companies, clusters, and organizations across technology, industry, agriculture, and forestry. The members have all committed to invest in collaborative research, development, and demonstration of biobased technologies within the PPP. The BIC membership also consists of associate members (Research and Technology Organizations, universities, associations, and technology platforms).

In July 2014, EU and industry leaders have launched a new European Joint Undertaking on Biobased Industries.[31] The aim is to trigger investments and create a competitive market for biobased products and materials sourced locally and made in Europe, tackling some of Europe's biggest societal challenges.

€3.7 billion will be injected into the European economy between 2014 and 2024—€975 million from the European Commission and €2.7 billion from the Bio-based Industries Consortium (BIC)—to develop an emerging bioeconomy sector.[32] Through financing of research and innovation projects, the BBI will create new and novel partnerships across sectors, such as agriculture, agro-food, technology providers, forestry/pulp and paper, chemicals, and energy.

### 11.6.2 Bio-Based Industries' Vision

BBI's vision is for a competitive, innovative, and sustainable Europe: leading the transition toward a postpetroleum society while decoupling economic growth from resource depletion and environmental impact.

Together with pan-European and cross-sector industries/SMEs, research organizations, universities, regions, and EU Member States, BIC will develop an economy that

- Sources domestic renewable raw materials.
- Produces food, feed, chemicals, materials, and fuels locally.
- Creates jobs in a broad range of sectors in Europe, triggering rural growth across regions.
- Places sustainability, smart, and efficient use of resources at the heart of industrial, business, and social activities.

In this vision, the biobased industries will optimize land use and food security through a sustainable, resource-efficient, and largely waste-free utilization of Europe's renewable raw materials for industrial processing into a wide range of bio-based products such as advanced transportation fuels, chemicals, materials, food and feed, and energy (Figure 1.1).[32,33]

In doing so, biobased industries will play an important role in spurring sustainable growth and boosting Europe's competitiveness by reindustrializing and revitalizing rural areas, thus, providing tens of thousands of high-skilled research, development, and production jobs over the next decade.

### 11.6.3 BIOBASED INDUSTRIES CONSORTIUMS' WORK

BIC's work is carried out along three stages:

1. Reinforce innovation and extend current infrastructure across the economy
2. Build and strengthen value chains across industry sectors
3. Realize a connected biobased economy from research to end consumer[34]

The BBI will optimize and create new value chains from primary production to consumer markets. These are

- *Value Chain 1 (VC1)*: From lignocellulosic feedstocks to advanced biofuels, biobased chemicals, and biomaterials.
- *Value Chain 2 (VC2)*: The next generation forest-based value chains.
- *Value Chain 3 (VC3)*: The next generation agro-based value chains.
- *Value Chain 4 (VC4)*: Emergence of new value chains from (organic) waste.
- *Value Chain 5 (VC5)*: The integrated energy, pulp, and chemicals biorefineries.[35]

The BBI vision is that of a competitive, innovative, and sustainable Europe leading the transition toward a postpetroleum industry while decoupling economic growth from resource depletion and environmental impact.[35] In this vision, the biobased industries will optimize land use and food security through a sustainable, resource-efficient, and largely waste-free utilization of Europe's renewable raw materials for industrial processing into a wide array of biobased products including advanced transportation fuels, chemicals, materials, food ingredients and feed, and energy.

The Bio-based Industries Joint Undertaking's first call for proposals in 2014 included topics for four of the five value chains: VC1, VC2, VC3, and VC4.

The work plan is organized around two types of actions:

- Research and innovation actions
- Innovation actions including demonstration actions and flagship actions

In addition, supporting actions, addressing cross cutting, and nontechnological issues are foreseen.

Priority in this first work plan is given to

1. Initiatives that have a high potential to deliver results on the short and medium terms.
2. Challenges to ensure medium- and long-term sustainable biorefinery approaches.
3. High impact and complex long-term issues.


## 11.7  BUILDING BLOCKS FROM BIOMASS

Major chemical, oil, and consumer goods companies are heavily investing in the bio-chemical sector. An example of this investment is the partnership of Coca-Cola with Virent, Gevo, and Avantium to accelerate the commercialization of a 100% biobased PET or PEF for its next-generation plant-based bottles.[36] In June 2015, Coca-Cola introduced world's first 100% biobased PET bottle (Table 9.5). Virent Bioformpx® paraxylene was used for this first PET entirely renewable bottle.[37]

Wertz[38] recently published a synthesis report devoted to biobased building blocks derived from carbohydrates through biorefineries. This work identified about 30 of the most promising biobased building blocks. Many of these mole-cules can lead to polymers that are commonly present in high volume plastics and rubbers. Besides building blocks able to generate mass products, some of them can lead to the production of biobased high added value products. This report has been made in the context of promoting biomass value chains to primarily nonenergy products. It should support regional strategies for the development of a biobased economy.

These building blocks that are able to generate a whole range of products are pro-duced in a biorefinery. These biorefineries will progressively replace the petroleum refineries to cope with the decrease in petroleum reserves and with the increase in $CO_2$ emissions. The future biorefineries will have to convert all the components of vegetal biomass into bioenergy and biobased products while maximizing the pro-duced added value.

Today, a plenty of building blocks originate from glucose as raw material. Glucose is easily accessible in sugar plants or starch plants, that is, food plants. However, glucose is also present but in a less accessible way, in nonfood ligno-cellulosic biomass. Currently, research is performed everywhere in the world to pass from one glucose to the other and, thus, from first to second generation technologies. The first lignocellulosic units are being built in the world but at a low speed.

The advent of biorefineries that convert cellulosic biomass into liquid transporta-tion fuels will generate substantially more lignin than necessary to power the opera-tion, and therefore, efforts are underway to transform it to value-added products.[21] Research on lignin deconstruction has recently become the center of interest for scien-tists worldwide, racing toward harvesting fossil fuel-like aromatic compounds.[39] The natural complexity and high stability of lignin bonds make lignin depolymerization a highly challenging task. Several efforts have been directed toward a more profound understanding of the structure and composition of lignin to devise pathways to break down the polymer into useful compounds.

## 11.8  THE SUSTAINABILITY CHALLENGE

In 1989, the World Commission on Environment and Development (Brundtman Commission) has defined sustainable development as "development which meets the needs of current generations without compromising the ability of future generations to meet their own needs."[40] The concept supports strong economic and social development, in particular, for people with a low standard of living. At the same time, it underlines the importance of protecting the natural resource base and the environment. Economic and social wellbeing cannot be improved with measures that destroy the environment. Intergenerational solidarity is also crucial: all development has to take into account its impact on the opportunities for future generations.

Maintaining the services provided by natural resources, promoting economic development, and providing conditions that support human and societal health are all critical components of a sustainable development. The sustainability efforts are organized around environmental, social, and economic, the three core pillars of sustainability (Figure 11.4).

Most biobased production routes to fuels, chemicals, and power can save energy in production processes and significantly reduce $CO_2$ emissions.[43–45] Vegetable solvents emit few or no volatile organic compounds, and biobased chemicals offer potential to reduce the generation of toxic wastes.[46] The social impacts primarily relate to employment creation and the potential for rural development. The economic benefits derive from a growing biobased chemicals market and the macroeconomic savings of biobased chemicals, when compared with petrochemical based chemicals.

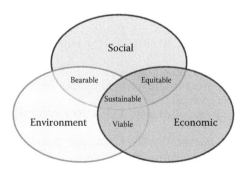

**FIGURE 11.4**  The three pillars of sustainability. (Adapted from http://www.environ.ie/en/Environment/SustainableDevelopment/PublicationsDocuments/FileDownLoad,30454,en.pdf; http://www.epa.gov/sustainability/docs/framework-for-sustainability-indicators-at-epa.pdf.)

## REFERENCES

1. https://www.vlaanderen.be/nl/publicaties/detail/bioeconomy-in-flanders.
2. https://ec.europa.eu/jrc/en/news/new-eu-bioeconomy-observatory.
3. http://poet-dsm.com/pr/first-commercial-scale-cellulosic-plant.
4. http://www.upmbiofuels.com/biofuel-production/biorefinery/Pages/Default.aspx.
5. http://www.fuelsamerica.org/blog/entry/four-commercial-scale-cellulosic-ethanol-biorefineries-to-enter-production.
6. http://biofuels.dupont.com/cellulosic-ethanol/nevada-site-ce-facility/.
7. http://www.storaenso.com/rethink/new-kind-of-gold-from-nordic-forests.
8. http://www.matrica.it/.
9. J.S. Golden and R.B. Handfield, Opportunities in the emerging bioeconomy, why biobased? 2014. http://www.biopreferred.gov/files/WhyBiobased.pdf and http://www.agri-pulse.com/USDA-report-outlines-opportunities-in-the-emerging-bioeconomy-10072014.asp.
10. Bioenergy Technologies Office, Energy Efficiency & Renewable Energy, DOE, multi-year program plan, 2013. http://www1.eere.energy.gov/bioenergy/pdfs/mypp_may_2013.pdf.
11. http://genomicscience.energy.gov/program/index.shtml.
12. http://genomicscience.energy.gov/research/.
13. http://genomicscience.energy.gov/centers/.
14. U.S. Department of Science Genomic Science Program, DOE Joint BioEnergy Institute, 2014. http://genomicscience.energy.gov/centers/jbei.shtml.
15. http://bioenergycenter.org/besc/research/characterization.cfm.
16. https://www.glbrc.org/about.
17. https://www.glbrc.org/research.
18. http://genomicscience.energy.gov/centers/BRCs2014LR.pdf.
19. J.G. Linger, D.R. Vardon, M.T. Guarnieri, E.M. Karp, G.B. Hunsinger, M.A. Franden, C.W. Johnson et al., *Proc. Natl. Acad. Sci. USA* 111, 12013, 2014. http://www.pnas.org/content/111/33/12013.full.
20. A.J. Ragauskas, G.T. Beckham, M.J. Biddy, R. Chandra, F. Chen, M.F. Davis, B.H. Davison et al., *Science* 344, 1246843, 2014. doi:10.1126/science.1246843.
21. Phys.org, Bionic liquids from lignin: New results pave the way for closed loop biofuel refineries, August 18, 2014. http://phys.org/news/2014-08-bionic-liquids-lignin-results-pave.html.
22. L. Yarris, Berkeley Lab, News Release, Bionic liquids from lignin, 486–5375, August 18, 2014. http://newscenter.lbl.gov/2014/08/18/bionic-liquids-from-lignin/.
23. A.M. Socha, R. Parthasarathi, J. Shi, S. Pattathil, D. Whyte, M. Bergeron, A. George et al., *Proc. Natl. Acad. Sci.* 111, E 3587, 2014. http://www.pnas.org/content/111/35/E3587.full.
24. European Commission, Bio-based economy in Europe: State of play and future potential Part 2, 2011. http://ec.europa.eu/research/bioeconomy/pdf/biobasedeconomyforeuropepart2_allbrochure_web.pdf.
25. http://ec.europa.eu/research/bioeconomy/pdf/2013_bioeconomy_stakeholders_conference_report_en.pdf.
26. https://biobs.jrc.ec.europa.eu/.
27. http://ec.europa.eu/research/bioeconomy/pdf/official-strategy_en.pdf.
28. https://ec.europa.eu/jrc/en/publication/eur-scientific-and-technical-research-reports/structural-patterns-bioeconomy-eu-member-states-sam-approach.
29. https://biobased.vito.be/node/282.
30. http://biconsortium.eu/about/about-bic-bbi.

31. http://www.bbi-europe.eu/news/eu-and-industry-partners-launch-%E2%82%AC37-billion-investments-renewable-bio-based-economy.

32. http://biconsortium.eu/sites/biconsortium.eu/files/documents/Biobased_Industries_position_EU_CircularEconomyPackage_July2015.pdf.

33. http://biconsortium.eu/about/our-vision.

34. http://biconsortium.eu/about/our-work.

35. http://ec.europa.eu/research/participants/data/ref/h2020/other/legal/jtis/bbi-sira_en.pdf.

36. Biofuelsdigest, The Jesus molecule, December 16, 2011. http://www.biofuelsdigest.com/bdigest/2011/12/16/the-jesus-molecule.

37. http://www.virent.com/news/virent-bioformpx-paraxylene-used-for-worlds-first-pet-plastic-bottle-made-entirely-from-plant-based-material/.

38. J.L. Wertz, Biobased building blocks derived from carbohydrates, 2014. http://www.valbiom.be/publications/molecules-plateformes-biobasees-issues-de-glucides.htm#.VDvcZhbpVI0.

39. C. Xu, R.A.D. Arancon, J. Labidi and R. Luque, *Chem. Soc. Rev.* 2014. doi:10.1039/C4CS00235K.

40. United Nations Economic Commission for Europe, UNECE in 2004-2005, Sustainable development-concept and action. http://www.unece.org/oes/nutshell/2004-2005/focus_sustainable_development.html.

41. http://www.environ.ie/en/Environment/SustainableDevelopment/Publications Documents/FileDownLoad,30454,en.pdf.

42. http://www.epa.gov/sustainability/docs/framework-for-sustainability-indicators-at-epa.pdf.

43. Organisation for Economic Co-operation and Development, Industrial biotechnology and climate change, 2011. http://www.oecd.org/dataoecd/15/54/49024032.pdf.

44. Organisation for Economic Co-operation and Development, The application of biotechnology to sustainability, 2002. http://www.oecd.org/dataoecd/61/13/1947629.pdf.

45. EuropaBio,Howindustrialbiotechcantackleclimatechange,2009.http://www.europabio.org/industrial/positions/how-industrial-biotechnology-can-tackle-climate-change.

46. Europe Innova, Eco-Innovation, Biochem, 2010. http://www.europe-innova.eu/c/document_library/get_file?folderId=177014&name=DLFE-11072.pdf.

# Glossary

## A

**(α/α)₆ barrel:** fold consisting of six inner and six outer α-helices forming a barrel-like structure

**α/β hydrolase fold:** fold characterized by a β-sheet core of 5–8 strands connected by α-helices to form a α/ β/α sandwich

**α-helix:** right- or left-handed coil conformation, resembling a spring, in which every backbone group donates a hydrogen bond to the backbone of the CO group four residues earlier

**α-glycosidase:** glycosidase that acts on 1,4-α bonds

**acetogen bacteria:** organisms capable to reduce $CO_2$ to acetate through the acetyl coenzyme A (acetyl-CoA) or Wood–Ljungdahl pathway

**actin:** globular, roughly 42-kDa protein found in most eukaryotic cells

**active site:** in biology, the active site is the small portion of an enzyme in which substrate molecules bind and undergo a chemical reaction. This chemical reaction occurs when a substrate collides with and slots into the active site of an enzyme

**aerobic organism:** organism that can survive and grow in an oxygenated environment

**aglycone:** nonsugar group of a glycoside

**allomorph:** any of the alternative crystalline forms of a substance, such as native cellulose

**alpha-complementation:** the *lac-Z* gene product (β-galactosidase) is a tetramer, and each monomer is made of two parts—lacZ-alpha and lacZ-omega. If the alpha fragment was deleted, the omega fragment is nonfunctional; however, alpha fragment functionality can be restored in-trans through the plasmid

**amylopectin:** branched α-linked-D-glucan

**amylose:** linear α-1,4-D-glucan

**anaerobic digestion:** series of processes in which microorganisms break down bio-degradable material in the absence of oxygen

**anaerobic organism:** organism that does not require oxygen for growth and may even die in its presence

**angiosperm:** flowering plant in which the seeds are enclosed in an ovary

**anhydroglucopyranose unit:** cyclic six-membered glucose residue resulting from the reaction of an alcohol and a hemiacetal to form an acetal and water

**anomer:** one of the two diastereoisomers (designated α and β) generated by hemiacetal ring formation

**anomeric center (anomeric carbon):** asymmetric carbon atom at C1 generated by hemiacetal ring formation

**antiparallel:** parallel but oriented in opposite directions; a chain packing is antiparallel when the chains running side by side alternate in direction

**apiogalacturonan:** pectic polysaccharide consisting of an α-1,4-linked D-galacturonic acid backbone with apiose and apiobiose side chains through O2 or O3 links

**apoplast:** free diffusional space outside the plasma membrane; structurally, the apoplast is formed by the continuum of cell walls of adjacent cells as well as the extracellular spaces, forming a tissue level compartment comparable to the symplast, which is the inner side of the plasma membrane

**aquasolv:** pretreatment of biomass with hot water or saturated steam

**aqueous phase reforming:** production of hydrogen from biomass-derived oxygenated compounds such as sugars, sugar alcohols, and glycerol

**arabinoxylan:** water-soluble hemicellulose consisting of β-1,4-xylan backbone with arabinose side chains

**asymmetric unit:** smallest portion of a crystal structure to which crystallographic symmetry can be applied in order to generate the complete unit cell

**autohydrolysis:** pretreatment of biomass by exposure to compressed liquid hot water or steam

**axial:** perpendicular to the mean plane of the ring

# B

**β barrel:** large β-sheet that twists and coils to form a closed structure in which the first strand is hydrogen bonded to the last

**(β/α)$_8$ barrel:** fold consisting of 8 repeating units of β/α module, in which the 8 β strands form an inner parallel β sheet arranged in a barrel structure, which is surrounded by the 8 α helices

**β-1,3;1,4 glucans:** hemicellulose polysaccharides consisting of β-1,4-linked glucans with interspersed single β-1,3-linkages

**β-glycosidase:** glycosidase that acts on β-1,4-bond

**β jelly roll:** variant of a Greek key topology with both ends of a sandwich or a barrel fold being crossed by two interstrand connections

**β-propeller:** type of all β protein architecture characterized by 4–8 blade-shaped β-sheets arranged toroidally around a central axis

**β sandwich:** fold consisting of two β-sheets that pack together, face-to-face, in a layered arrangement

**β-sheet:** assembly of β strands that are hydrogen-bonded to each other

**β strand:** a stretch of amino acids typically 5–10 amino acids long whose peptide backbones are almost fully extended

**bacterium:** group of unicellular microorganisms

**basidiomycetes (basidiomycota):** one of two large phyla that, together with the Ascomycetes, comprise the subkingdom Dikarya, often referred to as the higher fungi, within the kingdom Fungi

**biochemical conversion:** technological platform involving essentially three basic steps: (1) biomass pretreatment, (2) conversion of pretreated biomass into sugars, and (3) sugar fermentation

**biocrude:** crude-oil substitute made from biomass

**biofuel:** solid, liquid, or gas fuel derived from biomass

**biological carbon cycle:** one of the major biogeochemical cycles in which carbon moves from the atmosphere, through the Earth's ecosystems (the biosphere), and back to the atmosphere

**biomass:** material of biological origin, excluding the material embedded in geologic formation and/or fossilized

**biomass-to-liquids (BtL):** multistep process to produce liquid fuels (such as gasoline and diesel) from biomass through a thermochemical route; the term BTL is also applied to the fuels made through the BTL process

**bio-oil:** a liquid fuel produced by the pyrolysis of biomass

**biorefinery:** facility that integrates biomass conversion processes and equipment to produce energy (fuels, power, and heat), and chemicals and materials from biomass

**black liquor:** by-product from the Kraft process when digesting pulpwood into paper pulp

**building block:** basic component for organic synthesis

# C

**capnophilic:** $CO_2$-loving

**carbohydrate-active enzyme:** enzyme that degrades, modifies, or creates glycosidic bonds

**carbohydrate-binding module:** contiguous amino acid sequence within a carbohydrate-active enzyme with a discreet fold having carbohydrate-binding activity

**carbonization:** conversion of an organic substance into carbon or a carbon-containing residue through pyrolysis or destructive distillation (fossil fuels, in general, are the products of the carbonization of vegetable matter)

**catabolism:** phase that consists of the disintegration of complex organic compound to release energy

**catalytic subunit (module):** contiguous amino acid sequence within a carbohydrate-active enzyme with a discreet fold having catalytic activity

**cellobiohydrolase:** exoglucanase, which cleaves the $\beta$-1,4-glucosidic bond at one of the ends of the polysaccharide

**cellulase:** class of enzymes produced mainly by fungi, bacteria, and protozoans that catalyze the hydrolysis of cellulose

**cellulose synthase:** glycosyltransferase that catalyzes the chemical reaction: UDP-glucose + ($\beta$-1,4-D-glucosyl)$_n$ → UDP + ($\beta$-1,4-D-glucosyl)$_{n+1}$

**cellulose synthase complex (CSC):** cellulose-synthesizing complex also called terminal complex

**cellulose synthase-like genes:** superfamily of plant genes whose amino acid sequences are related to the *CesA* (cellulose synthase catalytic subunit) genes

**cellulosome:** multienzyme complex produced by many cellulolytic microorganisms

**center chain:** chain passing through the center of the unit cell

**center sheet:** sheet formed by the hydrogen bonding between the adjacent center chains

**charophytes:** a division of green algae, including the closest relatives of the land plants

**chiral (or handed):** a object or system is called chiral if it differs from its mirror image and its mirror image cannot be superposed on the original object; a chiral molecule and its mirror image are called enantiomers

**chlorophytes:** a division of green algae

**cholesteric:** describing a type of liquid crystal with a helical structure and which is therefore chiral

**cinnamic acid:** 3-phenylacrylic acid ($C_6H$-CH=CHCOOH)

**cinnamyl alcohol dehydrogenase (CAD):** enzyme responsible for the last step in lignin monomer biosynthesis, catalyzing the conversion of coniferaldehyde, sinapaldehyde, and $p$-coumaraldehyde to coniferyl alcohol, sinapyl alcohol, or $p$-coumaryl alcohol, respectively

**circular economy:** a circular economy is one that is restorative and regenerative by design, and which aims to keep products, components, and materials at their highest utility and value at all times, distinguishing between technical and biological cycles

**cladogram:** tree-like diagram showing evolutionary relationships

**clan:** group of glycoside families sharing a fold and catalytic machinery

**cohesin:** reiterated domain on scaffoldin, which interacts with the docker in domain on each enzymatic subunit

**combinatorial synthesis:** a process to prepare large sets of organic compounds by combining sets of building blocks

**commelinoid (or commelinid) monocots:** monophyletic group comprising the palms, the spiderworts, the graminaceous plants, and the gingers

**condensation reaction:** chemical reaction in which two molecules combine to form one single molecule, with loss of a small molecule such as water

**conformation:** spatial arrangement of a molecule

**conformation of the hydroxymethyl group (cellulose):** conformation expressed by two letters, the first referring to the torsion angle $\chi$ (O5-C5-C6-O6), that is, the relative orientation of the C6-O6 bond to the C5-O5 bond about the C5-C6 bond, and the second to the torsion angle $\chi'$ (C4-C5-C6-O6)

**coniferin:** glucoside of coniferyl alcohol

**conserved domain:** recurring unit in polypeptide chains (sequence and structure motif), determined by comparative analysis; molecular evolution uses such domains as building blocks and these may be recombined in different arrangements to make different proteins with different functions

**conserved plant-specific region:** one of the two plant-specific regions that is relatively conserved

**convergent evolution:** acquisition of the same biological trait in unrelated lineages

**cotyledon:** a significant part of the embryo within the seed of a plant

**cyanobacteria (blue–green algae):** phylum of bacteria that obtain their energy through photosynthesis, using water as an electron donor and producing oxygen

**cytoplasm:** gel-like substance residing between the cell membrane holding all the cell's internal substructures, except for the nucleus

# D

**designer cellulosome:** artificial cellulosome comprised of recombinant chimeric scaffoldin constructs and selected dockerin-containing enzyme hybrids

**dicotyledon (Dicot):** flowering plant whose seed has two embryonic leaves or cotyledons

**directed evolution:** protein engineering strategy in which random mutagenesis is applied to a protein, and a selection regime is used to pick out variants that have the desired properties

**dockerin:** protein domain, contained as a component part of each cellulosomal enzyme subunit, which functions to bind the enzyme to a cohesin domain within the scaffoldin protein

# E

**embryophytes:** land plants

**EMP pathway:** a chain of ten enzyme-catalyzed reactions that oxidize glucose into pyruvate

**endoderm:** one of the three primary germ cell layers in the very early embryo

**endoglycanase:** glycosyl hydrolase that cleaves internal linkages

**endoglucanase:** endo-$\beta$-1,4-D-glucan glucanohydrolase that usually causes a random breach of $\beta$-glucosidic bonds

**endosperm:** tissue produced within the seeds of most flowering plants around the time of fertilization; it surrounds and nourishes the embryo

**endoxyloglucanase:** specific endoglucanases that depolymerize high molecular mass xyloglucan into xyloglucan oligosaccharides

**endwise degradation:** peeling-off reaction

**endwise polymerization:** polymerization in which, after a certain initial period, the monomer precursors are joined to the ends of the growing polymer instead of combining with each other

**equatorial (lattice plane):** parallel to the chain axis

**equatorial (sugar):** in the mean plane of the ring

**exocytosis:** durable process by which a cell directs the contents of secretory vesicles out of the cell membrane

**exoglycanase:** glycoside hydrolase that cleaves a substrate at the end

**exoglucanase:** $\beta$-1,4-D-glucan cellobiohydrolase, which cleaves the $\beta$-glucosidic bond at one of the ends of the polysaccharide

**exon:** DNA region in a gene that codes for the protein

**exoxyloglucanase:** enzyme that hydrolyzes $\beta$-1,4 glucosidic linkages in xyloglucans, so as to successively remove oligosaccharides from the chain end

**expression cloning:** a technique in DNA cloning that uses expression vectors to generate a library of clones, with each clone expressing one protein. This expression library is then screened for the property of interest

# F

**FAME:** fatty acid methyl esters that constitute biodiesel

**fast pyrolysis:** pyrolysis characterized by moderate temperatures (400°C–600°C) and rapid heating rates

**Fenton's reagent:** solution of hydrogen peroxide and ferrous ions resulting in the production of hydroxyl radicals; the solution is used to oxidize contaminants or waste waters

**fermentation:** process of extracting energy from the oxidation of organic compounds, such as carbohydrates, using an endogenous electron acceptor, which is usually an organic compound; fermentation is caused by enzymes that are produced by microorganisms

**ferulate 5-hydroxylase (F5H):** enzyme responsible for the last hydroxylation of the syringyl-type lignin precursors

**ferulic acid or coniferic acid:** 4-hydroxy-3-methoxycinnamic acid, which is an abundant phenolic phytochemical found in plant cell wall components such as arabinoxylans as side chains. It is related to trans-cinnamic acid. As a component of lignin, ferulic acid is a precursor

**fiber (cell):** one of the two groups of sclerenchyma cells, which are the principal supporting cells in plant tissues that have ceased elongation

**first-generation biobased product:** product that has been derived from edible biomass

**first-generation biorefinery:** biorefinery that uses edible biomass as feedstock

**Fischer–Tropsch:** catalyzed chemical reaction in which synthesis gas is converted into liquid hydrocarbons of various forms

**fungus:** eukaryotic organism that is a member of the kingdom Fungi

**furanose:** simple sugar that contains a five-membered ring consisting of four carbon atoms and one oxygen atom

**furfural:** aromatic aldehyde ($OC_4H_3CHO$), also named furan-2-carbaldehyde

# G

**galactan:** polysaccharide consisting of polymerized galactose

**galactoglucomannan:** hemicellulose polysaccharide consisting of a $\beta$-1,4-linked backbone of D-glucose and D-mannose residues, with $\alpha$-1,6-linked D-galactose side groups

**galactomannan:** hemicellulose polysaccharide consisting of a $\beta$-1,4-linked D-mannose backbone with $\alpha$-1,6-linked D-galactose side groups

**gasification:** process that converts organic- or fossil-based carbonaceous materials into carbon monoxide, hydrogen, carbon dioxide, and methane. This is achieved by reacting the material at high temperatures ($>700°C$), without combustion, with a controlled amount of oxygen and/or steam

**geologic carbon cycle:** one of the major biogeochemical cycles in which carbon moves between rocks and minerals, the oceans, and the atmosphere over large timescales

**glucan:** polyglucose

**glucanase:** any enzyme that breaks down a glucan

**glucomannan:** water-soluble hemicellulose consisting of $\beta$-1,4-linked D-glucose and D-mannose; it is mainly straight-chain polymers, with a small amount of branching; it is found in the roots of the Asian konjac plant and also as a hemicellulose in the wood of conifers and cotyledons

**glucosidase:** enzyme of the hydrolase class that hydrolyzes glucose residues from glucosides

**glucoside:** a molecule in which glucose is bound through its anomeric carbon to another group through a glucosidic bond

**glucosidic bond:** chemical linkage that is formed between the hemiacetal group of glucose and another group

**glucuronoarabinoxylans:** hemicellulose polysaccharides having a $\beta$-1,4-xylan backbone with glucuronic acid and arabinose side chains

**glucuronoxylans:** hemicellulose polysaccharides having a $\beta$-1,4-xylan backbone with glucuronic acid side chains

**glycan:** polysaccharide

**glycan synthase:** glycosyltransferase that catalyzes the synthesis of glycans

**glycoconjugates:** general classification for carbohydrates covalently linked with other chemical species

**glycolysis:** metabolic pathway that converts glucose into pyruvate; glycolysis takes place in the cytoplasm of the cell

**glycone:** sugar group of a glycoside

**glycopeptides:** peptides that contain carbohydrate moieties (glycans) covalently attached to the side chains of the amino acid residues that constitute the peptide. N-linked glycans derive their name from the fact that the glycan is attached to an asparagine (Asn, N) residue, and are among the most common linkages found in nature. O-linked glycans are formed by a linkage between an amino acid hydroxyl side chain (usually from serine or threonine) with the glycan. C-linked glycans are the least common compounds

**glycosidase:** glycoside hydrolase

**glycoside:** any molecule in which a sugar group is bonded through its anomeric carbon to another group through a glycosidic bond. Glycosides can be linked by an O- (an *O*-glycoside), N- (a glycosylamine), S- (a thioglycoside), or C- (a *C*-glycoside) glycosidic bond

**glycoside hydrolase (also called glycosidase):** enzyme that catalyzes the hydrolysis of the glycosidic linkage to release smaller sugars. The glycosidic bond may be between two or more carbohydrates or between a carbohydrate and a noncarbohydrate moiety

**glycoside hydrolase (GH) also known as glycosidase:** enzyme that hydrolyzes the glycosidic bond between two or more carbohydrates or between a carbohydrate and a noncarbohydrate moiety

**glycosidic bond:** chemical linkage formed between a hemiacetal group of a sugar molecule and an alcohol group. In the glycosidic bond, the sugar group is known as the glycone and the nonsugar group as the aglycone

**glycosyltransferase (GT):** enzyme that catalyzes the transfer of a monosaccharide residue from an activated donor substrate to an acceptor molecule, forming glycosidic bonds

**Golgi apparatus:** organelle found in most eukaryotic cells whose primary function is to process and package the macromolecules that are synthesized by the cell

**Greek key:** topology in which typically three antiparallel β strands connected by hairpins are followed by a longer connection to the fourth strand, which lies adjacent to the first

**green liquor:** dissolved smelt of sodium carbonate, sodium sulfide, and other compounds from the recovery boiler in the kraft process

**greenhouse gases (GHG):** greenhouse gases are the gases in the atmosphere that absorb and emit radiations within the thermal infrared range

**gymnosperm:** plant whose seeds are not enclosed in an ovary

# H

**heme:** prosthetic group (i.e., tightly bound cofactor) that consists of an iron atom contained in the center of a large heterocyclic organic ring called a porphyrin

**hemicellulases:** group of enzymes that hydrolyze hemicelluloses

**hemicelluloses:** branched glycans composed of 5-carbon and 6-carbon sugars. In plant cell walls, hemicelluloses bind to cellulose

**homogalacturonan:** linear pectin chain of α-1,4-linked-D-galacturonic acid residues in which some of the carboxyl groups are methyl esterified

**homolog:** a gene related to a second gene by descent from a common ancestral DNA sequence; the term may be applied to the relationship between genes separated by the event of speciation (ortholog) or to the relationship between genes separated by the event of genetic duplication (paralog)

**hydrogenation:** chemical reaction that typically constitutes the addition of pairs of hydrogen atoms to a molecule

**hydrogenolysis:** chemical reaction whereby a carbon–carbon or a carbon-heteroatom single bond is cleaved by hydrogen

**hydrothermal liquefaction:** thermochemical conversion process in which high temperatures and pressures are used to decompose wet biomass

**hydrothermal treatment:** processing of biomass that use water in its liquid state, including hot water extraction, pressurized hot water extraction, liquid hot water pretreatment, hydrothermal carbonization, and hydrothermal liquefaction

**hydrothermolysis:** hydrolysis of lignocellulosic materials using pure water at elevated temperatures

**hypervariable region:** one of the two plant-specific regions that represents highly divergent DNA sequences

# I

**integral membrane protein:** a protein that is permanently attached to the biological membrane

**inversion (mechanism):** single nucleophilic displacement that leads to an inversion of the anomeric configuration

**inverting glycosyltransferase:** enzyme that catalyzes the transfer of sugar moieties with an inversion of the anomeric configuration

**ionic liquid:** a salt in the liquid state

**isoenzymes:** enzymes that differ in amino acid sequence but catalyze the same chemical reaction

**isoform (protein):** any of the several different forms of the same protein formed because of single nucleotide polymorphisms

# J

**jelly roll:** variant of Greek key topology with both ends of a sandwich or a barrel fold being crossed by two interstrand connections

# L

**laccases:** multicopper oxidases of wide specificity that carry out one-electron oxidation of phenolic and related compounds, and reduce $O_2$ to water

**levelling-off degree of polymerization (LODP):** when the degree of polymerization of a cellulose is plotted against time of hydrolysis, the degree of polymerization initially decreases sharply and then levels off at a value characteristic of the cellulose called LODP

**lignans:** polyphenolic substances derived from phenylalanine through dimerization of monolignols to a dibenzylbutane skeleton

**lignification:** the process that includes the biosynthesis of monolignols, their transport to the cell wall, and polymerization into the final molecule

**lignin:** generic term for a large group of aromatic polymers resulting from the oxidative combinatorial coupling of 4-hydroxyphenylpropanoids

**lignin peroxidase (LiP):** hemoprotein that catalyzes the oxidative cleavage of C–C bonds and oxidizes benzyl alcohols, such as veratryl alcohol, to aldehydes or ketones; LiP attacks both phenolic and nonphenolic lignin substrates by a one-electron oxidation reaction to generate unstable aryl radical cations. It requires hydrogen peroxide as an electron acceptor

**ligninase:** lignin-degrading enzyme

**lignocellulosic biomass (or simply biomass):** plant biomass that is composed of cellulose, hemicelluloses, and lignin

**lyase:** enzyme that catalyzes the breaking of various chemical bonds by means other than hydrolysis and oxidation

**lycophyte:** oldest living vascular plant division, including some of the most primitive extant species

**lyocell fiber:** regenerated cellulose fiber produced through the direct dissolution of cellulose in organic solvents such as *N*-methylmorpholine-*N*-oxide

# M

**manganese peroxidase (MnP):** hemoprotein involved in the oxidative degradation of lignin in white-rot basidiomycetes; it catalyzes the oxidation of Mn(II) to Mn(III) by hydrogen peroxide; the highly reactive Mn(III) is stabilized through chelation in the presence of dicarboxylic acids

**mannan:** polymannose

**mercerization:** treatment for cotton fabric and thread that gives the fabric a lustrous appearance

**metabolism:** set of enzyme-catalyzed reactions that convert substrates that are external to the cell into various internal products, that is, the set of reactions that occur in living organisms to maintain life. The metabolism consists of two basic types of interdependent phases: the phase in which the energy is freed (catabolism) and the phase in which energy is captured (anabolism)

**metabolite:** metabolites are the intermediates or products of metabolism. A primary metabolite is directly involved in normal growth, development, and reproduction. A secondary metabolite is not directly involved in those processes, but usually has an important ecological function. Ethanol and lactic acid are primary metabolites produced by industrial microbiology

**microfibril (cellulose):** fibrous crystalline aggregate of cellulose molecules

**microtubule:** one of the components of the cytoskeleton; they have a diameter of 25 nm and length varying from 200 nm to 25 μm

**mixed-linked glucan:** hemicellulosic polysaccharide consisting of β-D(1-3) and β-D(1-4)-linked glucosyl residues

**module (enzymes):** independently folding, structurally and functionally discrete unit of modular enzymes

**molecular screening (high throughput):** automated, simultaneous testing of thousands of distinct chemical compounds in models of biological mechanisms

**monoclinic:** one of the seven lattice point groups; in the monoclinic system, the crystal is described by three vectors of unequal length that form a rectangular prism with a parallelogram as its base

**monocotyledon (monocot):** flowering plant whose seed has a single embryonic leaf or cotyledon

**monolignols (or hydroxycinnamyl alcohols):** lignin monomers, the most abundant of which are *p*-coumaryl (4-hydroxycinnamyl) alcohol, coniferyl (3-methoxy 4-hydroxycinnamyl) alcohol, and sinapyl (3,5-dimethoxy 4-hydroxycinnamyl) alcohol

**monooxygenase:** enzyme that incorporates one hydroxyl group into substrates in many metabolic pathways

**morphology:** study of form, comprising shape, size, and structure

# N

**N-glycosylation or N-linked glycosylation:** attachment of oligosaccharides—a process known as glycosylation—to a nitrogen atom in an amino acid residue (usually the N4 of asparagine residues) in a protein

**non-commelinoid monocot:** APG (Angiosperm Phylogeny Group) clade comprising orchids and lilies

**non-processive glycosyltransferase:** enzyme that catalyzes the addition of only one sugar residue

**nucleotide:** structural unit of DNA (deoxyribonucleic acid) and RNA (ribonucleic acid) that consists of a nitrogenous base, a sugar, and a phosphate group

**nucleotide sugar (or sugar nucleotide):** activated monosaccharides that act as donors of sugar residues in nucleotide sugar metabolism; they are substrates for glycosyltransferases

# O

**olefin metathesis or transalkylidenation:** a chemical reaction in which the bonds between different atoms are broken and new bonds are formed; in olefin metathesis, the carbon bonds change places

**open reading frame:** portion of an organism's genome, which contains a sequence of bases that could potentially encode a protein

**Operon:** functioning unit of key nucleotide sequences, including an operator, a common promoter, and one or more structural genes, which is controlled as a unit to produce messenger RNA (mRNA), in the process of protein transcription

**organosolv:** pulping technique that uses an organic solvent to solubilize lignin and hemicelluloses

**origin chain:** chain passing through the origin of the unit cell

**origin sheet:** sheet formed by hydrogen bonding between adjacent origin chains

**orthologs:** genes in different species that are similar to each other because they originated from a common ancestor; orthologs retain the same function in the course of evolution

**oxidative combinatorial coupling (lignin polymerization):** oxidative radicalization of phenols, followed by combinatorial radical coupling (see combinatorial synthesis)

# P

**parallel down:** sense of the polymer chain when the $z$ coordinate of O5 is smaller than that of C5

**parallel up:** sense of the polymer chain when the $z$ coordinate of O5 is greater than that of C5

**paralogs:** homologs whose evolution reflects gene duplication events; paralogs evolve new functions

**peak oil:** the point in time when the maximum rate of global petroleum extraction is reached, after which the rate of production enters terminal decline

**pectin:** any of several heterogeneous polysaccharides that are composed primarily of $\alpha$-1,4-D-galacturonic acid units or their methyl esters; in plant primary cell walls, pectins form a gel phase in which the cellulose–hemicelluloses network is embedded

**perisperm:** nutritive tissue of a seed derived from the nucellus and deposited external to the embryo

**peroxidase:** hemoprotein catalyzing the oxidation by hydrogen peroxide of a number of substrates

**phenylpropanoids:** a class of organic compounds that are biosynthesized from the amino acid phenylalanine; their name is derived from the phenyl group and the propene tail of cinnamic acid

**photoautotroph:** autotrophic organism that carries out photosynthesis to acquire energy and fix carbon

**photoheterotroph:** heterotrophic organism that uses light for energy, but cannot use $CO_2$ as their sole carbon source; consequently, they use organic compounds to satisfy their carbon requirements

**photophosphorylation:** production of ATP using the energy of sunlight

**photosynthesis:** metabolic pathway that converts light energy into chemical energy

**plasma membrane:** selectively permeable lipid bilayer found in all cells. A plasma membrane has two surfaces, one toward the outside of the cell and termed the exoplasmic (outer) surface (ES) and one adjacent to the cytoplasm and termed the protoplasmic (inner) surface (PS)

**plastids:** major organelles found in the cells of plants and algae; they are responsible for photosynthesis, storage of products such as starch and for the synthesis of many classes of molecules

**polarity (cellulose):** cellulose polarity refers to the chemical difference of the two ends of the molecule, inducing possible parallel and antiparallel chain packings

**polymorph:** any of the different crystalline forms of the same chemical compound

**prebiotic (nutrition):** prebiotics are nondigestible fiber compounds that pass undigested through the upper part of the gastrointestinal tract and stimulate the growth and/or activity of advantageous bacteria that colonize the large bowel by acting as a substrate for them

**primary cell wall (plant):** cell wall of growing cells, which are relatively thin, and only semirigid to accommodate future cell growth

**primary scaffoldin:** enzyme-binding scaffoldin

**primitive lattice:** a crystal lattice in which there are lattice points only on the cell corners

**processive glycosyltransferase:** enzyme that catalyzes the addition of multiple sugar residues

**processivity:** ability of an enzyme to repetitively continue its catalytic function without dissociating from its substrate

**promoter:** region of DNA that facilitates the transcription of a particular gene

**protein engineering:** the design and construction of new proteins with novel or desired functions, through the modification of amino acid sequences using recombinant DNA technology

**protists:** group of eukaryotic microorganisms

**pyranose:** simple sugar that contain a six-membered ring consisting of five carbon atoms and one oxygen atom

**pyrolysis:** thermochemical decomposition of organic material at elevated temperatures in the absence of oxygen. Pyrolysis typically occurs under pressure and at operating temperatures above 430°C

**pyrolytic lignin:** water-insoluble fraction of bio-oil

# R

**recalcitrance:** robustness to deconstruction

**recombinant DNA:** a form of synthetic DNA that is engineered through the combination or insertion of one or more DNA strands, thereby combining DNA sequences that would not normally occur together

**reducing end (cellulose):** chain-end containing an unsubstituted hemiacetal, the nonreducing end containing an additional hydroxyl group at C4

**repeat unit:** a part of a polymer chain whose repetition would produce the complete polymer (except for the ends) by linking the repeating units together successively along the chain

**respiration:** set of the metabolic reactions and processes that take place in organisms' cells to convert biochemical energy from nutrients into adenosine triphosphate (ATP), and then release waste products. The reactions involved in respiration are catabolic reactions that involve the oxidation of one molecule and the reduction of another

**retaining glyciosyltransferase:** enzyme that catalyzes the transfer of sugar moieties with a retention of the anomeric configuration

**retention (mechanism):** double displacement mechanism that leads to the retention of the anomeric configuration

**rhamnogalacturonan I (RG I):** pectic polysaccharide containing a backbone of the repeating galacturonic acid–rhamnose disaccharide, in which the rhamnosyl residues can be substituted with arabinan, galactan, and/or arabinogalactans

**rhamnogalacturonan II (RG II):** highly branched pectic polysaccharide containing an $\alpha$-1,4-linked galacturonic acid backbone with side chains comprising a wide and unusual range of sugar residues

**rosette:** a structure of cellulose-synthesizing terminal complexes

**Rossmann fold:** protein structural motif found in proteins that bind nucleotides. The structure with two repeats is composed of six parallel $\beta$ strands linked to two pairs of $\alpha$-helices in the topological order $\beta$-$\alpha$-$\beta$-$\alpha$-$\beta$

**RuBisCO (Ribulose-1,5-bisphosphate carboxylase/oxygenase):** enzyme involved in the Calvin cycle, that catalyzes the first major step of carbon fixation

# S

**scaffoldin:** multifunctional integrating subunit of the cellulosome

**sclereids:** one of the two groups of sclerenchyma cells, which are the principal supporting cells in plant tissues that have ceased elongation

**second generation biobased product:** product that has been derived from lignocellulosic biomass

**second-generation biorefinery:** biorefinery that uses nonedible biomass as feedstock

**secondary cell wall (plant):** structure found in many plant cells located between the primary cell wall and the plasma membrane; the cell starts producing the secondary cell wall only after the primary cell wall is completed and the cell has stopped growing; it is thicker and stronger than the primary cell wall

**shikimic acid:** an important biochemical intermediate in plants and microorganisms, better known under its anionic form shikimate (3,4,5-trihydroxycyclohex-1-ene-1-carboxylate); the shikimate pathway is the biosynthetic sequence employed by plants and bacteria to generate the aromatic amino acids

**sinapyl alcohol:** 4-(3-hydroxy-1-propenyl)-2,6-dimethoxyplenol; a unit in the lignin polymer

**solanoceae:** family of flowering plants that contains a number of important agricultural crops such as potato, tomato, and aubergine, as well as many toxic plants

**solvolysis:** special type of nucleophilic substitution or elimination in which the nucleophile is a solvent molecule

**stepwise polymerization:** polymerization in which any two monomers present in the reaction mixture can link together at any time and growth of the polymer is not confined to chains that are already forming

**sugar nucleotide (or nucleotide sugar):** activated monosaccharides that act as donors of sugar residues in nucleotide sugar metabolism; they are substrates for glycosyltransferases

**supercritical (fluid):** substance above its critical temperature and critical pressure, where distinct liquid and gas phases do not exist

**synthase:** enzyme that catalyzes a synthesis process

**synthesis gas (or syngas):** gas mixture that contains varying amounts of carbon monoxide and hydrogen

**syringin:** glucoside of sinapyl alcohol

# T

**tars:** condensable organic compounds with boiling point between 80°C and 350°C that contain a high degree of aromatic rings

**tautomerization:** chemical reaction resulting in the migration of a proton and the shifting of bonding electrons; the concept of tautomers that are interconvertible by tautomerization is called tautomerism

**terminal complex:** cellulose-synthesizing complex that contain a number of cellulose synthases organized in spinnerets at the cell membrane

**thermochemical conversion:** technological platform involving generally biomass gasification or pyrolysis processes

**thermotropic:** exhibiting different phases at different temperatures; a liquid crystal is thermotropic if the order of its components is determined or changed by temperature

**transcription factor:** protein that binds to specific DNA sequences, thereby controlling the flow (or transcription) of genetic information from DNA to mRNA; transcription factors bind to either enhancer or promoter regions of DNA adjacent to the genes that they regulate; depending on the transcription factor, the transcription of the adjacent gene is either up- or down-regulated

**transglycosidase:** glycosidase that is also able to form glycosidic bonds

**transverse microfibril:** microfibril that is transverse to the main cell axis

**triclinic:** one of the seven lattice point groups; in the triclinic system, the crystal is described by three vectors of unequal length that are not mutually orthogonal

**tubulin:** one of the several members of a family of globular proteins; the most common members of the tubulin family are $\alpha$-tubulin and $\beta$-tubulin, the proteins that make up microtubules

**type-I cohesin:** domain of the scaffoldin that binds the cognate dockerin domain of each enzymatic component

**type-II cohesin:** domain on cell-surface anchoring proteins, which interacts with a type II dockerin domain

**type-I dockerin:** domain on the enzymes that interacts with the type-I cohesin domains

**type-II dockerin:** domain on scaffoldin that interacts with the type II cohesin domains contained on cell-surface anchoring proteins

# U

**uniplanar orientation:** preferential orientation of cellulose microfibrils to the cell wall surface (film surface)

**unit cell:** smallest repeating unit that can generate the whole crystal with only translation operations

**uridine diphospho-glucose (UDPGlc):** a compound in which $\alpha$-glucopyranose is esterified at C1 with the terminal phosphate group of uridine-5′-pyrophosphate

**uronic acid:** sugar acid with both a carbonyl and a carboxylic acid function; it can be thought of as a sugar in which the terminal carbon's hydroxyl function has been oxidized to a carboxylic acid; the names of uronic acids are generally based on their parent sugars (e.g., glucuronic acid)

# V

**vascular plant:** green plant having a vascular tissue system (e.g., ferns, gymnosperms, and angiosperms)

**versatile peroxidase (VP):** hemoprotein, combining catalytic properties of manganese peroxidase (oxidation of [MnII] to Mn[III]), lignin peroxidase (oxidation of benzyl alcohols to aldehydes or ketones) and plant peroxidase (oxidation of hydroquinones and substituted phenols)

**viscose fiber:** regenerated cellulose fiber produced through the dissolution of a cellulose dithiocarbonate ester called xanthate in an aqueous solution of caustic soda

# W

**whisker:** short single crystal fiber or filament used as a reinforcement in a matrix

**white liquor:** strongly alkaline solution mainly of sodium hydroxide and sodium sulfide. It is used in the first stage of the kraft process in which lignin and hemicelluloses are separated from cellulose

# X

**xylan:** polyxylose

**xylanase:** class of enzymes that hydrolyze the $\beta$-1,4 linkage in the xylan backbone, yielding short xylooligomers. Most known xylanases belong to families GH 10 and 11

**xylogalacturonan:** pectic polysaccharide consisting of an α-1,4-linked D-galacturonic acid backbone with β-1,3-linked xylose side chains

**xyloglucan endotransglycosylase/hydrolases (XTH):** enzymes that are able to cleave and to reattach xyloglucans polymers in plant cell walls

**xyloglucan endotransglycosylases (XET):** enzymes in plants that can cut and rejoin xyloglucan chains

**xyloglucan:** hemicellulose polysaccharide with a linear β-1,4-glucan backbone substituted at O6 with mono-, di-, and triglycosyl side chains

# Z

**ZSM-5 (Zeolite Sieve of Molecular porosity):** aluminosilicate zeolite mineral belonging to the pentasil (a pentasil unit consists of eight five-membered rings) family of zeolites; it is widely used in the petroleum industry as a heterogeneous catalyst for hydrocarbon isomerization reactions such as the isomerization of metaxylene to paraxylene

# Index

**Note:** Page numbers followed by f and t refer to figures and tables, respectively.

Printed and bound by CPI Group (UK) Ltd, Croydon, CR0 4YY

01/11/2024

01782622-0007